Advances in Engineered Wood Products and Timber Structures

Advances in Engineered Wood Products and Timber Structures

Guest Editors

Almudena Majano Majano
Antonio José Lara Bocanegra
Francisco Arriaga

MDPI

Basel • Beijing • Wuhan • Barcelona • Belgrade • Novi Sad • Cluj • Manchester

Guest Editors

Almudena Majano Majano
Department of Building Structures and Physics
Universidad Politécnica de Madrid
Madrid
Spain

Antonio José Lara Bocanegra
Department of Building Structures and Physics
Universidad Politécnica de Madrid
Madrid
Spain

Francisco Arriaga
Department of Forestry and Environmental Engineering and Management
Universidad Politécnica de Madrid
Madrid
Spain

Editorial Office
MDPI AG
Grosspeteranlage 5
4052 Basel, Switzerland

This is a reprint of the Special Issue, published open access by the journal *Applied Sciences* (ISSN 2076-3417), freely accessible at: https://www.mdpi.com/journal/applsci/special_issues/Q7Y9BF9P04.

For citation purposes, cite each article independently as indicated on the article page online and as indicated below:

Lastname, A.A.; Lastname, B.B. Article Title. *Journal Name* **Year**, *Volume Number*, Page Range.

ISBN 978-3-7258-5519-3 (Hbk)
ISBN 978-3-7258-5520-9 (PDF)
https://doi.org/10.3390/books978-3-7258-5520-9

Cover image courtesy of Antonio José Lara Bocanegra

Contents

Preface

In the context of growing environmental concerns and the urgent demand for sustainable construction practices, timber has re-emerged as a key material in contemporary architecture and structural engineering. The papers included in this Special Issue represent the work of a diverse group of researchers actively engaged in advancing the field. Their contributions reflect a broad spectrum of perspectives, methodologies, and areas of expertise, encompassing advances in engineered wood products—including bio-based adhesives, fiber reinforcements, and hybrid solutions—as well as innovative design strategies that expand the potential of timber across diverse structural applications, from high-rise and multi-story modular buildings to spatial frame systems. We gratefully acknowledge all the authors for their valuable contributions, the reviewers for their constructive feedback, and the editorial team of *Applied Sciences* for their continued support and assistance throughout the publication process.

Almudena Majano Majano, Antonio José Lara Bocanegra, and Francisco Arriaga
Guest Editors

applied sciences

MDPI

Review

Contemporary Strategies for the Structural Design of Multi-Story Modular Timber Buildings: A Comprehensive Review

Marina Tenório [1,*], Rui Ferreira [2], Victor Belafonte [1], Filipe Sousa [2], Cláudio Meireis [2], Mafalda Fontes [2], Inês Vale [1], André Gomes [1], Rita Alves [2], Sandra M. Silva [1], Dinis Leitão [1], André Fontes [2], Carlos Maia [2], Aires Camões [1] and Jorge M. Branco [1]

[1] Department of Civil Engineering, Institute for Sustainability and Innovation in Engineering Structures (ISISE), University of Minho, Campus de Azurém, Av. da Universidade, 4800-058 Guimarães, Portugal; victorbelafonte@gmail.com (V.B.); ines.pvale@gmail.com (I.V.); andrelg1810@gmail.com (A.G.); sms@civil.uminho.pt (S.M.S.); dleitao@civil.uminho.pt (D.L.); aires@civil.uminho.pt (A.C.); jbranco@civil.uminho.pt (J.M.B.)

[2] Lab2PT, School of Architecture, Art and Design, University of Minho, Campus de Azurém, Av. da Universidade, 4800-058 Guimarães, Portugal; rui_ferreira_3@hotmail.com (R.F.); filipe_sousas@hotmail.com (F.S.); claudio13dias@gmail.com (C.M.); mafaldaflcfontes@gmail.com (M.F.); ritaalves.arq@gmail.com (R.A.); afontes@cfaarch.com (A.F.); ocarlosmaia@gmail.com (C.M.)

* Correspondence: tenorio.mcu@gmail.com

Abstract: Modular timber construction embodies a pioneering and eco-friendly methodology within the building sector. With the notable progress made in manufacturing technologies and the advent of engineered wood products, timber has evolved into a promising substitute for conventional materials such as concrete, masonry, and steel. Beyond its structural attributes, timber brings environmental advantages, including its inherent capacity for carbon sequestration and a reduced carbon footprint compared to conventional materials. Timber's lightweight nature, coupled with its versatility and efficiency in factory-based production, accelerates modular construction processes, providing a sustainable solution to the growing demands of the building industry. This work thoroughly explores contemporary modular construction using wood as the primary material. The investigation spans various aspects, from the fundamentals of modularity and the classification of modular timber solutions to considerations of layout design, structural systems, and stability at both the building and module levels. Moreover, inter-module joining techniques, MEP (mechanical, electrical, and plumbing) integration, and designs for disassembly are scrutinized. The investigation led to the conclusion that timber modular construction, drawing inspiration from the steel modular concept, consistently utilizes a structural approach based on linear members (timber frame, post-and-beam, etc.), incorporating stability configurations and diverse joint techniques. Despite the emphasis on modularization and prefabrication for adaptability, a significant portion of solutions still concentrate on the on-site linear assembly process of those linear members. Regarding modularity trends, the initial prevalence of 2D and 3D systems has given way to a recent surge in the utilization of post-and-beam structures, congruent with the ascending verticality of buildings. In contrast to avant-garde and bold trends, timber structures typically manifest as rectilinear, symmetric plans, characterized by regular and repetitive extrusions, demonstrating a proclivity for centrally located cores. This work aims to offer valuable insights into the current utilization of modular timber construction while identifying pivotal gaps for exploration. The delineation of these unexplored areas seeks to enable the advancement of modular timber projects and systems, fully leveraging the benefits provided by prefabrication and modularity.

Keywords: timber buildings; modular construction; timber structures; prefabrication; modularity; joining techniques; MEP installations; disassembly

1. Introduction

The alarming global climate scenario, which has rapidly worsened in recent decades, has highlighted the urgency of reversing the consumption patterns practiced by various economic sectors, demanding the establishment of increasingly ambitious collective climate and energy mitigation goals. As a growth strategy for a modern, competitive, resource-efficient economy which is, consequently, conducive to sustainable development, one of the priorities outlined by the European Union is achieving carbon neutrality by 2050 [1].

In this context, the prioritization of revising the traditional models replicated by the construction sector becomes crucial when confronted with data regarding their impact. The production and processing of materials for the sector hold the most significant share of energy consumption and greenhouse gas emissions [2], with cement and steel accounting for 4 to 7% [3] and 5% [4] of global CO_2 emissions per year, respectively.

Wood, then, emerges as a promising construction material as an alternative to concrete and steel, with the benefit of its inherent contribution to emission reduction through its carbon storage capacity—approximately 0.9 t CO_2/m^3 of material—and lower embodied energy [5].

The population growth, coupled with the urbanization trend that will lead more than 2/3 of the global population to live in urban areas by 2050 [6], requires the establishment of increasingly dense urban forms, which often translates into high-rise buildings. According to the UK Committee on Climate Change [7], by 2060, 230 billion square meters of new construction will be added, with 2 billion square meters per year between 2019 and 2025 [8], contributing to the release of approximately 415 $GtCO_2$ into the atmosphere over the next 40 years [9]. Thus, population and urban expansion turn cities into opportunities for sustainable development [10], highlighting the potential of the timber industry to provide carbon-neutral solutions [11].

Constructing tall buildings with timber under the current standards is a pioneering effort made possible by innovations in the timber-based industry, both at the product and process levels. The introduction of Engineered Wood Products (EWP), such as laminated veneer lumber (LVL), and, in particular, cross-laminated timber (CLT), has optimized the structural properties of timber and overcome many of its limitations [12], reinventing and propelling the industry with products that exhibit high stability, rigidity, and competitive mechanical properties compared to concrete and steel [13].

On the other hand, the development of CAD/CAM (Computer-Aided Design/Computer-Aided Manufacturing) technology combined with the use of CNC (Computer Numeric Control) equipment and the integrated approach of various specialties within the BIM methodology has increased the precision and quality of wood products. Produced in a controlled factory environment, these products minimize on-site activity, thereby reducing construction time, site costs, noise, pollution, waste, accident risks, the need for traditional construction equipment, and material deterioration due to exposure to moisture [14–18]. An example is the Forté building, located in Melbourne and inaugurated in 2013. At that time, it was considered the largest timber apartment complex in the world, with 10 floors and 30% less construction time than that estimated for a similar structure using conventional materials [19]. According to APA [20], using a case study analysis, opting for prefabricated wood panels over conventional materials can result in even more substantial assembly time savings, exceeding 50%.

The objectives of this state-of-the-art review on timber modular construction are multifaceted, aiming to present a comprehensive and innovative analysis of the current landscape of this modern and promising construction typology that integrates sustainable building practices and efficient resource utilization. The first specific objective is to investigate and map current trends in timber modular construction, considering technological advancements, market changes, and recent innovations. This includes exploring production methods and design techniques that can positively influence the efficiency and sustainability of modular construction. This approach will enable us to gain an understanding of the contemporary scenario and identify knowledge gaps. Subsequently, the review aims to identify the challenges faced by timber modular construction and examine solutions proposed in the recent literature. This will not only facilitate an understanding of existing limitations but also highlight innovative approaches to overcoming these challenges. Lastly, the review seeks to provide insights and future directions for research and practice in timber modular construction. This involves synthesizing our findings to formulate practical recommendations and suggestions for future investigations, aiming for continuous advancement and innovation in this dynamic and ever-evolving field.

2. Methodology

This state-of-the-art review employed a systematic approach to identify, analyze, and synthesize the relevant literature. The methodology involved the following steps:

- Scope definition: Clearly defined geographical, temporal, and thematic boundaries were established through a preliminary literature review, mapping the current landscape of modular timber construction, and identifying knowledge gaps. Documents published before the period of 1990–2023 were disregarded from the review;
- Source selection: Rigorous selection of information sources included academic databases, specialized journals, conferences, books, chapters, and technical reports, all written in English. The search engines used were Google Scholar and Google Books. These platforms provide a wide range of scholarly literature across disciplines, offering free accessibility and user-friendly interfaces for efficient navigation. Additionally, the inclusion of citation metrics assists in evaluating the impact of works, facilitating the identification of influential papers and authors. The literature search utilized specific keywords such as "timber", "wooden", "modular", "buildings", "construction", "structures", "systems", "prefabrication", "EWP", "modularity", "design", "guide", "layout", "multi-story", "module", "panel", "frame", "transport", "joining techniques", "connections", "inter-module", "MEP", "infrastructures", "installation", "stability", "performance", "disassembly", and "innovations". The keywords were consolidated using Boolean operators such as "and", indicating that all search terms must be present in the retrieved records, "or", indicating that any of the search terms can be present, and "not", instructing the database to disregard concepts that might be implicit in the search terms;
- Screening and classification: A screening process based on relevance and quality criteria was applied to the identified documents. Selected works were categorized thematically, facilitating an effective analysis of the diverse approaches and perspectives in the existing literature;
- Analysis and synthesis: A critical analysis of the selected articles identified trends, challenges, innovations, and gaps in current modular timber construction research. The information synthesis aimed to construct a cohesive narrative highlighting the present state of knowledge in the field, incorporating various perspectives and relevant findings;
- Article structuring: The article structure followed a logical sequence reflecting historical evolution, recent developments, and future projections for modular timber construction. Each section was carefully organized to provide a comprehensive and comprehensible overview of the state of the art, incorporating diverse insights and relevant results.

Figure 1 illustrates a flowchart identifying the main tasks performed in the development of this state-of-the-art review.

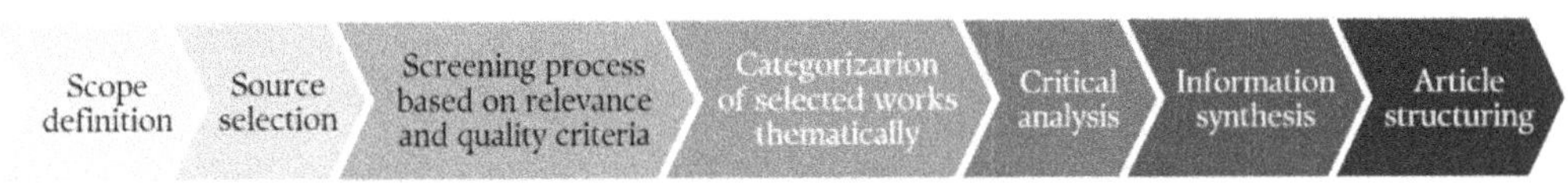

Figure 1. Flowchart describing the main steps of the research.

This methodology facilitated the compilation of a state-of-the-art article offering a comprehensive and updated perspective on modular timber construction. It presented key contributions from the existing literature and indicated directions for future research.

The examination of the existing literature resulted in 124 full-text records. Figure 2a displays the distribution patterns of the chosen records after undergoing filtering and the removal of duplicates. Figure 2b showcases the temporal distribution of the selected records.

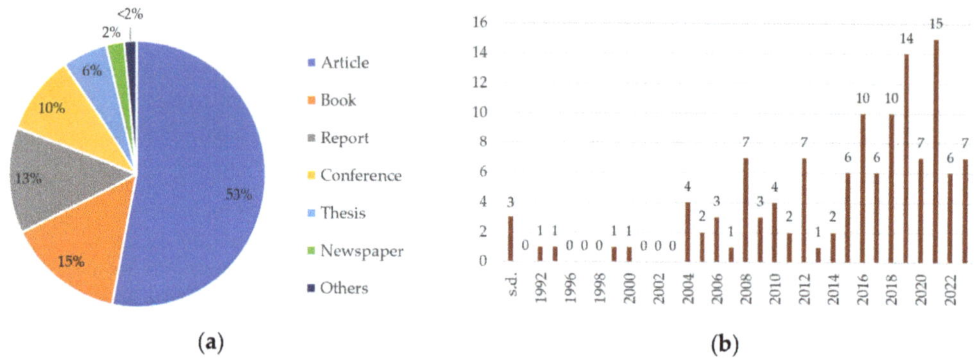

Figure 2. Distribution patterns of the selected records of the review (**a**) by record typology; (**b**) by year.

3. Wood as a Modern Construction Material for Modular Buildings

The significant role of wood in tall buildings has captured the attention of researchers, construction professionals, and governments worldwide in the past decade. The material's sustainable profile, coupled with its structural properties, such as favorable bending behavior and weight-resistance ratio, has boosted the growing interest in using wood for multi-story structures in the current context.

Addressing the question "Why should one believe it is possible to construct modern tall buildings from wood?", Smith and Frangi [21] conducted a study focused on the specifics of tall timber buildings to identify the key requirements and structural engineering challenges. Analyzing buildings between 10 and 20 stories, they concluded that wood and wood-based products meet the required structural strength functions similarly to concrete and steel. Moreover, wood has the potential to surpass the conservatism of building codes regarding fire safety, ensuring an adequate amount of time for evacuation and firefighting. They emphasized the need for prefabricated systems to ensure that the proper construction details for enhanced mechanical responses and expected functionality (occupant comfort and structural maintenance) can be met.

In recent years, the typology of multi-story timber buildings has gained momentum in European countries. The completion of pioneering projects, such as the Limnologen residential tower in Växjö (2008) and the Stadthaus in London (2009), the world's first nine-story all-timber residential building, indicates a growing confidence in this typology.

Historical large-scale fires played a crucial role in hindering the use of structural wood elements in tall buildings, evident in numerous regulatory barriers present in building codes worldwide. In the European context, despite regulatory requirements not distinguishing between materials, limitations on combustible materials make timber less competitive [22], confining most timber buildings in European countries to a restricted number of stories.

In addition to the standardization of national regulations, government incentives have also made a significant contribution to the consolidation of tall timber construction, especially in leading European countries. Whether to encourage the reduction of greenhouse gas emissions or to promote the timber industry, national economies, employment, or exports from countries with forest reserves, governments have encouraged the use of timber in buildings through environmental policies that promote wood as a building material over steel and concrete. In France, for example, a law was passed requiring the use of timber for new buildings, except in cases where evidence of incompatibility, usually related to safety, health, or building performance, is provided [23]. In British Columbia, Canada, the Wood First Act [24] aims to foster a wood culture by requiring it to be the first material considered in the early stages of all new provincially funded building projects. In Japan, the Wood First law highlights a preference for wood in the development of public buildings [25]. Programs like Nationella träbyggnadsstrategin and Trästad in Sweden aimed to promote tall wood building construction and were responsible for the rapid increase in interest in the typology in the country [26]. In certain locations, strategic targets were set for vertical wood building construction to reach 50% by 2020 [27]. In Finland, the Ministry of Employment and the Economy led the National Wood Construction Programme from 2011 to 2015, which not only identified projects for large-scale wood construction but also invested in knowledge about high-quality and energy-efficient wood construction, the implementation of new competitive operation models for the construction stage of wood buildings, and the improvement of wood construction training, enhancing efficiency in design, research, and development activities [28].

Currently, educational and professional training mechanisms related to modern wood applications in construction are not at a level comparable to those for structural steel and concrete materials. However, significant initiatives are underway to enhance educational resources and professional training programs, aiming to bridge this gap and foster a more comprehensive understanding of the potential and intricacies of contemporary timber construction practices.

Therefore, the widespread adoption of multi-story wood construction is a result of collaboration between construction and forestry industries, research institutes, and governments. Positive marketing of wood use produced by incentive programs and collaborative research demonstration projects, along with increased customer interest in modern sustainable buildings, has also played a crucial role in this scenario [29,30].

This successful approach is evident in examples such as the e3 building in Berlin (2008), the Stadthaus in London (2009), the Forté apartment building in Melbourne (2012), Treet in Bergen (2015), Brock Commons Tallwood House in Vancouver (2017), and the HoHo hybrid residential-commercial development in Vienna (2020).

It is evident that countries where the use of wood is traditionally widespread, such as Sweden, Canada, the United States, and Japan, adopt new technologies for tall building construction more rapidly due to their experience, confidence, and the availability of the material, along with the established organization of this market. In Sweden, a country with 69% of its territory covered by forests and approximately 90% of single-family homes built using wood as early as 1980, and where timber constructions above two stories were not allowed until 1994, the market share of wood in multi-story buildings increased from 1% in 2000 to 15% in 2012 [26]. In Finland, this increase went from 1% in 2010 to 10% by 2015 [27,31]. In Canada, after the approval for the construction of five to six-story wood buildings in British Columbia, more than 250 buildings were completed in the province [32].

Locations such as the United Kingdom, Germany, and the Netherlands, while needing to address concerns about fire safety and material durability, have shown growth in the widespread use of wood, even for low-rise buildings. In the United Kingdom, the TF2000 project opened new markets for timber-frame constructions up to seven stories high [33] and increased the market share of timber frame from 8% in 1998 to 25% in 2008 [34,35]. In Germany, in 2014, 16.2% of new buildings were built from wood [36]. The use of wood for multi-story buildings was permitted in the country in 2002, and since then, has represented approximately 2% of the multi-story building market [26].

In recent years, there has been a noticeable increase in studies focused on designing multi-story buildings. Most of the existing scientific literature on multi-story timber buildings addresses technical, acoustic [37–39], structural [40–44], energy-related [45], and sustainability aspects [46,47]. However, very few comprehensive and comparative studies have been conducted regarding the practical implementation of wood. Ilgin et al. [48] agree and state that there is a very limited number of studies focused on global trends and typologies regarding the architectural and structural design considerations of modular timber buildings.

4. Engineered Wood Products and Prefabrication

Erecting timber buildings is currently feasible and desirable due to the advantages that wood construction offers, especially when combined with a prefabrication processes.

As a building material, wood presents some deficiencies, such as a high degree of variability and anisotropy, both concerning strength and movements caused by moisture variations, dimensional instability with changes in humidity, and a limited availability of widths and lengths [49]. To overcome these deficiencies and fully leverage the capabilities of new manufacturing methods for a more effective use of the unique characteristics of wood, various products derived from the material were developed during the 20th century, either in the form of panels or beams, known as Engineered Wood Products (EWP).

According to Forestry Innovation Investment [50], EWP refers to value-added wood products made by bonding wood veneers, laminations, strands, or fibers, usually with an adhesive. This manufacturing process generates high-performance and dimensionally stable products for construction projects of different scales. It has benefited from developing and improving adhesive technologies, mechanical connections, and classification and manufacturing technologies [51].

Indeed, the development of EWP has expanded the application of wood in the construction sector. Produced through controlled industrial processes, these products exhibit a high level of quality control and facilitate and endorse the prefabrication process of timber buildings.

Prefabrication involves producing construction units in a controlled factory environment, followed by transporting and assembling these units on-site for the complete construction of the building [52]. The low weight of timber facilitates these steps compared to other structural materials [18]. Within this modern production philosophy, a holistic strategy is necessary, including 3D modelling of the building, well-defined technical systems for resolving complexities in the factory, the use of information communication technology, high levels of planning and process control, and stronger relationships between stakeholders [53]. Prefabrication also impacts the design process, which, in addition to discussions about material performance and structural optimization, must consider solutions regarding construction process efficiency and on-site safety management [18].

Concerning economic sustainability, the construction sector has shown widespread inefficiency and unpredictability in terms of both products and processes due to traditionally employed construction methods [54]. This results from the high degree of complexity in activities and phases, and the competencies of the actors involved, leading to frequent cost increases and delays. On average, the initially estimated time for a project is extended by 20%, resulting in cost overruns of up to 80% [55]. Regarding social sustainability, the sector has the highest rates of workplace accidents and the highest associated costs [56].

In this context, prefabrication has proven to have various benefits for the construction process. The technique, taking place in a factory environment and allowing strict quality controls over prefabricated components, promotes process predictability and enhances the quality and precision of products. A well-managed factory process and well-coordinated efforts on-site also result in improvements in schedules and productivity [57,58] and, consequently, in the reduction of overall project timelines and costs [59,60]. Lehmann [61] estimated a timeframe of three to four months for wood buildings up to nine stories high. According to Richard [62], there is evidence that construction costs can be reduced by 85% when the prefabrication plant is operating at full capacity. This is an economic benefit for builders, who bear lower on-site costs, and for investors, who benefit from the quick marketing of the property and timely delivery [63]. The reduction in on-site activities also contributes to minimizing interference from external conditions and the exposure of components to the elements, preserving their integrity and durability; on-site labor; construction traffic; noise impact and disturbances to the local community; accidents, promoting a controlled and safe construction site; the need for traditional construction equipment; wet processes; waste, as local factors like overordering, poor execution, and improper material storage are eliminated; and joints, gaps, and penetrations, supporting the creation of a comfortable indoor environment; among others [17,64–68].

The prefabrication construction method inherits concepts that integrate environmental aspects into the product development process which can potentially reduce the negative environmental impacts caused throughout its life cycle phases. Through modular design and dry production (Lean Construction), a wide variety of sustainability goals can be achieved [69]. By promoting the upgrade, adaptation, and modification of components to give them an extended lifespan and influencing component disassembly, prefabrication reduces the industry's environmental burden. Okodi-Iyah [70] proposed that prefabricated construction could save about 44% of the energy used in the construction phase and improve performance by more than 7% during the use phase.

The emergence of fully prefabricated systems has led to production lines consisting of numerous sequential processes, where each station is responsible for a specific function that will contribute to the overall quality and productivity of the assembly line. Traditionally, these processes are completed by conventional means; however, in recent years, an increasing number of these processes are being automated to process materials efficiently and accurately [71].

Automated building processes facilitating efficient in situ resource utilization hold the potential to streamline construction operations in remote areas while offering a carbon-reducing alternative to conventional building practices [72] by minimizing transport distances, costs, and associated pollutant emissions. Importantly, it ensures that the ecological value associated with timber usage remains intact, as it mitigates concerns arising from long-distance transportation. The proximity between the factory and the assembly site enhances operational efficiency and streamlines quality control processes by enabling direct supervision and communication throughout the production cycle. Moreover, this strategy fosters prompt responses to project changes, offering greater flexibility and agility in tailoring components to meet specific project requirements. Unforeseen alterations to the project can be accommodated without significant delays in component delivery, while reduced waiting times for prefabricated components translate to increased on-site productivity. Furthermore, this solution stimulates local economies through job creation or the support of local sawmills and can benefit from the use of native materials. This approach can be partially implemented by focusing solely on essential services completed at the on-site facility. For instance, assembling prefabricated panels into modules on-site helps circumvent logistical challenges and economic concerns associated with transporting fully assembled 3D modules.

There is significant market potential for using wood in various types of buildings using the combination of digital design and CNC (Computer Numeric Control) processing. Digital design and production using CAE (Computer-Aided Engineering), CAD (Computer-Aided Design), and CAM (Computer-Aided Manufacturing) have allowed wood construction to advance to new proportions. During the modeling phase, meticulous attention is given to defining the provisions and dimensions of each element and its components, ensuring compliance with structural requirements while respecting dimensional limitations identified in the design, production, transportation, and on-site application. Employing material optimization strategies is imperative to minimize cuts, thereby reducing production times and waste. This preparatory stage is pivotal for subsequent phases, requiring thorough planning to effectively manage technical and process performance. Following this, the CAD/CAM platform automates the generation of detailed instructions for CNC equipment, facilitating the precise cutting and drilling of elements. Further phases, including lifting, handling, positioning, assembling, and connecting the diverse components, are seamlessly executed by automated systems. This holistic integration optimizes construction processes, bolstering efficiency, consistency, accuracy, and safety, consequently streamlining workflows, and enhancing project outcomes. Innovative connections, modern wood-based materials, and advanced CNC techniques provide new possibilities and sculpt wood into striking forms [63].

In this context, the Building Information Model (BIM) methodology plays a prominent role, as it not only provides a digital model with parametric information about the building but also promotes the integration and coordination of different materials contained in the project and provides an efficient communication platform for sharing information among various stakeholders [73]. Moreover, Sacks et al. [74] studied the correlations between BIM functionalities and lean construction principles and found strong positive interactions. At present, BIM-enabled tools have widely matured for steel and concrete structures, but they are still not satisfactory for modeling wood structures [75].

Prefabrication represents a unique opportunity for wood technologies to lead the construction sector away from high-consumption, high-waste, and labor-intensive approaches towards a more industrialized and sustainable approach [18,76]. However, despite constant advances in terms of materials, prefabrication technologies, and multi-story timber building assembly methods, the design, as well as the installation procedures for construction services in multi-story timber buildings, lag significantly behind technological advancements observed for other construction materials [77].

5. Principles of Modularity in Timber Buildings

The use of modularization as a technique in construction and design aims to achieve design objectives with minimal variation in the building components. This approach is driven by the desire to reduce the number of distinct building elements in a project, promoting standardization and efficiency.

While in freeform construction buildings are designed with little or no adherence to regular geometrical shapes or standardized components, allowing for highly customized and innovative architectural designs but demanding more time and resources, by adopting the recent adaptative modular construction technique, designers and builders seek to balance standardization and adaptability, aiming for an efficient and versatile system. This approach aligns with broader trends in modular construction and prefabrication, emphasizing the benefits of standardized, repeatable elements in the built environment, such as quality, speed of construction, and cost-effectiveness.

The discrete design agenda has played a pivotal role in transitioning from strict modularity to adaptive modular construction. In this approach, fully functional and intricate buildings are assembled from serially repeating, rearrangeable sets of generic discrete elements, each treated as an individual entity with clear boundaries [78]. This approach diverges from the serialized production of identical units or generic solutions; instead, it relies on the combination and variation of purposefully designed parts to achieve customization and adaptability through scalable principles [79]. In this particular context, the kit-of-parts methodology emerges as an alternative strategy within sustainable design. This methodology revolves around the concept of a pre-designed collection of discrete building components that can be assembled in various ways into a finished building [80]. The kit's components are systematically generated using a combinatorial approach, affording a broad spectrum of design possibilities [81].

This design and construction methodology entails deconstructing a building or project into modular components, each conceptualized as integral elements of a standardized kit. These components are pre-designed, prefabricated, and mass-produced, facilitating seamless assembly, offering flexibility, and fostering efficiency in the construction process.

However, a notable challenge in expanding the application of the kit-of-parts approach lies in the absence of universally applicable toolkits and libraries of configuration tools. Consequently, manufacturers face the imperative to escalate their efforts to craft solutions that seamlessly integrate into the supply chain from design to production [82]. In the kit-of-parts approach, the overarching objective is to establish a system wherein these components can be amalgamated and reconfigured to cater to diverse needs, thereby offering a solution characterized by versatility and adaptability [81].

This approach has been employed in various architectural and construction projects, ranging from residential buildings to commercial structures. It aligns with modular construction and prefabrication principles, emphasizing standardization, efficiency, flexibility, and sustainability in the built environment.

6. Taxonomy of Modular Timber Products

Modular timber products' taxonomy involves categorizing them based on their characteristics, functionalities, and applications.

The fundamental load-bearing element in modular constructions is the modular unit. Different structural configurations of modules designed for modular high-rise buildings have been developed and can be broadly classified [83]. However, the literature does not consistently agree upon the structural categorization of mass timber buildings. As Salvadori [84] elaborates, Mass Timber Buildings (MTBs) can take shape through one-dimensional (1D) or two-dimensional (2D) structural elements (vertical and horizontal) or by employing three-dimensional (3D) modules, where the walls and floors are pre-assembled. This author established 32 categories that intricately combine and cross-reference the primary types of structures and materials used in various systems. In contrast, Svatoš-Ražnjevi'c et al. [85] introduced a classification system based on the predominant usage of 1D, 2D, or 3D timber elements in construction and their respective combinations. The established categories include the following (Figure 3):

- 1D frame structure;
- 2D bearing wall;
- 3D volumetric modules;
- Combination or hybrid.

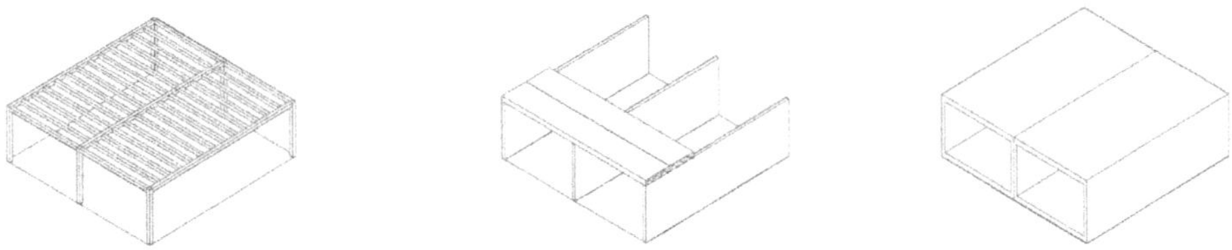

(a) (b) (c)

Figure 3. Modularity in (**a**) 1D elements; (**b**) 2D panels; and (**c**) 3D volumetric modules. Adapted from Kaufmann et al. [86].

According to Svatoš-Ražnjevi'c et al. [85] and Kuzmanovska et al. [87], panel and space module systems were the most common systems from 2000 to 2010, while from 2009 to 2011, there was a significant increase in frame structures.

The research findings of Svatoš-Ražnjevi'c et al. [85] establish that panel and volumetric module systems are predominantly employed in mid-rise projects, typically reaching up to 10 stories in height. Although taller structures exist within panel and volumetric systems, combination systems emerge as the preferred choice for projects exceeding 15 stories, including notable specimens ranging from 29 to 80 stories in height [85].

This trend aligns with the observation that a significant majority of the proposed but unbuilt projects considered by Svatoš-Ražnjevi'c et al. [85] opt for frame structures. These unbuilt projects encompass some of the tallest Modular and Sustainable Tall Buildings (MSTBs). However, it is crucial to acknowledge that this prevalence of frame structures in unbuilt projects may be influenced by the data sources and the limited visibility of modular projects in mainstream literature and timber databases.

6.1. One-Dimensional/1D Frame Structure/Linear Systems

One-dimensional (1D) frame structures, also known as linear systems, play a crucial role in timber buildings, offering a versatile and effective approach to structural design.

They are relatively simple in their assembly, making them practical for various building designs. Moreover, these structures offer versatility in terms of configuration, allowing architects and engineers to adapt them to different architectural styles and functional requirements.

One-dimensional elements form beam–column structures, slab–column structures, and exoskeleton structures, where the vertical supports are often limited to the exterior and fixed to a core, contributing to structural stability and architectural aesthetics. According to Svatoš-Ražnjevi'c et al. [85], one-dimensional element structures are primarily used for five to eight-story projects. Nearly one-third of the projects comprise buildings with more than nine floors (7% with more than twenty floors).

Although structures with one-dimensional elements may consist of just an articulated assembly of linear timber pieces, additional bracing systems, such as shear walls, diagonal EWPs, transverse metal bracing, and metal beams, can be added to the structures to ensure their lateral stability.

For these structures, various combinations of EWPs can be employed, such CLT slabs, ribbed slabs, or composite floors (CLT–concrete or glulam–concrete), to create floor slabs [85].

Using traditional column and beam structural forms in 1D frame structures may be considered effective and aligned with the structural design criteria in mass timber projects. This approach optimizes the use of timber elements and meets the sustainability goals associated with modern construction practices. On the other hand, in modern timber modules, adopting the conventional column and beam structural form may be considered excessive and is seldom utilized [81].

6.2. Two-Dimensional/2D Bearing Wall/Panelized Systems

A 2D panelized solution resembles the flat-pack assembly approach commonly employed in home furniture [88]. These systems involve the use of prefabricated panels that serve as both vertical load-bearing walls and horizontal floors or roofs, manufactured offsite and then transported to the construction site for assembly. This approach offers efficiency, speed, and versatility in construction.

These panels can be categorized into various types based on their specific functions within the building structure. They typically consist of a framework with integrated sheathing and insulation. They are also often made from EWPs, in particular CLT or other prefabricated timber panels. Where necessary, panels include conduits for essential services such as heating, ventilation, air conditioning (HVAC), and plumbing. These conduits can be seamlessly linked using standard connectors [88]. Two-dimensional panel systems primarily consist of medium-height projects up to ten floors. Medium-height projects ranging from five to eight floors are the majority, followed by low-rise building projects of three to four floors [85].

A distinctive feature of the modular product lies in its flexible design. For example, components can be combined to accommodate windows and doors in various locations and sizes [81]. Two-dimensional panelized solutions provide greater flexibility than three-dimensional modules [88]. For instance, large open-plan offices may be unsuitable for single 3D modular elements. The relevance of 2D panels extends to high-end residential projects, including single-family homes and apartments. In such projects, where differentiation is crucial and the ratio of wet to dry areas is lower, 2D panels offer a more adaptable and suitable solution [88].

The onsite assembly process for flat-pack panels is simpler than traditional construction but more intricate than assembling 3D modules. It does require additional onsite assembly and internal finishing. However, transporting the panels is much more convenient than larger 3D modules. Using flat-pack panels allows materials covering a significantly larger floor area to be transported at once. As stated in [88], it costs approximately 8 USD per square meter of floor space to transport 2D panels around 250 km, whereas the cost is nearly 45 USD per square meter for the 3D equivalent.

In contemporary timber construction, 2D bearing wall systems have gained popularity due to their efficiency and sustainability. The prefabrication of panels allows for precise engineering and quality control, resulting in structures that meet modern performance and energy efficiency standards. Additionally, these systems contribute to the growing trend of modular and offsite construction practices in the timber building industry.

6.3. Three-Dimensional/3D Volumetric Module/Modular Systems

Modules are autonomous three-dimensional units or partially assembled sections. Their versatility allows for repetition through stacking or joining them side by side, facilitating the extension of spaces [89]. This approach is most suitable for projects characterized by a high level of repeatability and a significant ratio of wet to dry rooms. It is important to note that repeatability in this context does not imply uniformity in appearance. Instead, a range of standardized modules can be assembled in various configurations to achieve a customized and diverse result [88]. According to Svatoš-Ražnjevi'c et al. [85], only 6% to 10% of the volumetric panel and module projects exceed nine floors, with a maximum height of 19 and 15 floors, respectively.

In contrast to panelized or linear types of prefabrication, modular construction involves the integration of most interior and exterior finishes within the factory setting [83,90]. In fact, the module represents the most complete form of prefabrication, with approximately 95% of the essential kitchen and bathroom facilities, storage amenities, and living spaces being fully fitted in the factory and subsequently sealed for transportation. This approach is particularly advantageous for rooms requiring intricate finishing, especially wet rooms such as bathrooms and kitchens [88]. The primary advantage of modular building is that the structure becomes immediately ready for use once power and water facilities are connected [89].

In addition to maximizing factory production, modular construction enjoys the advantages of being unaffected by weather conditions, achieving precision through advanced tooling, maintaining high-quality control, streamlining assembly lines with semi-skilled labor or automation/robotics, flexible grouping geometries, and benefiting from bulk purchasing of components [91].

Nevertheless, it requires significant (though not prohibitive) transportation costs, mainly because costs are primarily associated with volume and secondarily with weight. This implies that transporting empty modules (such as living areas, bedrooms, etc.) incurs costs comparable to fully equipped modules with value-added content (kitchen, bathroom, services, etc.) and is subject to vertical transportation restrictions [91]. This either raises the transportation costs for larger units or constrains the dimensions of modules, making 3D volumetric solutions particularly well-suited for applications such as hotels, hostels, or affordable housing [88].

Modules are designed for ease of assembly, yet compared to other prefabrication types, this approach is deemed the most cumbersome and technically challenging [89]. Onsite assembly in modular construction entails lifting the modules into place using cranes, bolting them into position and connecting services such as electricity [88]. During the lifting process, the modules are subjected to structural forces, particularly vertical bending and shear, which differ from the forces they experience after installation. The necessity to reinforce the modules for lifting and the doubling of floor, roof, and wall frames at the connection points leads to increased quantities of structural material [89].

In a case study, construction using the volumetric modular technique was completed 16% faster than that using the 2D approach. Despite the apparent transportation difficulties in modular construction compared to the 2D method, the volumetric construction method saves more time without sacrificing quality [92]. When evaluating the difference between the three options (1D, 2D, and 3D) for an affordable housing unit with four floors, Bertram [88] found that a 2D solution could be 17% cheaper than a traditional approach, while a 2D and 3D hybrid solution lowers costs by 20%, and a 3D solution by 24%. This would vary by project, but these estimates indicate the scale of potential savings.

Three-dimensional (3D) volumetric module systems in timber buildings represent an advanced approach to construction, leveraging prefabricated modules encompassing complete building volumes. These systems provide a holistic and efficient solution for assembling entire structures with integrated components.

6.4. Hybrid Systems

Hybrid systems in timber buildings refer to the integration of different structural approaches within a single construction project, combining elements from one-dimensional (1D), two-dimensional (2D), and three-dimensional (3D) systems. This hybridization allows for a tailored and optimized solution, leveraging the strengths of each system to address specific structural and design requirements.

In hybrid structural systems, categories are combined in multiple ways. Projects, therefore, fall into one of the following categories [85]:

- The upper and lower sections of the building modules are constructed with different structural characteristics;
- Different zones of the building floor plan are constructed using distinct systems.

In modular construction, it is feasible to employ a combination of 3D modules and 2D panels within a project or even integrate these approaches with traditional onsite construction, especially for components like basements and the first floor of larger projects [83,88,90]. An example is the construction of areas resembling corridors, often seen in hotels, that can be achieved by repetitively utilizing 3D units. In contrast, the corridors can be produced as planar elements or as extensions to room modules [93].

Typically, wet areas, such as bathrooms, are manufactured as bathroom pods, while the rest of the building is constructed using 2D panels. This strategic approach optimizes the construction process for different building areas, enhancing productivity for bathroom spaces and providing maximum flexibility for other areas. However, it is important to note that the manufacturing process required to deliver both solutions becomes more intricate, as does the coordination of the overall supply chain [88].

According to Svatoš-Ražnjevi'c et al. [85], hybrid structural systems are primarily composed of combinations of 1D elements and 2D elements, followed by the combination of 3D modules and exoskeletons, timber frames and solid wood slabs, 2D panels and 3D modules, and finally, projects combined with additional external structures such as balconies. Based on Žegarac Leskovar and Premrov [94], aiming to increase the height of buildings, hybrid systems of beams and columns associated with solid panels (CLT slabs or walls) are commonly used.

In contemporary timber building projects, adopting hybrid systems reflects a holistic and pragmatic approach to construction. This integration of 1D, 2D, and 3D elements allows for innovative and efficient solutions that meet the demands of modern design, structural integrity, and sustainability goals.

7. Manufacturing, Transportation, and Installation Dimensions

Dimensional constraints play a crucial role in the manufacturing, transportation, and installation of prefabricated wood panels and modules. In the manufacturing process, limited dimensions may be determined by the factory's capacity and the equipment used. Transportation constraints are related to the dimensions allowed on public roads and transport regulations. Additionally, the choice of transportation modes, such as trucks or containers, may be influenced by the dimensions of prefabricated components. During installation, dimension constraints impact the logistics of the construction site, including access to the site, available space, and the ability to handle bulky components. Therefore, careful consideration of these constraints is essential to optimize the prefabricated wood construction process.

The dimensions of 2D panels are primarily constrained by production and transportation considerations, given that their geometry and lightweight nature do not pose significant challenges during the lifting and installation phases. Typically, the cutting machinery utilized allows for dimensions of 3.50 × 9.00 m^2. When considering a standard truck, the panels are confined to the following specifications:

- Length: 13.50 m;
- Width: 2.45–2.48 m;
- Height: 2.50–3.00 m;
- Load weight: 24.00–25.50 tons.

Transportation is a critical and design-defining phase in the construction of room modules. Special forms of transport, subject to approval, are typically utilized for this purpose. However, the maximum size of room modules is predominantly determined not only by the production facility and on-site conditions but also by the transport capacities offered by the truck.

Standard semi-trailer trucks, with a loading length of 13.50 m, are suitable for most projects, but longer vehicles can handle considerably larger lengths with manageable additional costs, reaching a critical point at around 15 m. For smaller dimensions, up to approximately 6.50 m in length, transporting two modules in one vehicle is feasible. Most roads recommend a 4.30 m maximum total transport height. With standard loading platform heights ranging from 0.90 to 1.10 m, a minimum height of 3.20 m is retained for the modules. Special low-bed semi-trailer trucks enable the transportation of room modules with a maximum height of 4.20 m, subject to specific limitations regarding the length of the cubicles. Maintaining a total width of 3.25 m mitigates transportation issues, facilitating navigation through unforeseen construction sites. For transport widths exceeding 4.00 m, commissioning a transport study to analyze the route is advisable. The average costs for a transport route spanning several hundred kilometers are estimated to be around 5% of the pure construction costs, with the complexity of transport (police escort, etc.) contributing to price fluctuations of up to 30% [95].

Timber room modules, typically weighing 350–400 kg/m^2 in standard sizes, may approach the capacity limit of common semi-trailer trucks when exceeding 20 tons. Room modules are generally lifted using mobile cranes. If the modules are sufficiently rigid, they can be directly mounted on the crane hook using steel cables at four points. For less rigid modules, such as those made of timber-frame elements, an additional lifting structure, often in the form of a robust steel frame, may be necessary to provide additional attachment points. This particularly applies to large modules or those with a lower rigidity than that of cross-laminated timber [95]. When choosing transportation via maritime containers, it is crucial to consider specific dimensions: a length ranging from 5.90 m to 12.03 m, a width of 2.35 m, and a height of 2.39 m. Additionally, it is important to consider door opening dimensions when applicable.

Li et al. [81] assessed design details for the dimensions, weight, and floor area of volumetric timber items by analyzing modular products made by eight suppliers and the modular products used in 60 modular timber projects.

The length of a typical modular unit is often from 6 to 12 m [96]. The significant variation in lengths is a result of the adaptable design. Shorter lengths are derived from functional modules, such as bathrooms, separate from residential units. Modules less than 5 m long are functional or flexibly designed modules. Modules longer than 10 m are generally made for apartments. The most variable building types in module length are apartments and social housing structures, followed by hotels. This can be attributed mainly to the combination of modules tailored to fulfill the diverse demands for different types of living spaces. Shorter lengths are typically fused to generate more expansive areas. Shorter lengths are typically merged to create larger areas. For student housing, nursing homes, and hospitals, the modules do not typically use a combination design method. Nevertheless, the dimensions of the module may still vary depending on the application;

for instance, certain modules may need to be configured without obstacles and with slightly increased lengths to facilitate wheelchair access [81].

Most timber volumetric modules have a width ranging from 2.50 to 4.00 m. The module's intended function has little significant influence on its width. However, the modules utilized in apartments and social housing encompass an extensive range. Hotel modules exhibited the highest degree of variability in terms of width. Transportation conditions primarily contribute to the concentration of width values between 2 and 5 m. Typically, modular rooms exceeding a width of 5 m are constructed by integrating multiple 3D modules, commonly found in offices and educational facilities requiring more space [81].

Ceiling height is also a significant architectural consideration in timber buildings. More than half of the modules that were investigated by Li et al. [81] had heights in the interval [3.00, 3.50) m. The MEP (mechanical, electrical, and plumbing) system area is positioned either above the ceiling or below the floor of the room module, depending on the construction technique. Ali and Armstrong [97] proposed a height range of 2.40 to 2.70 m from floor to ceiling (including finishes) for residential purposes, irrespective of the country or region. The Tall Buildings Statement of London [98] recommends a ceiling height of 3.20 m, reflecting the need to accommodate facilities and services. According to Ilgin et al. [48], measurements considering the distance between the floor of one level and the floor of the next range from 2.81 to 3.30 m, with an average of 3.00 m. Only two out of the thirteen buildings they analyzed exceeded the 3.20 m suggested by the Tall Buildings Statement of London [98].

The largest floor area of 50 m^2 is used for hotels and residential buildings. The standard floor space for volumetric modules ranges from 15 to 25 m^2 [81].

Timber modules can be divided into functional and room modules based on the difference in mass. Functional modules encompass various facilities, such as lavatories (including saunas) and equipment rooms. The mass of timber room modules falls into the range of 5 to 15 t, and the mass of an equipment room is less than 25 t [81].

The dimensions of a standard modular unit are constrained by transportation and lifting requirements. In addition, the maximum lifting capacity of most tower cranes used in building projects is around 20 tons, and the cost of a tower crane with a lifting limit of over 20 tons might increase by up to 60%. Hence, transportation and lifting requirements serve as the primary limitations in establishing both the module size and hoist weight. In the case of high-rise buildings, the modules at the base experience substantial vertical loads. This leads to an increase in their member sizes and hoisting weight, thereby reducing the modules' life span. However, using short modules will result in an increased number of inter-module joints, so the benefits of modular construction are not fully maximized [83].

8. Layout Design

The arrangement of a building is influenced by its purpose, structural design, seismic considerations, topographical constraints, and the specifications of various design alternatives [81].

Although most of the literature focuses on technical and structural aspects, the few architectural design aspects examined have been limited to the geometric description of buildings, lacking a comprehensive exploration of the variety of forms and spatial organizations in timber construction. According to Žegarac Leskovar and Premrov [94], early modern timber buildings had a relatively regular geometry with symmetrical and repetitive floor plans. However, contemporary buildings exhibit more geometric variations, including vertical irregularities due to variations in floor elements, distortion of the basic structural grid, and asymmetric configurations of facade openings such as windows and balconies [87]. These architectural expressions pose technical challenges and require customized solutions.

Contrary to current trends in concrete and steel buildings, timber structures show a prevalence of rectilinear plans and regular extrusions with a low degree of innovation, relying on the simple use of timber rather than its expression as the primary material—a marketing strategy. Exceptions are rare, and more complex projects often involve hybrid structures or small footprints adapted to specific geometric constraints in medium- to high-density urban locations.

Svatoš-Ražnjevi'c et al. [85] reported a continuous increase in building height. The first five-story building was introduced in 2004, followed by the first six-story building in 2006, seven and eight-story buildings in 2008, a nine-story building in 2009, a ten-story building in 2012, and so forth.

The analysis of floor plans and building silhouettes reveals that most projects are rectangular, with the only recurring alternative being the "L" shape. Projects using two-dimensional panels include examples that deviate the most from orthogonal shapes, followed by projects using one-dimensional elements. Volumetric projects have no examples of non-orthogonal forms. Three-dimensional module projects present the smallest range of typologies, with only rectangular and similar examples. According to Li et al. [81], there are currently 21.77 modules on average per floor.

Furthermore, over 80% of projects are symmetrical, with 3D module projects having the fewest non-symmetrical designs, while combined systems comprise the majority of this typology. As an example of symmetry, Ilgin and Karjalainen [99] highlight a preference for centrally located cores, followed by peripheral cores, mirroring the trends seen in concrete or steel buildings. The core's location, and consequently those of the elevator and stairs, is a crucial architectural parameter for multi-story buildings, following the classification defined by Ilgin et al. [48], as represented in Figure 4.

The benefits of a central core include aspects such as structural support, compactness (allowing openings in the facade), and improved safety performance for fire stairs (which may have contributed to the prevalence of this typology). However, Kuzmanovska et al. [87] conclude that most central circulation schemes are independent of the core's location. On the other hand, the disadvantages of peripheral core configurations include low space efficiency, such as longer corridors and longer evacuation distances in the case of fire. The same applies to external core configurations, for which there are no records of their use in multi-story residential timber buildings.

Furthermore, regular extrusions with a flat roof dominate all structural systems. There is less variation in volume in 3D module projects, which do not have irregular examples. In contrast, 1D and 2D element projects show much more volumetric variations. Even within regular extrusions, variations in the roof are more common.

Through the work of Ilgin and Karjalainen [99], it was concluded that prismatic forms with rectilinear plans and regular extrusions are the most common and occur in eight out of ten case studies. The reasons presented by the authors encompass ease of construction, practicality, and efficient use of interior space (especially in rectangular floor plans). These are also the most used forms in residential buildings made of concrete or steel, indicating that this trend is not dependent on the structural material (Figure 5).

The grid is the dominant structural ordering system, followed by the linear strategy or a combination of both. One-dimensional element projects generally follow grid systems, while two-dimensional panel and three-dimensional module projects are primarily ordered by linear arrays. Combined systems have two dominant strategies: grid and linear array (Figure 6) [85]. According to Hua et al. [100], the symmetry in a grid is essential to the modularity in design and construction. There are multiple choices in terms of the symmetry in an architectural floor plan, but these decisions determine different prefabrication and logistics conditions.

Balconies and facades are other important design elements worth considering in a project. Both elements can be constructed on-site or assembled using modules in an off-site factory. The selection of materials is not restricted to timber and can be accomplished aesthetically to hide the appearance of modular construction [81].

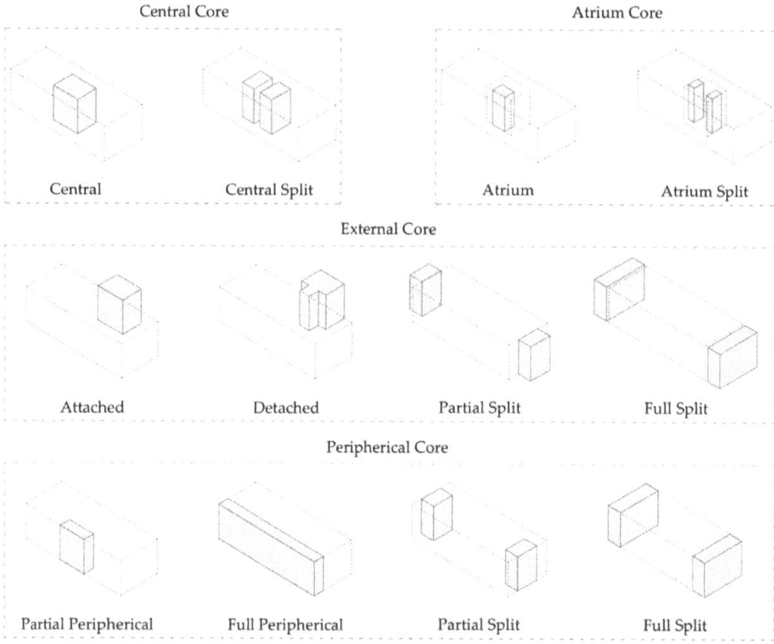

Figure 4. Classification of central arrangements for tall buildings. Adapted from Ilgin et al. [48].

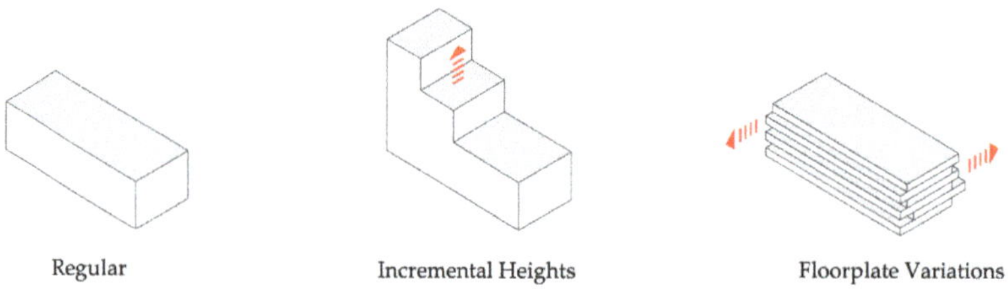

Figure 5. Examples of volumetric extrusions. Adapted from Svatoš-ražnjević et al. [85].

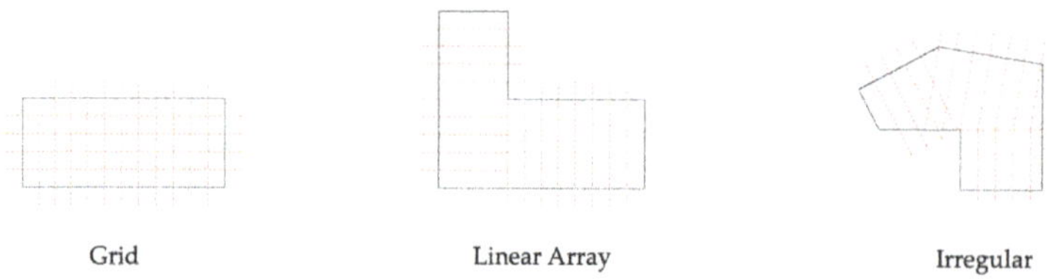

Figure 6. Examples of structural ordering systems. Adapted from Svatoš-ražnjević et al. [85].

The materials used for facade cladding encompass a range of options, such as timber battens, stainless steel, fiber cement board, glass panels, copper, photovoltaic panels (BIPV), and aluminum. Timber facades, where the timber elements lack protection, leading to an increased risk of degradation (use class 3, corresponding to service class 3 of Eurocode 5 [101]), should be constructed using hardwood species with inherent durability or treated appropriately. A greater diversity of facade forms can be provided using coated and corrugated aluminum panels. All the aforementioned facade designs are well-suited for on-site installation. Moreover, the building's facade can be selected from the external surfaces of the wood modules or chosen to impart a non-modular appearance to the modularly constructed buildings. In Li et al. [81], the term "non-modular façade" refers to a situation in which the material completely hides the modular connection details.

9. Building Occupancy and Use

The most common program in timber building construction is residential, comprising nearly half of all projects [85,99]. However, Kuzmanovska et al. [87] noted a shift away from entirely residential programs associated with a growing preference toward mixed-use programs. Currently, projects geared towards the development of modular timber systems are noticeable, catering not only to residential buildings but also extending to hotels, senior residences, student accommodations, hospitals, and schools.

Svatoš-Ražnjevi'c et al. [85] concluded that office buildings constitute the majority of one-dimensional systems. On the other hand, residential buildings (whether for social housing, student housing, or multifamily housing) comprise most of the panel systems. In 3D module construction, the even distribution between commercial and residential spaces results from a high number of hotels and hostels being counted as commercial spaces and a lower number of spaces being designated for offices. Most residential program projects are for affordable housing. Other available options include nursing homes, apartments, and various types of accommodations. School projects also have a significant presence in 3D module buildings. Residential programs constitute the majority of hybrid systems.

10. Structural Systems for Modular Buildings

10.1. Building Level

Structural systems for modular high-rise buildings are pivotal in ensuring stability, safety, and efficiency. Various approaches are employed to address the unique challenges posed by the vertical expansion of modular structures. Here we present the key considerations and structural systems commonly used in high-rise modular buildings.

Svatoš-Ražnjevi'c et al. [85] provided a complete overview of the sub-categories of structural systems that appeared during their project analysis (Table 1).

Ilgin et al. [48] identified six structural systems for tall buildings in their study (Figure 7).

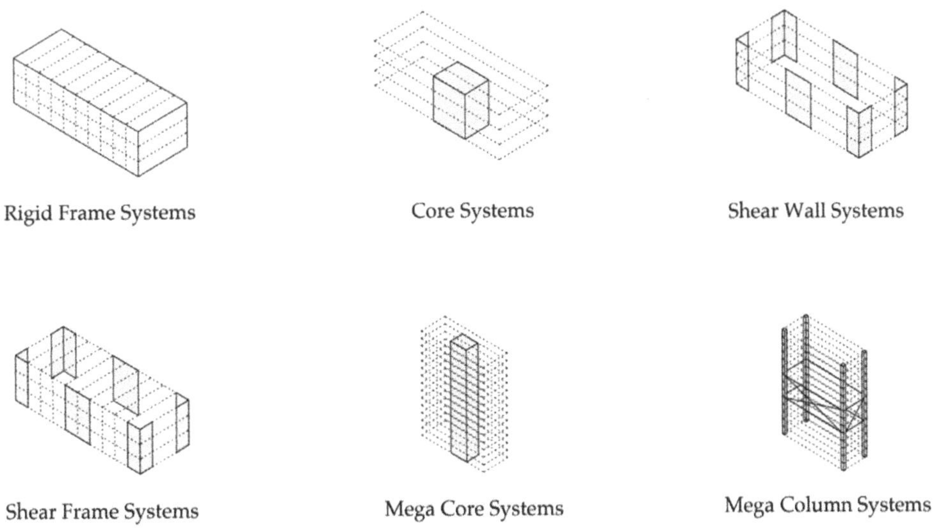

Figure 7. Structural systems for tall buildings. Adapted from Ilgin et al. [48].

Table 1. Sub-categories of structural systems. Adapted from Svatoš-ražnjević et al. [85].

Type of Structural System	Sub-System
One-dimensional Frame structure Linear systems	Exoskeleton Post-and-beam Post-and-beam with linear bracing Post-and-slab Post-and-slab-band
Two-dimensional Bearing wall Panelized systems	Crosswall and party wall Honeycomb Panel + beams Panel + box beams Panel + truss Panel + beams + columns Panel + columns Panel + external frame (balconies)
Three-dimensional Volumetric modules Modular systems	Space modules Space modules + external frame (balconies)
Combination Hybrid systems	Frame + panel Frame + space modules Exoskeleton + space modules Light frame + mass timber Panel + beams + external frame (balconies)

Within the domain of stacked Prefabricated Volumetric Modular Systems (PFVMS) buildings, a vertical augmentation of modular units facilitates the establishment of multi-story structures. This modular system optimizes vertical space utilization without encroaching upon ground space [102]. Particularly prevalent in low-to-mid-rise residential apartments, this solution combines modules in a linear or planar fashion, allowing for the pursuit of space efficiency [93]. This approach is applied similarly regardless of the chosen material. The modules in this stack are interconnected using rigid intermodular connections [103] (Figure 8a). Nevertheless, relying solely on intermodular connections may not effectively contribute to load resistance, necessitating the incorporation of stability technologies to withstand lateral forces in high-rise Modular Steel Buildings (MSBs). These stability systems may include steel braces, steel-plate shear walls, multi-column walls, base isolation systems, concrete cores, and moment frames. The development of these stabilization technologies can follow traditional or modern modular methodologies, utilizing precast, steel, composite, or steel–concrete hybrid materials [104].

Another option is a superstructure of beams and columns as the main structure into which prefabricated modules are inserted, avoiding direct overloading on the underlying elements (Figure 8b). This approach is common practice in modular constructions, where the support infrastructure is established using vertical (columns) and horizontal (beams) elements. Prefabricated modules, manufactured off-site, are then positioned, and secured within this framework. This configuration offers advantages as it allows for a more efficient assembly of modules, simplifies the construction process, and provides flexibility in the design of interior spaces. The beams-and-columns superstructure provides the necessary stability and the ability to support vertical loads, ensuring that modules can be integrated safely and effectively. This methodology is commonly employed in modular construction projects, where coordination between the superstructure and prefabricated modules is essential to ensure the structural and functional integrity of the building. However, this solution also presents challenges, including additional costs due to structural redundancy and complexities in the connections between the megastructure and the modules.

The third structural solution involves anchoring the modular units directly to the central core of the building, contributing to structural cohesion (Figure 8c). This approach enhances stability and ensures a robust connection between the prefabricated modules and the building core, providing a secure and integrated structural system.

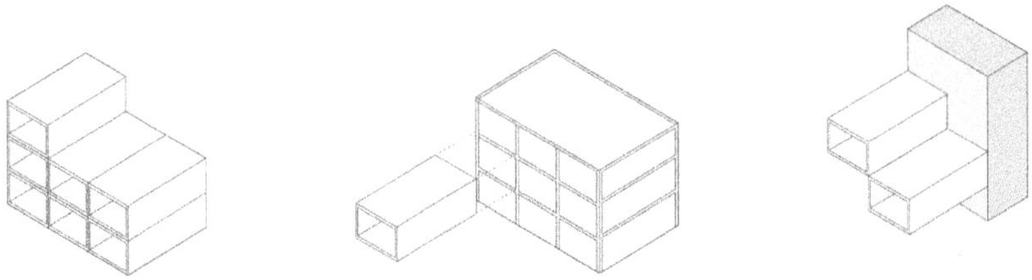

Module stacking

(a)

Module anchoring to an exoskeleton

(b)

Module anchoring to the core

(c)

Figure 8. Structural solutions to support volumetric modules: (**a**) module stacking; (**b**) module anchoring to an exoskeleton; (**c**) module anchoring to the core.

Timber has a strength (parallel to grain) that is similar to that of reinforced concrete; hardwood is slightly stronger, and softwood is slightly weaker. Nevertheless, timber cannot rival the compression strength of modern high-strength concrete. Timber is less stiff than concrete, and both materials are far less stiff or strong than steel. However, timber has a low density compared to these other conventional structural materials. Due to this, when considering strength-to-weight and modulus-to-weight ratios, softwood demonstrates a comparable performance to steel. This implies that timber is notably structurally efficient in constructions, or components of constructions, where a significant portion of the load to be withstood is the self-weight of the structure. Examples are roofs, some bridges, and tall buildings' gravity load-resisting systems [105].

In designing tall timber buildings under specific circumstances, such as in seismically active regions, an additional challenge arises beyond the effects of gravity and wind loading. It is related to the dissipation of seismic energy within the structure, as timber is a non-ductile material. In a full-timber construction, the dissipation of seismic energy occurs at the connecting points between the timber elements, such as nails, screws, bolts, hold-downs, etc. Structural timber members can be combined with other materials, such as steel and concrete, to cope with the disadvantages found within tall buildings' structural and seismic design that have been mentioned [106].

In multi-story timber construction, access cores (such as emergency exits and those used to brace buildings), firewalls, or entire reinforced concrete base stories are now often combined with timber structures, and steel-reinforced concrete frames or compartmentalized structures with a building envelope made of highly insulated timber panel elements are increasingly being used [86].

Combining concrete with timber in multi-story buildings offers several advantages over solid wood structures. This combination allows for longer spans, providing better performance in terms of vibration, deflection, and sagging due to increased structural stiffness. It enhances fire safety and soundproofing and minimizes vertical subsidence. Additionally, the concrete layer protects the underlying timber during construction and potential interior leaks. However, composite timber–concrete structures can be heavier if additional mass is not needed for soundproofing purposes [86].

Steel, known for its structural strength, is often used in situations that require support for long spans and heavy loads. By incorporating timber in this context, design flexibility is expanded, allowing for the creation of innovative, sustainable, and aesthetically pleasing structures. Structural efficiency is then optimized by leveraging the strength of steel against compression and tension forces, resulting in more robust and durable structures. A steel frame is also sometimes combined with a timber secondary structure to transfer heavy loads in multi-story timber buildings [86].

In general, most timber buildings exhibit a hybrid and complex structure, as mentioned by Salvadori [107]. With only 78 examples, fully timber constructions are uncommon in all of the structural categories considered, representing 22.6% of all projects, while the remaining structures are hybrids. Most hybrid structures in timber buildings (54%) combine timber and concrete. Timber–concrete–steel structures account for 13.1%, and timber–steel structures for 5.4%.

The classification also considers the presence of these materials in subsequent structural components [85]:

- A podium or plinth (only the ground floor or a set of floors in the lower part of the building), which can be made of concrete or sometimes steel;
- The core, which can be made of timber (CLT or LVL), concrete, or an articulated set of one-dimensional metallic elements;
- Floor slabs, which can be made from a range of EWP or in a composite solution (timber with a layer of concrete or embedded precast concrete or steel beams);
- Lateral bracing and vertical/horizontal structural elements, which can be made of timber or steel;
- Others, mostly comprising external structural elements such as stairs and circulation areas.

According to Kuzmanovska et al. [87], the use of concrete cores has increased over time (from 38% in 2009–2013 to 57% in 2018–2020) and is more prevalent in structures with more than 16 floors. Conversely, the use of CLT cores has decreased (from 63% to 43% over the same period). Additionally, their study revealed a discrepancy among nine-story structures: an increased use of CLT cores and a decline in concrete cores. However, there was a noticeable growth in the use of steel cores between 2017 and 2018, indicating a transition to hybrid models as systems evolve. The study's results make it challenging to determine whether the increased use of concrete cores results from taller building heights, increased beam–column system construction (which may require a core with more mass than CLT can provide), the trend of hybridization, or a combination of these factors.

Kuzmanovska et al. [87] also noted a growing trend in the use of podiums (25% in 2009–2013 to 60% in 2018–2020). Interestingly, this measure seems not to correlate with a building's height or its program.

According to Ilgin et al. [48], using concrete podium structures offers advantages, such as accommodating ground-level commercial zones, providing open areas and large openings, and creating fire-resistant enclosures for substantial mechanical/electrical services and equipment. The timber structure is also outside the range of water splashing and earth moisture, and level transitions between inside and out can be created while providing structural protection for the timber without complex detailing. This type of structure also enables planners to create emergency staircases leading outside and access routes for emergency services without having to add special separate fire safety measures [86].

On the other hand, the motivation for using concrete in the core may be based on promoting lateral stiffness and strength in the structure, benefiting from concrete's inherent fire resistance, and reducing wind-induced sway, a commonly reported issue in tall buildings. In the case of Brock Commons Tallwood House, it was found that using concrete in the core also simplified the project approval process [48]. Nevertheless, integrating various materials in vertical structural components may also lead to challenges or issues. Concrete staircases, lift shafts, or similar structural components must usually be built at an early stage of construction and drying times and formwork processes can significantly extend construction times. The dimensional accuracies that concrete and timber buildings require, and their subsidence, also differ substantially [86].

Pan et al. [108] present a hybrid timber–concrete building composed of two parts: a concrete core with concrete flat slabs on every third floor as the main structure and prefabricated light timber-frame modules as substructures. Two out of three concrete slabs are replaced by timber modules to create the livable spaces. The concrete slabs offer fire separation at each third level. Bolted connections were designed to connect the main structure and the substructures.

Most timber buildings have reinforced concrete foundations and basements if present. The exception is using timber piles or foundations with bolted metal cap beams.

Concrete cores and podiums with additional steel elements incorporated on the inside or outside the structure are more common. Some projects have entire blocks made of concrete or steel, and balconies and exterior circulation areas are often constructed with these materials.

10.2. Module Level

In the context of structural solutions for prefabricated modules, it is essential to note that there are numerous intersections among them, regardless of the type of material used, whether timber, concrete, or steel.

Three generic forms of modular construction exist: continuously supported, four-sided, or panelized/monolithic modules, where vertical loads are transmitted through the walls; open-sided, corner-supported or framed at the edge's modules, where vertical loads are transmitted through corner and intermediate posts; and non-load bearing modules, often called pods, that are supported by the floor or a separate structure. These three construction methods are employed in diverse applications, contingent on the need for cellular spaces, like hotel bedrooms, or open-plan areas [91,93]:

- Continuously supported: Continuously supported or four-sided modules are supported on their longitudinal sides, which bear on the walls of the modules below. Continuous-supported modules are divided into a lightweight load-bearing wall system and a massive load-bearing wall system based on the type of sidewall support system [109]. Monolithic load-bearing wall modules are commonly used in concrete buildings, in which the concrete walls are used to transfer gravity loads to the foundation and resist the lateral loads [110]. The end walls are usually highly perforated with large windows at one end and a door and service riser at the other. In some systems, the edge beams in the floor and ceiling provide an indirect means of load transfer by bearing on each other, but in most cases, the side walls provide direct load transfer. The indirect transfer method relies on the resistance of the edge beams to compression across their depth, so this type of modular construction can be more limited in height. The direct transfer of loads through the side walls relies on the compression resistance of the structural elements. In continuously supported modules, corner posts can be incorporated to serve as lifting points and connection points to other modules and structural elements. Stability is provided either by placing X-bracing in the walls of the modules, by the diaphragm action of sheathing boards, or by a separate bracing system [83,93];
- Open-sided: Corner-supported modules have posts at their corners and sometimes at intermediate points, and edge beams span between them. In this way, the modules may be designed with open sides, although infill walls can be used to form the cellular space. Because of that, this solution usually has more flexibility in architectural design and larger module sizes. This leads to fewer modules and connections [91,96]. They are generally made of steel in which the gravity loads are transferred to the slab, the edge beams and corner columns, and the foundations [110]. Therefore, frame-supported modular units are employed in high-rise MSBs due to the robust compression, torsional, lateral-torsional, and bending resistance of columns and beams [111]. However, because the beam-to-post connections are relatively weak in terms of their bending resistance, the stability of the group of modules is provided by additional bracing often located around the stair and lift core [93]. The lateral loads due to winds and earthquakes are transferred to shear wall cores and bracing systems via ceilings and floors [93,111]. These types of open-sided modules are often used in the health and educational sectors [93]. The corner-supported module is commonly used in high-rise buildings due to its high load-carrying capacity [110]. Another variation is the open-ended modules that can be manufactured so full-height glazing can be provided, and modules can also be combined along their length [93];

- Non-load bearing: The third type is non-load bearing modules, which can only support their weight and crane lifting forces. Simultaneously, relying on adjacent members or modules, such as wall support landings and half-landings of staircase modules, resist externally exerted forces. On the other hand, the bathroom, kitchen, and plant-room modules are installed on finished floors to transfer external loads [112].

Timber framing has been a prevalent construction method in the residential sector for modular housing since the 1960s, especially in the United States. Initially used in modular construction, particularly in temporary or relocatable buildings, this approach relies on prefabricated timber wall panels utilizing wall studs with a corresponding top and bottom track. Typically sheathed with plywood or oriented strand board (OSB), these wall panels incorporate one or two layers of plasterboard on the interior. The external walls are insulated externally with rigid insulation boards or with mineral wool placed between the timber studs. There is also a vapor barrier and a waterproof membrane. Floor and ceiling panels, designed with deeper joists, may include edge beams such as deep laminated beams, enabling up to 10 m spans between corner posts [93]. To optimize the strength and stability of the structure, it is a common practice to use varied stud spacings depending on the loads to which the modules are subjected: modules on lower floors have reduced stud spacing, while modules on upper floors, with less load, have larger stud spacings.

Over time, timber modular systems have evolved into versatile solutions suitable for both temporary and permanent constructions, accommodating a variety of functions. The construction techniques employed in timber buildings encompass timber framing for low-rise timber structures and post-beam and paneling systems for high-rise mass-timber projects, both strategically organized according to modular principles [81]. However, systematically dividing timber construction into more general methods such as panel, frame, and solid timber construction no longer seems reasonable in this context. By intelligently combining various building elements, customized solutions can be created in practice, giving planners the greatest possible design freedom. One frequent combination is timber panel exterior walls and ceilings and load-bearing interior walls made of dowel laminated timber or glued laminated timber, using the elements' thermal insulation, soundproofing, and fire safety advantages. Planar structural components have always been combined to brace structures and enclose spaces in frame structures [86].

The widespread practice of combining timber with other materials, such as reinforced concrete or steel, in construction has led to the creation of timber–concrete composite (TCC) and steel–timber composite (STC) components [106].

TCC solutions are commonly applied to beams, floors, and walls at the module level. TCC beams consist of a timber beam coupled with reinforced concrete in the transversal or longitudinal direction. TCC slabs, a well-known hybrid component, feature a slab-type EWP or timber beams in a joisted system connected to a reinforced concrete slab through shear connectors. TCC walls are created by connecting a massive timber or framed structure to a reinforced concrete slab with various materials such as solid timber, GLT (glued laminated timber), LVL, adhesive-free EWPs, regular concrete, lightweight concrete, or high-performance concrete. TCC fabrication can occur in wet–dry or dry–dry construction systems, involving processes where wet concrete is poured onto dry timber or precast concrete is connected to the timber part, respectively [106].

STC solutions are frequently employed in construction, especially when there is a need to distribute heavy loads across multiple points. Steel elements are used as connectors in frame construction, integrated into box or slab elements, and inserted in prestressed beams or frame structures through steel cables. This integration allows for longer spans, flush beams and joists, and more slender support cross-sections in modern timber buildings. Additionally, when steel elements are incorporated into timber, the surrounding timber provides fire safety, eliminating the need for a separate fire protection coating for the steel to ensure a sufficient duration of fire resistance [86].

In certain situations, edge timber modules on both longitudinal sides of the building incorporate a combination of steel and wood to resist lateral movement. Steel frames can be fully concealed within the timber cavities to maintain a seamless appearance and prevent external observers from discerning the difference [81].

11. Stability

11.1. Building Level

In high-rise buildings, lateral loads from winds and earthquakes are counteracted by an independent lateral stability system, such as a bracing truss or shear wall core. The lateral stability system can be constructed from steel, concrete, and hybrid steel–concrete structures using either traditional or modular construction methods [104]. While timber modular structures are primarily confined to low-rise applications, steel and precast modular units have established themselves as the industry standard for high-rise projects [109].

In contrast to the bracing truss, the shear wall core has the flexibility to accommodate elevators, stairs, and service risers. In current practice, four out of five of the world's tallest modular buildings constructed by the 3D volumetric method are based on concrete shear wall cores. Concrete cores can be constructed using either conventional methods, such as slipform and jumpform, or through the modular approach employing precast concrete systems. The composite steel–concrete–steel sandwich system is another option for the shear wall core [83]. Due to the discontinuity in the structural components among modules, the modules are designed to resist the gravity loads only whereas the core wall resists all the lateral loads. The result is a laterally stable building if the inter-module connection is designed correctly to transfer the horizontal load via axial and shear forces. The floor slab design in each module needs to be stiff enough so that each can act as a rigid plate [113].

While the use of pure timber structures for medium- and high-rise buildings remains a challenge, it is undoubtedly feasible up to a certain height. The Mjøstårnet tower in Norway, standing at 18 stories, exemplifies current possibilities and challenges. Additional mass, such as screed or concrete slabs without composite action, has been introduced to control sway and vibration. Although visible, the stabilizing system's diagonals sometimes impose limitations on functionality [114].

A typical 50–60 m tall timber high-rise building usually consists of a concrete core (to meet fire protection, acoustic, and structural requirements) in combination with timber or timber–concrete composite floors and possibly timber columns along the facade (at relatively short intervals). For taller buildings, additional bracing for sway vibration and comfort requirements or additional stability systems are usually also required [115].

11.2. Module Level

To facilitate the transfer of lateral loads from the module to the lateral stability system, supplementary bracing systems should be incorporated within the floors and ceilings of the module. Also, the inter-module connections should have sufficient shear strengths [83].

Bracing is often achieved by double-sided cladding of the framework with panels made of a processed timber construction material to enhance stability. Alternatively, internal bracing with plasterboard is employed, ensuring structural integrity, and contributing to the overall strength of the module [116].

When additional stabilizing elements, such as wall cladding, are to be avoided, the structure can be braced with rigid corner elements. However, the total allowable height of these structures is limited to three stories due to fire protection and structural constraints. For buildings exceeding three stories, non-structural timber construction modules can be integrated into load-bearing steel or reinforced-concrete frame structures [116].

The complexity of a structure rises proportionally with the overall height of the building. For low-rise buildings (three to five stories), where the forces to be resisted are relatively low, shear wall systems mainly made of CLT panels can be used. Alternatively, timber-frame systems arranged around the perimeter can be provided by bracing. Moment-bearing structures (rigid frames) rely on the interaction between beams and columns that transfer the moment caused by horizontal forces directly at the connection. Due to the typically low strength and stiffness of timber moment connections, this type of system is not very common [114].

12. Joining Techniques

The design of modular building systems is highly influenced by structural integrity against critical loading conditions. One of the main aspects is to ensure that all the assembled parts can, together, have a lateral load resistance, a compression capacity related to vertical forces and load-transfer, and a deformation capacity within the connections between the structural elements and modules.

In terms of connections, a modular building connection system can be divided into three different types: (i) inter-modular (IMC); (ii) intra-modular; and (iii) module to foundation (Figure 9).

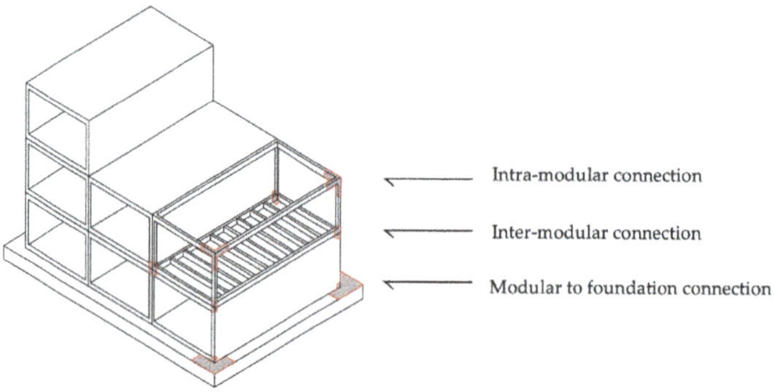

Figure 9. Types of modular building connections. Adapted from Rajanayagam et al. [117].

The inter-modular connections are crucial for determining the overall structural behavior of the system [118] since they are mainly responsible for the horizontal and vertical links within stacked modules. The inter-module connections between adjacent modules are pivotal for maintaining the integrity, stability, and robustness of modular building systems [119].

Intra-modular connections refer to those within a module, which are crucial for maintaining the integrity and stability of each individual module. These connections also play a vital role in ensuring that the module's structural behavior is properly supported and transferred to adjacent modules, all the way down to the foundation.

The module-to-foundation connection governs the sliding and overturning of entire modular building structures. An important aspect to consider in these connections is the fact that these failures, caused by lateral forces, are mostly due to how well attached the entire structure is to its base [117].

The Design for Manufacturing and Assembly (DfMA) principle has increasingly demanded that innovative connections be designed and tested so that performance can be directed and driven to fulfill requirements such as seismic design, robustness, and lateral load resistance.

12.1. Intra-Module

For the assembly of timber elements composed of panels and modules, screws and metal connectors are predominantly employed. Among the metal connectors available in the market, notable options include galvanized steel brackets, perforated plates, metal shoes, and extruded aluminum profiles. The array of metal connectors within the modules encompasses the following:

1. Self-drilling screws with partial thread for timber, wide-head;
2. Self-drilling screws with partial thread for timber, countersunk head;
3. Self-drilling screws with full thread for timber;
4. Ring-shank nails;
5. Hold-down angle plate;
6. Shear angle plate;
7. Perforated plates.

Numerous specifications are necessary for connectors to join elements, including type, size, quantity, and angle, which must be carefully designed on a case-by-case basis and can be consulted in catalogues from specialized suppliers.

Concerning the geometry of the union between elements, the method of joining walls often involves simple juxtaposition, unlike the junction between walls and floors. The latter may require different solutions depending on factors such as architecture, fire safety, aesthetics, and the function and location of the elements.

12.2. Inter-Module

In modular buildings, ensuring structural integrity, overall stability, and robustness relies heavily on the connections between modular units. The inter-modular connections entail horizontal connections in two plane directions from neighboring modules and a vertical connection within stacked modules.

The main aspect is to provide a path for load sharing and transmission between modular units which are the linking elements that allow stacked modules to transmit loads effectively to the base. Another great point is that they fulfil robustness requirements through load sharing by avoiding catastrophic incremental collapse due to local failure.

Differential movements between modules are also restricted by inter-module connections to avoid both serviceability failure (e.g., high drift ratio) and damage to the layers' protections coats.

In terms of seismic performance, despite the fact that the behavior of multiple stories of modular units is not adequately understood, the inter-module connections play a key role by dissipating the energy to maintain the structural behavior between the joined elements [119].

While welded connections offer rigidity, they are not preferred during on-site construction due to the need for highly skilled labor, substantial working space, and time-consuming post-weld inspections. Various inter-modular connection techniques have been developed for modular buildings, categorized into three types [83]:

- Rod based;
- Connector based;
- Bolt based.

12.3. Module-to-Foundation Connection

The module-to-foundation connection is another essential type for modular building systems and it becomes more critical as the height of the building rises. Typically, for a post-and-beam system these connections are usually made by using a corner fitting or embedded column. In addition, for timber–hybrid systems, hold-down connectors or anchor bolts are responsible for the foundation attachment to the frames or panels (e.g., CLT panelized system).

13. MEP Installations

There are several possible solutions regarding the location of the building's infrastructure. Pipes can run through the ceiling, walls, floor, or even the intersections of different planes. The study conducted by Monsberger et al. [120] highlights several crucial principles to consider when designing solutions involving piping in timber structures. A recommended approach is to concentrate pipes in specific and well-defined areas, facilitating inspections and access for maintenance. This organization not only simplifies the identification of damage to the timber structure in the case of leaks but also expedites the implementation of corrective measures. The strategic installation of maintenance doors is another important consideration, providing sporadic access to facilities and contributing to the ease of inspections. In inaccessible locations, it is advisable to take additional measures to protect pipes, such as including tapes or reinforcement tubes with larger diameters. However, it is important to note that this approach may increase the size of the cuts. Despite being underdeveloped, electrical humidity monitoring systems are also mentioned as a possible solution for such situations. The same author also mentions that an alternative could be using prefabricated cores as an autonomous module, which can be added to and adapted to the proposed solutions.

Incorporating "service cores" within structural 3D modules is a significant development in offsite construction. These prefabricated modules, equipped with ample structural capacity, include crucial components such as kitchens, bathrooms, mechanical–electrical shafts, stairs, and elevator shafts in housing projects. Upon on-site installation, these cores are strategically positioned to support slabs, partitions, and envelope panels, creating spacious and adaptable "served" areas like living rooms, dining rooms, and bedrooms [91].

Specialized pods designed for toilets and kitchens are commonly found in educational and healthcare buildings, commercial establishments, and hotels. Since the mid-1980s, large bathroom/toilet units have been utilized in city center commercial buildings [93].

Pods, as non-load-bearing modular units supported by the main building structure, play a crucial role in construction, especially in highly serviced areas. Typically measuring 2.00 m × 2.40 m in external dimensions, these units come in various materials. Specialized manufacturers produce pods in standard designs and can customize them for larger production runs. Some pods, designed without floors, aim to minimize construction depth and weight, addressing concerns about stepping into a bathroom module. Modules with open bases often feature ceilings that serve as the floor for the module above, a solution commonly used in prisons and secure accommodations. Innovations include under-floor heating systems with pipes embedded in the ceiling slab. Connection points on the exterior wall of these pods facilitate the integration of the main building's services and drainage systems [93]. An alternative could be prefabricating the utility core as an autonomous module, which can be added and adapted to the proposed solutions [120].

To optimize the utilization of shared service risers, bathroom pods are commonly positioned back-to-back around the service riser, allowing for the concentration of up to four pods in one area of the slab. While some situations may permit a step from the general floor level into the bathroom, in most cases, a thinner floor is necessary under the bathroom pod, or a layer of screed is applied over the adjacent floor to align it with the pod's floor level [93].

To enhance building design, there is a desire to reduce the construction depth and increase the headroom in modular structures. In contrast to conventional buildings, where a single beam supports both the ceiling of the lower story and the floor of the upper story, modular structures consist of separate units that are stacked. Each module has a ceiling beam supporting the MEP services and a floor beam supporting dead and imposed loads. When modules are stacked vertically, the two beams leave a small gap between the upper and lower stories, resulting in extra vertical space consumption compared to conventional buildings. If a more significant depth of the floor beam is required due to a long span or higher design-imposed load, the construction depth increases, further sacrificing headroom [96].

14. Design for Disassembly

The topic of flexibility in construction is not new. For Groák [121], the definition can be divided into two types of intervention. On the one hand, adaptability can be achieved by designing rooms or units that have many uses, mainly accomplished through arranging rooms, flow patterns, and space designations. On the other hand, flexibility is reflected in making changes to the building's physical structure, such as by expanding or adding rooms or units and utilizing sliding or folding walls and furnishings. Lehmann and Kremmer [122] also point out the necessity of identifying that there are two phases: pre-configuration in the design phase and reconfiguration during the use of the building. This means recognizing that the hybrid nature of timber construction potentially allows for the creation of systems with different configurations, which can even be combined with other structural systems. However, there are many other factors besides the construction itself. According to Askar et al. [123], it is necessary to define the term adaptability as the ability of buildings to change in response to different needs, including social and local factors (available materials or user demands), environmental issues (climate change, environmental disasters), technical requirements (technological developments), economic factors, and bureaucratic issues (regulations), among others. However, there is a consensus about the numerous interpretations: true adaptability lies in the ability to respond to the impossibility of predicting future changes.

Design for Disassembly (DfD) is a theme frequently used over time by industries to acknowledge the need to maintain and repair their products. Its main objective is to reduce the consumption, cost, and waste of materials, eliminating waste that cannot be reintroduced into production cycles and increasing their life cycle. In the building industry, this can be carried out by simplifying systems and connections so that buildings function as stores of replaceable materials over time. According to Bogue [124], the connection points must be made up of reversible connections with good accessibility and visibility so that they can be dismantled using only mechanical fixings. Despite the variation in opinions on this theme, concepts like circularity, which includes maintenance and repair, adaptation, relocation, reconfiguration, and reuse, are all based on DfD principles [125] (Figure 10).

Developments around these concepts have triggered greater investments in reusing systems and components in construction. While recent years have seen a paradigm shift towards transforming linear production systems into circular ones, the current focus is on reducing the flow of this circularity to allow for greater longevity in construction.

Within the scope of academic research, there are several ways of categorizing the different types of adaptability that serve as proof of the consequent exploration of the topic; however, the multitude of terms reveals that there is a problem in standardizing these concepts. One of the most common classifications in academic circles, which seeks to standardize the various existing concepts into sub-categories, is that of Schmidt II and Austin [126].

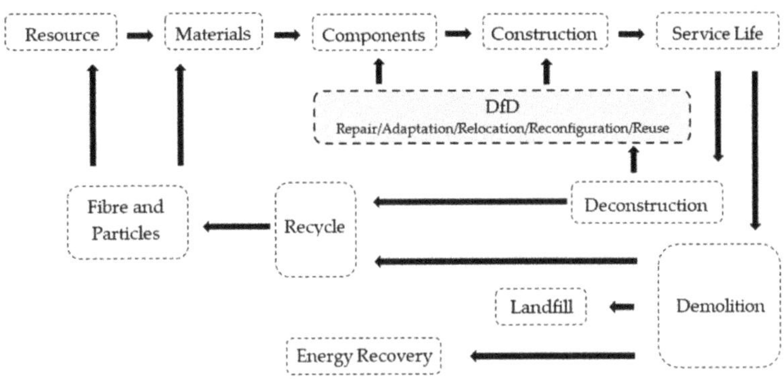

Figure 10. Reuse principles on timber life cycle. Adapted from Ottenhaus et al. [125].

The division was created to establish a standard for design models and their main fundamental characteristics to allow for adaptability in architecture, with a spectrum of options that allow for multiple changes over the lifetime of buildings [127]. In this sense, the proposed sub-categories are adjustable, versatile, refitable, convertible, scalable, and movable. Despite the unanimity of this categorization, there are other definitions that, despite grouping the same concepts and using them as a basis, try to systematize the information differently. Magdziak [128] organizes the different types of adaptability based on location, scale, and intended function, while Schneider and Till [129] distinguish between different parameters, from spatiality to technical construction issues.

As pointed out previously, one of the main challenges in this type of system is the bonding between materials. Wood is a composite of structured natural fibers, meaning loads that exceed the material's maximum elastic limit will cause it to break [125]. Demands for reduced cost and speed of execution mean that most market solutions opt for using chemical connections to link components, giving rise to monolithic solid constructions, which essentially focus on building a final product [130]. Even though this type of solution is effective, it has several limitations regarding future disassembly.

According to Dams et al. [131], the use of dry or mechanical connections should be favored, along with the use of screws, bolts, or staples, which are easily accessible and uniform. This type of element should also be applied using gravity (without momentum), thus making the assembly and disassembly process easier. Stainless steel is one of the main materials for making durable and reusable connections. The main solutions employed can thus be categorized as dry solutions, which require a single material (wood), or those that use metal fittings, ranging from simple conventional solutions (simple fasteners) to more sophisticated, market-based, and highly standardized options. Although over the last few years dry solutions have been progressively replaced by metallic solutions, the technological advances in manufacturing processes (such as CNC), which allow for the precise construction of large-scale parts, the construction of systems based on traditional logics has been the focus of some recent research [125]. One example is the SI-Modular system developed by Metsa [132], which enables houses to be constructed completely from timber without screws by simply using interlocking connections in the installation. However, these developments are only isolated cases without many repercussions from the rigidity of the timber and the need to take care when exposing it to variations in temperature and humidity, which make it difficult for the connections to function, meaning that metal solutions continue to be the focus in the development of solutions. Some studies have conducted tests to understand the applicability of metallic solutions to different types of timber systems.

Yan et al. [133] applied CLT panel fixing solutions to timber-frame systems to assess their behavior. Some of the main conclusions were surrounding the deformations in the materials, which question their future disassembly and reuse, and they concluded that there is a need for future research on the subject. Also, Arvaniti [134] points out that one of the main challenges of disassembling timber systems is deformation (in the holes, fasteners, or the wood elements themselves), which alters their characteristics and presents lower resistance properties compared to new materials. According to Bogue [124], these deformations are inherent to the physical behavior of the material, and they should be a fundamental factor to take into account in the design phase, designing solutions in which the changes resulting from traction, bending, and corrosion, among others, should occur mainly in the replaceable metal elements.

Another major concern for companies in the sector is the compatibility of MEP infrastructures with reversible systems. Given the ever-increasing volume of infrastructure required to meet today's energy demands and living conditions, the possibility of moving the planes where this type of pipework is usually located is proving to be a challenge. According to Nelson [135], although the experience of locating installations on the floor makes it possible to redesign the non-structural elements inside the modules, installation on the walls seems to be the most widely used solution on the market.

Finally, one of the fundamental issues for a building's adaptability is the variations in scale and the consequences for the structure. As seen in previous sections, the structure of volumetric module systems is based on partition walls with a structural function. Removing or changing one of these walls requires the reinforcement of the entire system if this has not been analyzed and previously defined in the design phase [134]. In this sense, it is essential to make the structural solution compatible with any possible changes the building may experience in the future. One example of the difficulty of making changes that have not been idealized in the design phase is the spacing of the substructure in lightweight systems, which conditions the future opening of doors between rooms. Since the metric used is a subdivision of the dimensions of the materials, ensuring that the smaller the spacing, the better the structural behavior of the wall, it is straightforward to use standard measurements of 60 cm spacing, which makes it impossible to open a future doorway. On the other hand, flexibility is facilitated in systems where the structure is separated from the rest of the building's components, as seen the study on the principles of Open Building and Shearing Layers, which provide total freedom in the definition of interior spaces since the dividing walls have no structural function.

All these challenges imply greater investment in the building design phase, which, according to Arvaniti [134], implies a significant increase in execution costs. However, according to a study conducted on a sample of 48 buildings in the US, the implementation of adaptability strategies resulted in only a 1% increase in the initial construction costs compared to conventional strategies [135].

15. Discussions

A comprehensive exploration of the current state of the art in modular timber construction has revealed a dynamic landscape marked by both notable achievements and inherent challenges. Our examination showed that structural design, transportation considerations, and material utilization underscore the evolving nature of this construction methodology. Insights from various studies shed light on the increasing interest in tall timber buildings, driven by the material's sustainability and unique structural properties. Despite advancements, the discussion also identified persistent hurdles, such as societal perceptions, regulatory hesitations, and gaps in knowledge. This overview sets the stage for a nuanced discussion of modular timber construction's development, challenges, opportunities, and knowledge gaps.

Timber, one of the oldest building materials and the dominant one in construction until the 20th century, saw its usage become limited due to public perception and restrictive regulations introduced for fire safety, contributing to the consolidation of the concrete industry in the market [63].

A decade ago, there was a notable absence of structural wood use in multi-story buildings globally, primarily limited to residential one- or two-story structures. However, a unique historical moment was also unfolding, suggesting that wood was transitioning from a material with a long history to one that would define construction for the next millennium. Significant potential awaits the expansion of modular timber constructions, especially within dense and sustainable urban landscapes.

Despite skepticism about the material's performance, stemming from years of its structural disuse and resistance to change due to our dependency on established standards, the recent resurgence of wood construction has primarily occurred through revisions to regulations, extra regulatory incentives from collaborative public–private sector entities promoting wood use, and cooperation between the construction and forestry industries and research institutes, aiming to enhance product refinement or even revive a preference for wood, making it the primary construction material of this century [136].

The construction of various timber buildings erected in recent years has showcased the multiple possibilities of wood and its technologies, restoring confidence in the material. This is reflected in the growing interest in such projects, whether motivated by ecological value, thermal qualities, structural attributes like advantageous bending behavior and a favorable weight-resistance ratio, or aesthetic appeal [29].

This was feasible mainly due to the introduction of Engineered Wood Products and the utilization of digital design and production through CAE (Computer-Aided Engineering), CAD (Computer-Aided Design), and CAM (Computer-Aided Manufacturing). Leading to innovative systems characterized by high quality and precision, these advancements have streamlined and enhanced the prefabrication process for timber buildings and its associated benefits.

Many countries, however, still experience hesitancy regarding the safety and durability of wood in tall construction, resulting in few and conservative projects being executed or in progress [137]. Moreover, due to the recent emergence of modern wood construction techniques, particularly for tall and modular buildings, this typology has not been scientifically studied in depth or widely accepted [95]. This gap is evident in the collective understanding of the material and poses a significant research limitation. Despite consumers and industry professionals recognizing wood as a natural, renewable material with excellent thermal properties, it has not been universally regarded as strong, durable, modern, and fire-resistant, despite extensive advancements and discourse surrounding timber construction and its applications. Consequently, the wood industries remain fragmented, consistently confronting potential threats to their competitiveness and development, leading to a notable decrease in the dissemination of information on the subject. The divergence of opinion among industry professionals and consumers regarding skepticism and rejection of wood usage indicates a lack of training and understanding of the typology among various construction actors, such as engineers and architects. According to Roos et al. [138], architects acknowledge wood's environmental and aesthetic value but perceive it as culturally dependent. On the other hand, engineers appreciate its strength-to-weight ratio and new wood products but remain concerned about fire safety, sound transmission, implications due to moisture variations, and a shortage of specialized labor, preferring more familiar and explored materials. Additionally, authorities and contractors view wood structures as being synonymous with higher commercial risks. Despite consumers associating the material with environmentally friendly construction practices and considering it an important factor in decision making for product purchases and service hiring, they indicated a lack of knowledge about wood's contribution to reducing greenhouse gas emissions through carbon storage. In a study focused on the acceptance of LVL, McGregor et al. [139] highlighted that, while consumers recognized the sustainable benefit of choosing this product, which has a lower carbon footprint than steel, this factor did not significantly impact material selection.

It is conceivable that consumer perceptions regarding fire behavior, moisture, and biological agents result from instinctive motivations based on experience [19]. However, promotional campaigns for wood usage implemented, for example, in the United Kingdom and Austria have demonstrated that perceptions about the material can be altered among consumers and construction professionals [140]. Understanding markets and spreading knowledge about the negative perceptions associated with the typology are fundamental components for its progression, enabling the development and proposal of solutions capable of mitigating common misconceptions and consolidating the typology as an alternative construction method [19]. Social acceptance becomes vital for successfully implementing any technological innovation, especially those that are not competitively cost effective, such as wood structures compared to concrete and steel.

Beyond the challenge of social acceptance, many countries still exhibit a constrained market for wood construction, grounded in the lack of specific regulations and the scarcity of specialized professionals in the sector. Efforts to provide specific education for civil engineers and architects in the field of wood structures are rare and recent, contributing to a limited workforce [137]. Additionally, often only a fraction of native wood resources is utilized, with wood imports being prevalent for this typology, which represents a significant missed opportunity for economic, social, and environmental gains [141].

Furthermore, knowledge gaps are the primary obstacles to using structural wood [63]. The dissemination of tall timber buildings in countries with harsh climates has promoted the accumulation of knowledge and the proposal of suitable solutions for this specific scenario, which may not directly apply to other contexts. Therefore, there is a need to develop knowledge to adapt this typology to different situations, considering variables such as climate conditions favoring biological deterioration; temperature ranges; urban regulations for new construction; national normative requirements for functional (thermal and acoustic) and fire safety; the diversity of available forests for supply and the species and products offered by the forestry sector; widely used construction materials in terms of quality and cost, whose application and maintenance are well understood by technicians and users; prefabrication technologies available to the wood construction sector; traffic regulations and limiting conditions for material transportation; and architectural and aesthetic preferences. This would enable the proposal of suitable, reliable, efficient, and durable construction systems that fully leverage the benefits of the material.

16. Conclusions

Within the domain of prefabrication and modularity in timber systems, a list of topics remains underexplored and warrants attention. For instance, further research about which solution (1D, 2D, or 3D) should be used in particular project scenarios is essential. Factors such as transport, the solution's applicability considering site issues, the flexibility of the resulting spaces, or the local producer's manufacturing conditions influence the choice made. An extended study of various case studies is necessary to standardize solutions and narrow the options according to the specific need.

Even in light of recent developments regarding a specific standard value in terms of module length and width, there is a need for a more in-depth studies that look at issues such as the different types of transport, the regulatory dimensions, and the less important issue of which layouts (in module or overlapping) allow for the best percentage of return on transportable space; the selection of the materials used, their direct relationship with the metrics and repercussions on the final solution; or even the requirements and dimensions of the different programs, taking into account the regulations governing the use of space.

There is also a need to develop BIM tools in the timber construction sector in the same way as they have been for the concrete and steel sectors, in which are quite advanced and even more interconnected. Much of the software used today is due to the ease with which parts can be prepared (including the design of fittings and holes) and direct communication with CAM manufacturing processes. However, there are some problems when it comes to cross-referencing with the main programs used in the construction sector, and it is necessary to analyze ways of interconnecting communication to facilitate the development of solutions in the future.

Contrary to current trends in concrete and steel buildings, timber structures show a prevalence of rectilinear, symmetric plans and regular, repetitive extrusions with a low degree of innovation, possibly due to the lack of development of solutions that allow for greater layout variability. In this sense, there is a need for further analysis and cataloguing of the different market solutions for connecting planes to promote future research that tries to develop versatile solutions that are reflected in different space configurations. This will also contribute to the modularization technique, balancing standardization and adaptability.

Within the domain of structural solutions, timber modular construction derives inspiration from the steel modular concept, incorporating its very linear member-based construction method, stability arrangements, and joint techniques (screws, welding, or adhesives). Nevertheless, steel modular construction holds a distinct advantage in technical advancement, boasting a more extensive array of solutions that enjoy greater diversity and public acceptance.

In terms of adaptability, despite the focus on modularization and prefabrication, most of the solutions still focus on the linear assembly process, focusing on developing a final product. This option makes it difficult not only to make changes to the period of use of the building but also to dismantle it at the end of its life cycle. In this way, it is important to explore research into the creation of new models that allow for the redesign, alteration, and dismantling of buildings, ensuring greater freedom both in the design phase and during the period of use of the building. Among the main challenges are developing reversible connections between the various components, implementing MEP, and ensuring a structural system that allows greater spatial freedom. It is also necessary to explore the reuse of materials, given the progressive requirements for certification, which differ according to the regulations in each country. Reusing components could be challenging, and further research into the limitations of and protocols for its feasibility is essential (performance tests, certifications, among others).

Table 2 summarizes the main information about each topic covered in this state-of-the-art review.

Table 2. Overview of the main topics explored in this state-of-the-art review.

<table>
<tr><th>Section</th><th colspan="2">State-of-the-Art Review</th></tr>
<tr><td>3. Wood as a modern construction material for modular buildings</td><td colspan="2">• Initiative to promote wood building construction (France, Canada Japan, Sweden, or Finland);
• Collaboration between construction and forestry industries, research institutes, and governments;
• Positive marketing of wood use produced by incentive programs and collaborative research demonstration projects.</td></tr>
<tr><td>4. Engineered Wood Products and prefabrication</td><td colspan="2">• Reduction in project timelines and costs (85%);
• Saving energy in the construction phase (44%);
• Performance during the use phase (by more than 7%);
• Avoid construction traffic, noise impact, disturbances to the local community, accidents, the need for traditional construction equipment, wet processes, waste, poor execution, improper material storage are eliminated and joints, gaps, and penetrations minimized;
• Lags significantly behind technological advancements observed for other construction materials.</td></tr>
<tr><td>5. Principles of modularity in timber buildings</td><td colspan="2">• Escalate efforts to craft solutions that seamlessly integrate the supply chain from design to production;
• Establish a system wherein the components can be amalgamated and reconfigured to cater to diverse needs.</td></tr>
<tr><td rowspan="5">6. Taxonomy of modular timber products</td><td colspan="2">• Panel and space module systems were the most common systems (2000–2010) predominantly employed in mid-rise projects (10 stories);
• From 2009, there is a significant increase in frame structures.</td></tr>
<tr><td>1D Linear Systems</td><td>• Five to eight-story buildings;
• Combinations of EWPs, such as CLT (cross-laminated timber) slabs, ribbed slabs, or composite floors.</td></tr>
<tr><td>2D Panelized Systems</td><td>• Medium-height (5–10 floors) and low-rise building projects (3–4 floors);
• Assembly process is simpler than traditional construction but more intricate than assembling 3D modules;
• Provide greater flexibility and more convenient transporting.</td></tr>
<tr><td>3D Modular Systems</td><td>• 6% to 10% exceed nine floors (max. nineteen floors);
• High level of repeatability and a significant ratio of wet to dry rooms;
• Most complete prefabrication (95%);
• Faster than the 2D approach (16%);</td></tr>
<tr><td>Hybrid Systems</td><td>• Enhancing productivity for bathroom and providing maximum flexibility for other areas;
• Deliver both solutions becomes more intricate (coordination of the overall supply chain);</td></tr>
<tr><td>7. Manufacturing, transportation, and installation dimensions</td><td colspan="2">• Typically, the cutting machinery utilized allows for dimensions of 3.50 m × 9.00 m;
• A standard truck, the panels are confined to the following specifications: Length: 13.50 m; width: 2.45–2.48 m; height: 2.50–3.00 m; load weight: 24.00–25.50 tons;
• Length of a typical modular unit: from 6 to 12 m (modules less than 5 m long: often functional or flexibly designed modules; modules longer than 10 m: generally made for apartments);
• Width of a typical modular unit: from 2.50 to 4.00 m (transportation conditions contribute to a range between 2 and 5 m);
• Height of a typical modular unit: from 2.80 to 3.50 m;
• The standard floor space for volumetric modules ranges from 15 to 25 m^2;
• The mass of timber modules ranges from 5 to 15 t, and the mass of equipment is less than 25 t.</td></tr>
<tr><td>8. Layout design</td><td colspan="2">• Rectangular (alternative "L" shape);
• Two-dimensional system include most examples that deviate most from orthogonal shapes;
• Regular extrusions with a flat roof dominate all structural systems (1D/2D show more volumetric variations);
• One-dimensional element projects generally follow grid systems (dominant);
• Two-dimensional panel and 3D three-dimensional ordered by linear arrays.</td></tr>
</table>

Table 2. *Cont.*

Section	State-of-the-Art Review	
9. Building occupancy and use	• The most common is residential (half of all projects); • Hotels, senior residences, student accommodations, hospitals, and schools have appeared more recently (repeatability); • 1D—office buildings; 2D—residential buildings (whether for social housing, student housing, or multifamily housing); 3D—hotels and hostels.	
10. Structural systems for modular buildings	Building level	• Structural systems for tall buildings: rigid frame; core systems; shear wall systems; shear frame systems; mega core systems; mega column systems; • Structural solutions to support volumetric modules: module stacking; module anchored to an exoskeleton; module anchored to the core; • Structural components often made with concrete/steel when in timber-hybrid buildings: podium or plinth; core; floor slab; lateral bracing; external structural elements (such as stairs and circulation areas);
	Module level	• Generic forms of modular construction: Continuously supported (or four-sided modules); Open-sided (or corner-supported or framed at the edge's modules); Non-load bearing modules (often called pods); • TCC solutions: beams; floors; walls; • STC components: connectors; box or slab elements; cables;
11. Stability	Building level	• Four out of five of the world's tallest modular buildings constructed using the 3D volumetric method are based on concrete shear wall cores; • A typical 50–60 m tall timber high-rise building usually consists of a concrete core in combination with timber or timber-concrete composite floors;
	Module level	• Bracing is often achieved by double-sided cladding of the framework with panels made of processed timber construction material. Alternatively, internal bracing with plasterboard is employed; • When additional stabilizing elements, such as wall cladding, are to be avoided, the structure can be braced with rigid corner elements;
12. Joining techniques	Intra-module	• Self-drilling screws with partial thread for timber, wide-head or countersunk head; self-drilling screws with full thread for timber; ring-shank nails; hold-down angle plate; shear angle plate; perforated plates.
	Inter-module	• Rod-based; connector-based; bolt-based.
	Module to foundation	• Corner fitting or embedded column (1D); Hold-down connectors or anchor bolts (hybrid systems).
13. MEP (mechanical, electrical, and plumbing) installations	• A recommended approach is to concentrate pipes in specific and well-defined areas, facilitating inspections and access for maintenance; • Incorporating "service cores" within structural 3D modules; • Specialized pods (2.00 m × 2.40 m) are designed for toilets and kitchens (educational, healthcare, commercial and hotels.	
14. Design for disassembly	• There is a need for future research on the fixing solutions between different systems; • Concern in the compatibility of MEP with reversible systems; • 1% increase in initial construction costs compared to conventional strategies.	

Author Contributions: Conceptualization, M.T. and J.M.B.; methodology, M.T.; data curation, M.T., R.F. and V.B.; writing—original draft preparation, M.T., R.F., V.B., F.S., C.M. (Cláudio Meireis), M.F., I.V., A.G. and R.A.; writing—review and editing, M.T., R.F., V.B., F.S., C.M. (Cláudio Meireis), M.F., I.V., A.G., R.A., S.M.S. and J.M.B.; supervision, S.M.S., D.L., A.F., C.M. (Carlos Maia), A.C. and J.M.B. All authors have read and agreed to the published version of the manuscript.

Funding: This research was supported by the doctoral grant PRT/BD/152841/2021 financed by the Portuguese Foundation for Science and Technology (FCT), with funds from the State Budget and the community budget through the European Social Fund (ESF), under MIT Portugal Program; and by the R&D Project "R2U Technologies | modular systems", with reference C644876810-00000019, funded by PRR—Plano de Recuperação e Resiliência—and by the European Funds Next Generation EU, under the incentive system "Agendas para a Inovação Empresarial". This work was partly financed by FCT/MCTES through national funds (PIDDAC) under the R&D Unit Institute for Sustainability and Innovation in Structural Engineering (ISISE), under reference UIDB/04029/2020 (doi.org/10.54499/UIDB/04029/2020), and under the Associate Laboratory Advanced Production and Intelligent Systems ARISE under reference LA/P/0112/2020.

Conflicts of Interest: The authors declare no conflicts of interest.

References

1. *European Green Deal*; European Council: Brussels, Belgium, 2019.
2. D'amico, D.; Pomponi, F.; Hart, J. Global potential for material substitution in building construction: The case of cross laminated timber. *J. Clean. Prod.* **2021**, *279*, 123487. [CrossRef]
3. Shanks, W.; Dunant, C.; Drewniok, M.P.; Lupton, R.; Serrenho, A.; Allwood, J.M. How much cement can we do without? lessons from cement material flows in the UK. *Resour. Conserv. Recycl.* **2019**, *141*, 441–454. [CrossRef]
4. Davis, S.J.; Lewis, N.S.; Shaner, M.; Aggarwal, S.; Arent, D.; Azevedo, I.L.; Benson, S.M.; Bradley, T.; Brouwer, J.; Chiang, Y.M.; et al. Net-zero emissions energy systems. *Science* **2018**, *360*, 6396. [CrossRef]
5. Jeffree, M. *Wood: Building the Bioeconomy*; The European Confederation of Woodworking Industries: Brussels, Belgium, 2019.
6. *World Population Prospects: The 2017 Revision, Methodology of the United Nations Population Estimates and Projections*; Working Paper; United Nations, Department of Economic and Social Affairs, Population Division: New York, NY, USA, 2017.
7. *Reducing UK Emissions: Progress Report to Parliament*; Committee on Climate Change: London, UK, 2018.
8. Snelson, T. Seven Perspectives on the Use of Timber in Building Design and Construction. ARUP. 2019. Available online: www.rethinkingtimberbuildings.pdf (accessed on 15 January 2024).
9. Abergel, T.; Dean, B.; Dulac, J. *Towards a Zero-Emission, Efficient, and Resilient Buildings and Construction Sector: Global Status Report 2017*; UN Environment and International Energy Agency: Paris, France, 2017.
10. *Sustainable Urbanization Strategy: UNDP's Support to Sustainable, Inclusive and Resilient Cities in the Developing World*; United Nations Development Programme (UNDP): New York, NY, USA, 2016.
11. *EU Forest-Based Industries 2050: A Vision of Sustainable Choices in a Climate-Friendly Future*; The European Confederation of Woodworking Industries: Brussels, Belgium, 2019.
12. Riggio, M.; Alhariri, N.; Hansen, E. Paths of innovation and knowledge management in timber construction in North America: A focus on water control design strategies in CLT building enclosures. *Archit. Eng. Des. Manag.* **2019**, *16*, 1–26. [CrossRef]
13. Santi, S.; Pierobon, F.; Corradini, G.; Cavalli, R.; Zanetti, M. Massive wood material for sustainable building design: The Massiv–Holz–Mauer wall system. *J. Wood Sci.* **2016**, *62*, 416–428. [CrossRef]
14. Evison, D.; Kremer, P.; Guiver, J. Mass Timber Construction in Australia & New Zealand—Status, and Economic and Environmental Influences on Adoption. *Wood Fiber Sci.* **2018**, *50*, 128–138.
15. Carvalho, L.F.; Jorge, L.F.C.; Jerónimo, R. Plug-and-Play Multistory Mass Timber Buildings: Achievements and Potentials. *J. Archit. Eng.* **2020**, *26*, 2. [CrossRef]
16. Smith, R.E.; Griffin, G.; Rice, T.; Hagehofer-Daniell, B. Mass timber: Evaluating construction performance. *Archit. Eng. Des. Manag.* **2018**, *14*, 127–138. [CrossRef]
17. Chen, Z.; Gu, H.; Bergman, R.; Liang, S. Comparative Life-Cycle Assessment of a High-Rise Mass Timber Building with an Equivalent Reinforced Concrete Alternative Using the Athena Impact Estimator for Buildings. *Sustainability* **2020**, *12*, 4708. [CrossRef]
18. Gasparri, E.; Aitchison, M. Unitised timber envelopes. A novel approach to the design of prefabricated mass timber envelopes for multi-storey buildings. *J. Build. Eng.* **2019**, *26*, 100898. [CrossRef]
19. Kremer, P.D.; Symmons, M.A. Overcoming Psychological Barriers to Widespread Acceptance of Mass Timber Construction in Australia. *For. Wood Prod.* **2016**, *1*, 1–40.
20. *Case Study: Mass Timber has Banks Seeing Green: First United Bank Invests in First Mass Timber Buildings in Texas*; APA: Tulsa, OK, USA, 2019.
21. Smith, I.; Frangi, A. Tall Timber Buildings: Introduction. *J. Int. Assoc. Bridge Struct. Eng.* **2008**, *18*, 2. [CrossRef]
22. Silva, C. Tall Buildings Using CLT: An Integrated Design Considering Moisture Induced Effects. Ph.D. Thesis, Universidade do Minho–Escola de Engenharia, Guimarães, Portugal, 2020.
23. DÉCRET n°8 2010-273. L'utilisation du Bois Dans Certaines Constructions. Available online: https://www.legifrance.gouv.fr/loda/id/JORFTEXT000021979658 (accessed on 26 March 2024).
24. Wood First Act, SBC 2009, c 18. Available online: https://canlii.ca/t/52j7x (accessed on 21 January 2024).
25. Forestry Innovation Investment—Assessing the Impact of the Japanese Wood First Act on Demand for PFC in the Public Non-residential Market in Japan. Available online: www.bcfii.ca (accessed on 21 January 2024).
26. Mahapatra, K.; Gustavsson, L.; Hemström, K. Multi-storey wood-frame buildings in Germany, Sweden and the UK. *Constr. Innov.* **2012**, *12*, 62–85. [CrossRef]
27. Toppinen, A.; Röhr, A.; Pätäri, S.; Lähtinen, K.; Toivonen, R. The future of wooden multistory construction in the forest bioeconomy—A Delphi study from Finland and Sweden. *J. For. Econ.* **2018**, *31*, 3–10. [CrossRef]

28. *From Forests to Pioneering Bioeconomy—Final Report on the Strategic Programme for the Forest Sector (MSO)*; Ministry of Employment and the Economy: Helsinki, Finland, 2015.
29. Shmuelly-Kagami, T.; Matsumura, S. Progress of Timber Multi-Storey Apartment Building in Europe and Japan. In Proceedings of the 10th World Conference on Timber Engineering, Miyazaki, Japan, 2–5 June 2008.
30. Mahapatra, K.; Gustavsson, L. Multi-storey Timber Buildings: Breaking Industry Path Dependency. *Build. Res. Inf.* **2008**, *36*, 638–648. [CrossRef]
31. Franzini, F. Wooden Multistory Construction in Finland: Perceptions of Municipality Civil Servants. Ph.D. Thesis, Helsingin Yliopisto Helsingfors Universitet, Helsinki, Finland, 2018.
32. CMHC. *Canada Mortgage and Housing Corporation—Housing Observer 2016. Mid-Rise Residential Wood Construction n° 1*; CMHC: Toronto, ON, Canada, 2016.
33. Grantham, R. *Design Guidance for Multi-Storey TIMBER frame Buildings*; Building Research Establishment: Warford, UK, 2004.
34. Wahl, A.; Lavoie, P.; Brooks, D. Wood Market Trends in Europe. *FPInnovations* **2008**, *49*, 1–47.
35. Mahapatra, K.; Gustavsson, L. General Conditions for Construction of Multi-storey Wooden Buildings in Western Europe. In *CONCERTO*; Växjö University, School of Technology and Design: Växjö, Sweden, 2009.
36. Steele, M. Timber Buildings Reach Market Share of more than 16%. Timber Forum. Available online: http://www.gtf-info.com/news/new-timber-opportunities-and-green-building/timber-buildings-reach-market-share-of-more-than-16/ (accessed on 21 January 2024).
37. Caniato, M.; Bettarello, F.; Ferluga, A.; Marsich, L.; Schmid, C.; Fausti, P. Acoustic of Lightweight Timber Buildings: A Review. *Renew. Sustain. Energy Rev.* **2017**, *80*, 585–596. [CrossRef]
38. Caniato, M.; Marzi, A.; Monteiro da Silva, S.; Gasparella, A. A Review of the Thermal and Acoustic Properties of Materials for Timber Building Construction. *J. Build. Eng.* **2021**, *43*, 103066. [CrossRef]
39. Jayalath, A.; Navaratnam, S.; Gunawardena, T.; Mendis, P.; Aye, L. Airborne and Impact Sound Performance of Modern Lightweight Timber Buildings in the Australian Construction Industry. *Case Stud. Constr. Mater.* **2021**, *15*, e00632. [CrossRef]
40. Demirci, C.; Málaga-Chuquitaype, C.; Macorini, L. Seismic Shear and Acceleration Demands in Multi-Storey Cross-Laminated Timber Buildings. *Eng. Struct.* **2019**, *198*, 109467. [CrossRef]
41. Ferdous, W.; Bai, Y.; Ngo, T.D.; Manalo, A.; Mendis, P. New Advancements, Challenges and Opportunities of Multi-Storey Modular Buildings—A State-of-the-Art Review. *Eng. Struct.* **2019**, *183*, 883–893. [CrossRef]
42. Huber, J.A.J.; Ekevad, M.; Girhammar, U.A.; Berg, S. Structural Robustness and Timber Buildings—A Review. *Wood Mater. Sci. Eng.* **2019**, *14*, 107–128. [CrossRef]
43. Stepinac, M.; Šušteršič, I.; Gavrić, I.; Rajčić, V. Seismic Design of Timber Buildings: Highlighted Challenges and Future Trends. *Appl. Sci.* **2020**, *10*, 1380. [CrossRef]
44. Voulpiotis, K.; Köhler, J.; Jockwer, R.; Frangi, A. A Holistic Framework for Designing for Structural Robustness in Tall Timber Buildings. *Eng. Struct.* **2021**, *227*, 111432. [CrossRef]
45. Bruno, R.; Bevilacqua, P.; Cuconati, T.; Arcuri, N. Energy Evaluations of an Innovative Multi-Storey Wooden near Zero Energy Building Designed for Mediterranean Areas. *Appl. Energy* **2019**, *238*, 929–941. [CrossRef]
46. Li, J.; Rismanchi, B.; Ngo, T. Feasibility Study to Estimate the Environmental Benefits of Utilising Timber to Construct High-Rise Buildings in Australia. *Build. Environ.* **2019**, *147*, 108–120. [CrossRef]
47. Skullestad, J.L.; Bohne, R.A.; Lohne, J. High-Rise Timber Buildings as a Climate Change Mitigation Measure—A Comparative LCA of Structural System Alternatives. *Energy Procedia* **2016**, *96*, 112–123. [CrossRef]
48. Ilgin, H.E.; Karjalainen, M.; Pelsmakers, S. Contemporary tall residential timber buildings: What are the main architectural and structural design considerations? *Emerg. Glob. Trends* **2022**, *41*, 26–46. [CrossRef]
49. Dinwoodie, J.M. *Timber: Its Nature and Behaviour*, 2nd ed.; CRC Press: New York, NY, USA, 2000.
50. Forestry Innovation Investment—Naturally Wood—Information Resource. Available online: www.bcfii.ca (accessed on 21 January 2024).
51. Canadian Wood Council. *The Design Possibilities of Engineered Wood Products*; International Building Series: Toronto, ON, Canada, 2006.
52. Mapston, M.; Westbrook, C. Prefabricated Building Units and Modern Methods of Construction (MMC). *Mater. Energy Effic. Therm. Comf. Build* **2010**, *1*, 427–454.
53. Lessing, J.; Stehn, L.; Ekholm, A. Industrialised Housing: Definition and Categorization of The Concept. In Proceedings of the 13th International Group for Lean Construction Conference Proceedings, Sydney, Australia, 19–21 July 2005; pp. 471–480.
54. Abanda, F.; Tah, J.; Cheung, F. BIM in Off-Site Manufacturing for Buildings. *J. Build. Eng* **2017**, *14*, 89–102. [CrossRef]
55. Agarwal, R.; Chandrasekaran, S.; Sridhar, M. *Imagining Construction's Digital Future*; Mckinsey Global Institute: New York, NY, USA, 2016; p. 14.
56. Schwatka, N.V.; Butler, L.M.; Rosecrance, J.R. An Aging Workforce and Injury in the Construction Industry. *Epidemiology* **2012**, *34*, 156–167. [CrossRef] [PubMed]
57. Woetzel, J.; Mischke, J.; Barbosa, F.; Ribeirinho, J.; Sridhar, M.; Parsons, M.; Bertram, N.; Brown, S. *Reinventing Construction: A Route to Higher Productivity*; McKinsey Global Institute: New York, NY, USA, 2017.
58. Aitchison, M.; Bashirzadeh, T.; Beim, A.; Couper, R.; Doe, R.; Engstrom, D.; Kuzmanovska, I.; Lidelow, H.; Maxwell, D.; Macarthur, J.; et al. *Prefab Housing and the Future of Building Product to Process*; Lund Humphries: London, UK, 2018.

59. Bonev, M.; Wörösch, M.; Hvam, L. Utilizing Platforms In Industrializes Construction. *Constr. Innov.* **2015**, *15*, 84–106. [CrossRef]
60. Ni, C.; Popovski, M. *Mid-Rise Wood-Frame Construction Handbook*; FPInnovations Special Publication SP-57E: Pointe-Claire, QC, Canada, 2015.
61. Lehmann, S. *Sustainable Infill Development Using Low Carbon CLT Prefabrication: Adaptation for the South Australian Context*; Zero Waste SA Research Centre for Sustainable Design and Behaviour & Barbara Hardy Institute, University of South Australia: Adelaide, Australia, 2011.
62. Richard, R.B. Industrialised Building Systems: Reproduction Before Automation and Robotics. *Autom. Constr.* **2005**, *14*, 442–451. [CrossRef]
63. Kuzman, M.K.; Sandberg, D. A new era for multi-storey timber buildings in Europe. In Proceedings of the 70th Forest Products Society International Convention, Portland, OR, USA, 26–29 June 2016.
64. Dujič, B.; Yates, M.; Linegar, M. An 8-storey residential building in London made from cross laminated solid timber panels. In Proceedings of the 19th International Scientific Conference on Wood Is Good—Properties, Technology, Valorization, Application, Zagreb, Croatia, 17 October 2008.
65. Lehmann, S. Sustainable construction for urban infill development using engineered massive wood panel systems. *Sustainability* **2012**, *4*, 2707–2742. [CrossRef]
66. Kremer, P.D.; Symmons, M.A. Mass timber construction as an alternative to concrete and steel in the Australia building industry: A PESTEL evaluation of the potential. *Int. Wood Prod. J.* **2015**, *6*, 138–147. [CrossRef]
67. Ji, Y.; Zhu, F.; Li, H.; Al-Hussain, M. Construction industrialization in China: Current profile and the prediction. *Appl. Sci.* **2017**, *7*, 180. [CrossRef]
68. Gallo, P.; Romano, R.; Belardi, E. Smart Green Prefabrication: Sustainability Performances of Industrialized Building Technologies. *Sustainability* **2021**, *13*, 4701. [CrossRef]
69. Halstenberg, F.A.; Buchert, T.; Bonvoisin, J.; Lindow, K.; Stark, R. Target-oriented modularization: Addressing sustainability goals in product modularization. *Procedia CIRP* **2015**, *29*, 603–608. [CrossRef]
70. Okodi-Iyah, E.Y. A Comparative Evaluation of the Environment Impact of Prefabrication Versus Conventional Construction in UAE's Construction Industry. Ph.D. Thesis, The British University in Dubai (BUiD), Dubai, United Arab Emirates, 2012.
71. Orlowski, K. Automated manufacturing for timber-based panelised wall systems. *Autom. Constr.* **2020**, *109*, 102988. [CrossRef]
72. Johns, R.L.; Wermelinger, M.; Mascaro, R.; Jud, D.; Hurkxkens, I.; Vasey, L.; Chli, M.; Gramazio, F.; Kohler, M.; Hutter, M. A framework for robotic excavation and dry-stone construction using on-site materials. *Sci. Robot.* **2023**, *8*, 84. [CrossRef]
73. Liao, F.; Pan, H.; Zhang, J. Application od BIM technology in the design of prefabricated architecture. *IOP Conf. Ser. Earth Environ. Sci.* **2021**, *760*, 012005. [CrossRef]
74. Sacks, R.; Dave, B.; Koskela, L.; Owen, R. Analysis Framework for the Interaction Between Lean Construction and Building Information Modelling. In Proceedings of the 17th IGLC: International Group for Lean Construction, Taipei, Taiwan, 18–19 July 2009; pp. 221–234.
75. Nawari, N. BIM Standardization and Wood Structures. *Comput. Civ. Eng.* **2012**, *1*, 293–300.
76. Gan, X.; Chang, R.; Zuo, J.; Wen, T.; Zillante, G. Barriers to the transition towards off-site construction in China: An interpretative structural modeling approach. *J. Clean. Prod.* **2018**, *197*, 8–18. [CrossRef]
77. Koppelhuber, D.; Magg, A. Building services in timber construction—Investigation on difficulties and reliefs during the installation procedures in multi-storey timber residential buildings. *IOP Conf. Ser. Mater. Sci. Eng.* **2020**, *800*, 012031. [CrossRef]
78. Retsin, G. Bits and Pieces: Digital Assemblies: From Craft to Automation. *Archit. Des.* **2019**, *89*, 38–45. [CrossRef]
79. Sanchez, J. Architecture for the Commons: Participatory Systems in the Age of Platforms. *Archit. Des.* **2019**, *89*, 22–29. [CrossRef]
80. Howe, S.; Ishii, I.; Yoshida, T. Kit-of-parts: A review of object-oriented construction techniques. In Proceedings of the ISARC'99: International Symposium on Automation and Robotics in Construction, Madrid, Spain, 22–24 September 1999; pp. 165–171.
81. Li, J.; Andersen, L.V.; Hudert, M.M. The Potential Contribution of Modular Volumetric Timber Buildings to Circular Construction: A State-of-the-Art Review Based on Literature and 60 Case Studies. *Sustainability* **2023**, *15*, 16203. [CrossRef]
82. Cao, J.; Bucher, D.; Hall, D.; Lessing, J. Cross-phase product configurator for modular buildings using kit-of-parts. *Autom. Constr.* **2021**, *123*, 103437. [CrossRef]
83. Tai Thai, H.; Ngo, T.; Uy, B. A review on modular construction for high-rise buildings. *Structures* **2020**, *28*, 1265–1290. [CrossRef]
84. Salvadori, V. Worldwide Structural Survey of 197 Multi-Storey Timber-Based Buildings from 5 to 24 Storeys. In Proceedings of the Conference: WCTE 2021—World Conference on Timber Engineering, Santiago del Chile, Chile, 9–12 August 2021.
85. Svatoš-ražnjević, H.; Orozco, L.; Menges, A. Advanced Timber Construction Industry: A Review of 350 Multi-Storey Timber Projects from 2000–2021. *Buildings* **2022**, *12*, 404. [CrossRef]
86. Kaufmann, H.; Krötsch, S.; Winter, S. *Manual of Multistorey Timber Construction: Principles—Constructions—Examples*, 2nd ed.; DETAIL: München, Germany, 2018.
87. Kuzmanovska, I.; Gasparri, E.; Tapias Monné, D.; Aitchison, M. Tall Timber Buildings: Emerging trends and typologies. In Proceedings of the 2018 World Conference on Timber Engineering, Seoul, Republic of Korea, 20–23 August 2018.
88. Bertram, N.; Fuchs, S.; Mischke, J.; Palter, R.; Strube, G.; Woetzel, J. Modular Construction: From Projects to Products. McKinsey & Company. 2019. Available online: https://www.mckinsey.com/~/media/mckinsey/business%20functions/operations/our%20insights/modular%20construction%20from%20projects%20to%20products%20new/modular-construction-from-projects-to-products-full-report-new.pdf (accessed on 24 October 2023).

89. Knaack, U.; Chung-Klatte, S.; Hasselbach, R. *Prefabricated Systems: Principles of Construction*; Birkhäuser: Berlin, Germany, 2012.
90. Boafo, F.E.; Hee Kim, J.; Tae Kim, J. Performance of Modular Prefabricated Architecture: Case Study-Based Review and Future Pathways. *Sustainability* **2016**, *6*, 558. [CrossRef]
91. Smith, R.E.; Quale, J.D. *Offsite Architecture: Constructing the Future*, 1st ed.; Routledge: London, UK, 2017.
92. *The Amazing Race: Volumetric Modular, vs. Kit-Of-Parts*; MARK III Construction: Sacramento, CA, USA, 2023.
93. Lawson, M.; Ogden, R.; Goodier, C. *Design in Modular Construction*, 1st ed.; CRC Press: London, UK, 2014.
94. Žegarac Leskovar, V.; Premrov, M. A Review of Architectural and Structural Design Typologies of Multi-Storey Timber Buildings in Europe. *Forests* **2021**, *12*, 757. [CrossRef]
95. Huß, W.; Kaufmann, M.; Merz, K. *Building in Timber—Room Modules*, 1st ed.; DETAIL: Munich, Germany, 2019.
96. Liew, J.Y.R.; Chua, Y.S.; Dai, Z. Steel concrete composite systems for modular construction of high-rise buildings. *Structures* **2019**, *21*, 135–149. [CrossRef]
97. Ali, M.; Armstrong, P. *Architecture of Tall Buildings*; McGraw-Hill Book Company: New York, NY, USA, 1995.
98. Tall Buildings Statement. *OPDC, Old Oak and Park Royal Development Corporation, Local Plan, Supporting Study*; Tall Buildings Statement: London, UK, 2018.
99. Ilgin, H.E.; Karjalainen, M. Tallest Timber Buildings: Main Architectural and Structural Design Considerations. In *Wood Industry—Past, Present and Future*; IntechOpen: London, UK, 2022.
100. Hua, H.; Hovestadt, L.; Li, B. Reconfigurable Modular System of Prefabricated Timber Grids. *Comput. Aided Des.* **2022**, *146*, 103230. [CrossRef]
101. *EN 1995:2004*; Eurocode 5: Design of Timber Structures. European Committee for Standardization: Brussels, Belgium, 2004.
102. Brown, R. *Fabulous Pre-Fab-Applying Modular Construction to Multifamily Residential Projects in Washington, DC*; Hickok Cole Architects: Washington, DC, USA, 2014.
103. Teribele, A.; Turkienicz, B. Generative model and fixing guidelines for modular volumetric architecture. *Rev. La Constr.* **2019**, *17*, 517–530. [CrossRef]
104. Mills, S.; Grove, D.; Egan, M. Breaking the pre-fabricated ceiling: Challenging the limits for modular high-rise. In Proceedings of the 2015 New York Conference Proceedings, CTBUH, New York, NY, USA, 26–30 October 2015; pp. 416–425.
105. Ramage, M.; Burridge, H.; Busse-Wicher, M.; Feredey, G.; Reynolds, T.; Shah, D.; Wu, G.; Yu, L.; Fleming, P.; DEnsley-Tingley, D.; et al. The wood from the trees: The use of timber in construction. *Renew. Sustain. Energy Rev.* **2016**, *68*, 333–359. [CrossRef]
106. Pastori, S.; Mazzucchelli, E.; Wallhagen, M. Hybrid timber-based structures: A state of the art review. *Constr. Build. Mater.* **2022**, *359*, 129505. [CrossRef]
107. Salvadori, V. Multi-Storey Timber-Based Buildings: An International Survey of Case-Studies with Five or More Storeys Over the Last Twenty Years. Ph.D. Thesis, Technische Universität Wien, Vienna, Austria, 2021.
108. Pan, Y.; Tannert, T.; Kaushik, K.; Xiong, H.; Ventura, E. Seismic performance of a proposed wood-concrete hybrid system for high-rise buildings. *Eng. Struct.* **2021**, *238*, 112194. [CrossRef]
109. Chen, Z.; Khan, K.D.R.; Khan, A.; Javed, K.; Liu, J. Exploration of the multidirectional stability and response of prefabricated volumetric modular steel structures. *J. Constr. Steel Res.* **2021**, *184*, 106826. [CrossRef]
110. Lawson, M.; Ogden, R.; Bergin, R. Application of modular construction in highrise buildings. *J. Archit. Eng.* **2012**, *18*, 148–154. [CrossRef]
111. Lawson, M.; Richards, J. Modular design for high-rise buildings. *Proc. Inst. Civ. Eng. Struct. Build.* **2010**, *163*, 151–164. [CrossRef]
112. Lawson, M. Light steel modular construction high-rise modular building. *SCI Steel Knowl.* **2016**, *1*, 1–8.
113. Ramaji, J.; Memari, A. Identification of structural issues in design and construction of multi-story modular buildings. In Proceedings of the 1st Residential Building Design and Construction Conference, Bethlehem, PA, USA, 20–21 February 2013; pp. 294–303.
114. Nieri, G.; Hein, C.; Smith, S. Hybrid structures in high-rise buildings: The use of appropriate materials. In Proceedings of the World Conference on Timber Engineering, Oslo, Norway, 19–22 June 2023; pp. 1017–1026.
115. *Building Systems: 3–8 Storey Modular Element Buildings*; Stora Enso: Helsinki, Finland, 2016.
116. Staib, G.; Dörrhöfer, A.; Rosenthal, M. *Components and Systems: Modular Construction—Design, Structure, New Technologies*; Birkhäuser: Munich, Germany, 2008.
117. Rajanayagam, H.; Poologanathan, K.; Gatheeshgar, P.; Varelis, G.; Sherlock, P.; Nagaratnam, B.; Hackney, P. A-State-Of-The-Art review on modular building connections. *Structures* **2021**, *34*, 1903–1922. [CrossRef]
118. Lacey, W.; Chen, W.; Hao, H.; Bi, K. Structural response of modular buildings—An overview. *J. Build. Eng.* **2018**, *16*, 45–56. [CrossRef]
119. Nadeem, G.; Safiee, A.; Bakar, A.; Karim, A.; Nasir, A. Connection design in modular steel construction: A review. *Structures* **2021**, *33*, 3239–3256. [CrossRef]
120. Monsberger, M.; Schickhofer, G.; Magg, A.K.; Ganster, K.; Neumayer, M.; Bretterklieber, T.; Kaufmann, C. Strategies for a Safe Integration of Water-Bearing MEP Systems in Prefabricated CLT Room Modules. In Proceedings of the Modular and Offsite Construction (MOC) Summit Proceedings, Banff, AL, Canada, 21–24 May 2019; pp. 397–404.
121. Groak, S. *The Idea of Building: Thought and Action in the Design and Production of Buildings*, 1st ed.; Taylor & Francis: London, UK, 1992.

122. Lehmann, S.; Kremer, P. Filling the Knowledge Gaps in Mass Timber Construction. *Mass Timber Constr. J.* **2023**, *6*, 1–10.
123. Askar, R.; Bragança, L.; Gervásio, H. Adaptability of Buildings: A Critical Review on the Concept Evolution. *Appl. Sci.* **2021**, *11*, 4483. [CrossRef]
124. Bogue, R. Design for Disassembly: A Critical Twenty-first Century Discipline. *Assem. Autom.* **2007**, *27*, 285–289. [CrossRef]
125. Ottenhaus, L.; Yan, Z.; Brandner, R.; Leardini, P.; Fink, G.; Jockwer, R. Design for adaptability, disassembly and reuse—A review of reversible timber connection systems. *Constr. Build. Mater.* **2023**, *400*, 132823. [CrossRef]
126. Schmidt, R., III; Austin, S. *Adaptable Architecture: Theory and Practice*, 1st ed.; Routledge: London, UK, 2016.
127. Chmielarski, P. Spatial Adaptability in The Context of Multifunctional Timber High-Rise Buildings. Master's Thesis, Delf University of Technology, Delft, The Netherlands, 2023.
128. Magdziak, M. Flexibility and Adaptability of the Living Space to the Changing Needs of Residents. *IOP Conf. Ser. Mater. Sci. Eng.* **2019**, *471*, 072011. [CrossRef]
129. Schneider, T.; Till, J. *Flexible Housing*, 1st ed.; Routledge: London, UK, 2008.
130. Durmisevic, E. Transformable Building Structures. Design for Dissassembly as a Way to Introduce Sustainable Engineering to Building Design & Construction. Ph.D. Thesis, TU Delft, Delft, The Netherland, 2006.
131. Dams, B.; Maskell, D.; Shea, A.; Allen, S.; Driesser, M.; Kretschmann, T.; Walker, P.; Emmitt, S. A circular construction evaluation framework to promote designing for disassembly and adaptability. *J. Clean. Prod.* **2021**, *316*, 128122. [CrossRef]
132. Stell, H.L. *Si-Modular Wood Framing System*; Metsa: Kreuzau, Germany, 2017.
133. Yan, Z.; Ottenhaus, L.; Leardini, P.; Jockwer, R. Performance of reversible timber connections in Australian light timber framed panelised construction. *J. Build. Eng.* **2022**, *61*, 105244. [CrossRef]
134. Arvaniti, E. Ensuring Adaptability in Modular Construction. Master's Thesis, Aalborg University, School of Architecture, Design and Planning, Aalborg, Denmark, 2020.
135. Nelson, A. *Toward a New Metropolis: The Opportunity to Rebuild America, The Brokings Institution Metropolitan Policy Program*; Brookings Institute: Washington, DC, USA, 2004.
136. Wood Construction. *Invest in Finland—A National Organization, Finland*; Wood Construction: Helsinki, Finland, 2006.
137. Hamburg, P.; Lellep, K.; Kiisa, M.; Vahtramäe, E.; Kivil, L.; Jalas, R.; Amado, M.; Guerreiro, L.; Kiisa, M.; Vahtramäe, E. International Study on Best Practices and Knowledge Gaps for Construction of High-Rise Timber Buildings. 2018. Available online: https://www.researchgate.net/publication/349899395 (accessed on 21 January 2024).
138. Roos, A.; Woxblom, L.; Mccluskey, D. The Influence of Architects and Structural Engineers on Timber Construction: Perceptions and Roles. *Silva Fenn.* **2010**, *44*, 871–884. [CrossRef]
139. Mcgregor, R.M.; Evison, D.; Ozanne, L. A pilot survey of structural engineers' perceptions of laminated veneer lumber in single-storey industrial buildings. *SESOC J.* **2011**, *24*, 25–28.
140. *CEI-Bois Roadmap 2010: Summary of Work Packages 1.1, 1.2 and 5.1*; INDUFOR, European Confederation of Woodworking Industries: Brussels, Belgium, 2004.
141. Arano, I.M.; Topi, C.; Pettenella, D.; Secco, L.; Masiero, M.; Follesa, M.; Fragiacomo, M.; Carnus, J.; Lefèvre, F.; Rigolot, E.; et al. *A Forest-Based Circular Bioeconomy for Southern Europe: Visions, Opportunities and Challenges: Reflections on the Bioeconomy*; European Forest Institute: Joensuu, Finland, 2018.

applied sciences

MDPI

Article

Analysis of Space Efficiency in High-Rise Timber Residential Towers

Hüseyin Emre Ilgın [1,*] and Özlem Nur Aslantamer [2]

1 School of Architecture, Faculty of Built Environment, Tampere University, P.O. Box 600, FI-33014 Tampere, Finland
2 Department of Interior Architecture and Environmental Design, Faculty of Art, Design and Architecture, Atılım University, Ankara 06830, Turkey; ozlem.aslantamer@atilim.edu.tr
* Correspondence: emre.ilgin@tuni.fi

Abstract: High-rise timber residential towers (≥eight-stories) represent a burgeoning and auspicious sector, predominantly due to their capability to provide significant ecological and financial advantages throughout their lifecycle. Like numerous other building types, spatial optimization in high-rise timber residential structures stands as a pivotal design factor essential for project viability. Presently, there exists no comprehensive investigation on space efficiency in such towers. This study analyzed data from 51 case studies to enhance understanding of the design considerations influencing space efficiency in high-rise timber residential towers. Key findings included (1) the average space efficiency within the examined cases was recorded at 83%, exhibiting variances ranging from 70% to 93% across different cases, (2) the average percentage of core area to gross floor area (GFA) was calculated at 10%, demonstrating fluctuations within the range of 4% to 21% across diverse scenarios, and (3) no notable distinction was observed in the effect of various core planning strategies on spatial efficiency when properly designed, and similar conclusions were drawn regarding building forms and structural materials. This research will aid in formulating design guidelines tailored for various stakeholders such as architectural designers involved in high-rise residential timber building developments.

Keywords: space efficiency; timber; high rise; residential building; gross floor area (GFA); net floor area (NFA)

1. Introduction

The global urban landscape is witnessing a significant paradigm shift in architectural planning and building practices, driven by imperatives of sustainability, resilience, and innovation. In this context, high-rise timber residential towers have emerged as a novel and promising solution, challenging conventional perceptions of skyscraper construction [1]. As the demand for urban housing continues to rise alongside concerns about climate change and resource depletion, exploring alternative building materials and methods becomes imperative [2].

Timber, once primarily associated with low-rise and traditional structures, is experiencing a renaissance in contemporary architectural discourse [3]. Advancements in engineered timber products, such as cross-laminated timber (CLT) have expanded the possibilities for timber construction to unprecedented heights [4,5], exemplified by projects like the 73 m tall HAUT in the Netherlands [6], and the 85 m tall Mjøstårnet in Norway [7]. The utilization of timber offers several compelling advantages over conventional building materials like concrete and steel, including lower carbon emissions, faster construction times, and enhanced thermal performance [8–13]. The integration of timber within interior spaces fosters esthetically pleasing settings, thereby upholding individual physiological health, affective disposition, and habitation quality [14,15].

The inherent hygroscopic nature of wood, particularly prominent in tropical and subtropical climates, presents significant challenges for mass timber construction, necessitating meticulous moisture management strategies integrating both passive and active solutions. Effectively safeguarding against moisture throughout storage, transportation, and operation entails the application of weather-resistant membranes to the exteriors of engineered wood products like CLT panels during the manufacturing process, as evidenced by prior research [16]. Moreover, ensuring the protection of exposed timber end-grains, particularly in vertical orientations vulnerable to direct rainfall, emerges as a critical consideration during the construction phase. While studies [17] have demonstrated the commendable performance of wooden structures in hot and humid environments like those prevalent in Australia, exemplified by cases such as Monterey, it is paramount to meticulously address climate-specific factors throughout the phases of design, construction, and maintenance to mitigate potential degradation risks.

The concept of high-rise timber residential towers represents a convergence of multiple disciplines, including architecture, engineering, materials science, and environmental sustainability [18]. This interdisciplinary approach is essential for comprehensively assessing the feasibility, safety, and long-term viability of timber skyscrapers [19]. Key considerations include structural integrity, fire resistance, acoustic performance, moisture management, and life-cycle analysis, among others [20]. As this high-tech and research-oriented progress advances, wooden building configurations will continuously challenge the limits of height, unveiling new prospects for vertical urbanization.

High-rise timber residential towers hold the potential to catalyze a shift towards more sustainable and resilient urban environments [21,22]. By sequestering carbon dioxide within their structural elements, timber buildings contribute to mitigating climate change while offering esthetically pleasing and biophilic living spaces [23]. Additionally, the use of prefabricated timber components can streamline construction processes, reduce waste generation, and minimize disruption to surrounding communities [24].

Research on high-rise timber residential towers is scientifically significant as it has the potential to transform the construction industry towards sustainability, combat climate change, guide urban planning, foster engineering innovation, and improve occupants' health and well-being. Timber's renewable properties and lower carbon footprint offer a viable substitute for traditional construction materials like concrete, aiding in reducing greenhouse gas emissions and serving as a carbon sink. Furthermore, the compact design of high-rise timber towers helps address urban sprawl issues, while innovative engineering solutions ensure their safety and durability. Additionally, the esthetic and natural benefits of timber construction can positively impact the mental and physical health of residents, making this research crucial for comprehensive urban development.

Space efficiency is a paramount consideration in architectural design, particularly within residential structures aiming to optimize usable area to enhance their financial appeal [25,26]. Despite its importance, academic literature offers limited exploration into the spatial efficiency of buildings [27]. Moreover, while comprehensive research exists on the technological, environmental, and commercial aspects of timber constructions (e.g., [28–32]), there is a notable absence of examination specifically focusing on spatial efficiency within this domain. This gap underscores the need for further investigation into how timber constructions can maximize space utilization while balancing other architectural, economic, and environmental considerations. Such research would not only contribute to advancing knowledge in the field but also provide valuable insights for architects, engineers, and developers seeking to design more efficient and sustainable timber buildings.

Among studies focusing on space efficiency in buildings, Tuure and Ilgın [33] examined spatial effectiveness in fifty-five mid-rise timber dwellings in the Finnish context. The study found space efficiency ranging from 78% to 88%. Ilgın [25–27,34–36] investigated space efficiency in various types of skyscrapers, including office, residential, and mixed-use towers. Common findings across these studies included the frequent use of outriggered frame systems and a central core, and an inverse relationship between tower height and space efficiency. Okbaz and Sev [37] developed a spatial effectiveness model for 11 tall office structures with free-form configurations. The study revealed that building structure significantly impacts spatial effectiveness, with conical configurations demonstrating the highest efficiency ratio. Ibrahimy et al. [38] evaluated space utilization efficiency in residential buildings in Kabul City, noting deviations from regulations due to insufficient attention to interior design and governmental directives. Goessler and Kaluarachchi [39] explored the impact of intelligent technologies on compact urban dwellings, suggesting a two- to threefold improvement in space efficiency compared to conventional configurations. Hamid et al. [40] found that corner placement maximizes land utilization efficiency in detached residences in Sudan. Suga [41] analyzed spatial effectiveness in hotels, highlighting the positive outcomes of prioritizing spatial efficiency. Arslan Kılınç [42] studied elements affecting core and load-bearing structures in box-shaped towers, noting an increase in core area with building height and no correlation between construction material and spatial efficiency. Von Both [43] introduced a stakeholder analysis approach to enhance area and space efficiency in planning processes. Höjer and Mjörnell [44] proposed a four-phase construction guideline for digitalization's impact on interior space demand and supply. Nam and Shim [45] investigated lease duration and corner configurations in high-rise buildings, finding a negligible impact of corner cuts on spatial efficiency and notable effects of lease duration. Zhang et al. [46] developed a free-form configuration to enhance solar radiation absorption in cold regions, resulting in improved radiation acquisition with minimal decline in spatial efficiency. Sev and Özgen [47] explored spatial effectiveness in high-rise offices, highlighting the influence of structural systems and core configuration. Saari et al. [48] studied the interplay between spatial optimization and the comprehensive expenditure of towering office edifices. Their findings unveiled a significant influence of enhanced spatial efficiency on attaining the targeted benchmarks of indoor environmental comfort. Kim and Elnimeiri [49] assessed spatial efficiency ratios in mixed-use high-rise structures, highlighting the integration of optimal structural systems and forms to enhance spatial effectiveness.

As evident from the literature review conducted earlier, the current scholarly literature lacks sufficient exploration of spatial efficiency within the domain of high-rise timber residential developments, which constitute a notable subclass of tall wooden structures. To fill this significant knowledge gap and gain a deeper understanding of global trends, this article aims to systematically map, aggregate, and analyze data pertaining to spatial efficiency in high-rise timber residential buildings. Through an extensive examination of 51 cases worldwide, the study scrutinizes critical architectural and structural attributes, seeking to elucidate patterns and insights in this underexplored domain.

To fathom spatial optimization alongside pertinent parameters for the blueprinting and erection of these complexes, the research inquiries are delineated as follows: (1) What delineates the spatial effectiveness within the sampled group of buildings, and what spectrum encapsulates these proportions? (2) What ramifications do architectural and structural planning factors impart on the spatial efficiency of these cases? (3) What actionable recommendations may be proffered to enhance spatial efficiency?

The main goal of this study is to carry out a rigorous and systematic examination aimed at advancing the comprehension of spatial efficiency within high-rise timber residential towers. Through a comprehensive analysis of data extracted from 51 case studies, the study seeks to elucidate critical design variables impacting spatial optimization in these structures. By scrutinizing core arrangements, building forms, structural material and system selections, and spatial efficiency indicators, the research endeavors to offer empirically driven insights and develop methodological frameworks tailored to inform the practices of architectural designers and relevant stakeholders engaged in the development of high-rise timber residential structures.

Concerning the description of tall or high-rise structures, a universally accepted criterion regarding their elevation and story count is yet to be established, and even the categorization of 'tall' or 'high-rise' in the realm of wood constructions remains a subject of continuing discourse. In this exposition, a 'high-rise timber building' is explicitly characterized as an edifice comprising eight stories or more [50].

It is worth noting that the hygroscopicity of wood, referring to its ability to absorb and release moisture from the surrounding environment, plays a crucial role in the space efficiency of timber-framed buildings [51–53]. Wood has a natural tendency to swell and shrink in response to changes in humidity levels, which can affect the dimensional stability of structural elements and finishes within a building. When exposed to high humidity, wood absorbs moisture, causing it to swell, potentially leading to warping, buckling, or distortion of building components. Conversely, in dry conditions, wood releases moisture and may shrink, leading to gaps or cracks. Understanding and managing the hygroscopic behavior of wood is essential for ensuring the long-term performance and durability of wooden structures, as well as optimizing space efficiency by minimizing the impact of dimensional changes on building integrity and esthetics. Proper design, construction techniques, and moisture control measures, such as adequate ventilation, vapor barriers, and moisture-resistant finishes, are essential for mitigating the effects of wood hygroscopicity and maximizing space efficiency in buildings.

The ensuing sections of this article are organized in the following manner: Initially, an exposition of the utilized research method is provided. Subsequently, the results derived from an assessment of 51 meticulously examined case study structures are delineated, succeeded by the discussion incorporating potential future studies. Lastly, conclusions are outlined.

2. Researching Space Efficiency in High-Rise Timber Residential Towers

Researching space efficiency in high-rise timber residential structures is crucial for following reasons, encompassing environmental sustainability, urban planning, resource management, and human well-being:

1. Environmental sustainability [54–56]: Timber, as a renewable and carbon-sequestering material, offers significant potential for reducing the carbon footprint of buildings. High-rise timber residential buildings have a lower embodied carbon footprint compared to traditional structures. Space efficiency studies in such buildings enable the quantification of the environmental benefits achieved through the optimized use of timber resources, including reduced energy consumption, greenhouse gas emissions, and resource depletion.
2. Urban planning and density [57–59]: Urban areas face challenges related to population growth and limited land availability. High-rise timber buildings provide a sustainable solution by offering vertical density while minimizing land use. Space efficiency analyses help urban planners determine the optimal distribution of residential units within a building footprint, maximizing population density without compromising livability or architectural integrity. This approach considers factors such as building height, floor area ratio, and population density to inform urban development policies and strategies.
3. Resource management [60–62]: Timber is known for its lightweight nature and low energy requirements during manufacturing and construction. Space efficiency studies in high-rise timber residential buildings focus on optimizing material usage, minimizing waste, and maximizing resource utilization. Scientific methods, such as life cycle assessments and material flow analyses, quantify the environmental impacts associated with different space configurations, guiding decision-making processes towards more sustainable construction practices.
4. Human well-being and comfort [63–65]: Research in environmental psychology and building science emphasizes the importance of indoor environmental quality for occupants' health and well-being. Space-efficient designs in high-rise timber buildings aim to enhance natural lighting, ventilation, and thermal comfort while promoting access to outdoor spaces and views. Scientific measurements, such as daylighting analysis, indoor air quality monitoring, and occupant satisfaction surveys, provide empirical evidence of the positive effects of space efficiency on human health, productivity, and satisfaction.

5. Structural integrity and safety [66–68]: High-rise timber buildings must meet stringent structural requirements to ensure occupant safety and building resilience. Space efficiency studies consider structural design parameters, such as load-bearing capacity, lateral stability, and fire resistance, in conjunction with spatial layouts. Advanced computational modeling and simulation techniques, including finite element analysis and performance-based design, assess the structural performance of different space configurations under various loading and environmental conditions, informing design decisions and regulatory compliance.
6. Economic viability [69–71]: Space efficiency directly impacts the economic feasibility and profitability of high-rise timber residential projects. Scientific cost–benefit analyses evaluate the financial implications of different space configurations, taking into account construction costs, operational expenses, rental income, and market demand. Optimization algorithms and decision support tools help developers and investors identify the most economically viable design solutions that maximize returns on investment while meeting sustainability objectives.

The authors underscore the significance of space efficiency in boosting financial returns, occupant well-being, and environmental sustainability within vertical urban settings, albeit without delving into sustainable planning aspects due to data constraints. It intentionally overlooks certain factors like circulation flow, space flexibility, and adaptability, as these warrant separate research endeavors. Additionally, this study focuses on proportional evaluations of critical building areas such as the core, rather than floor area dimensions and structural elements such as columns and pillars.

3. Research Method

Case studies were employed to classify and consolidate data regarding contemporary high-rise residential wooden structures, thereby enabling a systematic examination and evaluation of their architectural and structural attributes (Figure 1). The adoption of case study analyses stands as a widespread approach in evaluations linked with the built environment [72–74]. This investigation contained a thorough scrutiny, embracing a collective of 51 [out of a total of 56 buildings in the Council on Tall Buildings and Urban Habitat (CTBUH) database] [50] tower case studies. These towers were either in completion or under construction, with a distinct focus on structures primarily composed of timber. The criteria for selection comprised buildings comprising eight stories or more, with data drawn from documentation sources by (CTBUH).

CTBUH holds significant acclaim for its authoritative role in defining the heights of skyscrapers worldwide, earning the esteemed title of 'The World's Tallest Building'. Beyond this, CTBUH spearheads the 'Buildings of Distinction' initiative, which celebrates exceptional architectural designs by installing public signboards and plaques. Operating on a universal scale, CTBUH works as a pivotal stage for disseminating leading-edge knowledge and facilitating commercial interactions within the realm of tall building construction and urban development. Through its multifaceted endeavors, CTBUH not only establishes standards but also fosters innovation, collaboration, and recognition within the architectural and urban planning communities, contributing to the ongoing advancement of the built environment.

The analyzed structures in this paper were geographically distributed, encompassing thirty-three locations across Europe, as seen in Figure 2 (with nine from Norway, nine from Sweden, five from France, five from Finland, two from the Netherlands, two from Germany, and one from Italy), seven from North America (comprising four from Canada and three from the USA), and seven from the UK, besides four from Australia, as outlined in Appendix A.

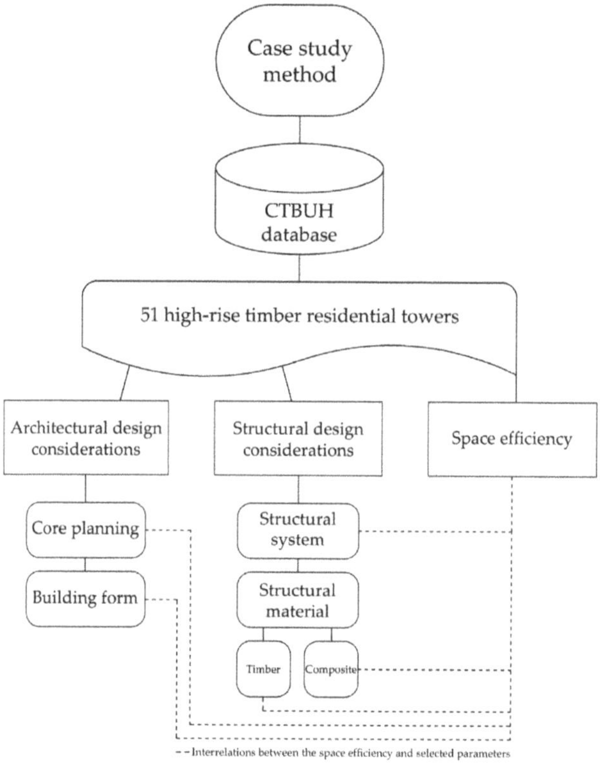

Figure 1. Methodology flowchart (image created by authors).

In the realm of wooden residential constructions, decision-making chiefly adheres to architectural and structural necessity-driven requisites, alongside the primary purpose of the edifice. The principal attributes encompass the subsequent factors:

Architecturally:

- Core arrangement impacting the configuration of vertical movement and the allocation of shafts.
- Building form influencing sizes and pattern of the floor slabs.

Structurally:

- Structural system influencing disposition and sizes of the load-bearing constituents.
- Structural material affecting the proportions of the load-bearing members.

The core planning scheme advocated by [33] was selected for implementation, owing to its comprehensive structure, encapsulating the following groupings: (1) central core, (2) atrium core, (3) external core, and (4) peripheral core. Building forms are classified as follows [33]: (1) prismatic, (2) setback, (3) tapered, (4) twisted, (5) tilted, and (6) free forms, as outlined in Appendix B.

Structural materials could be classified into two primary clusters: (a) 'timber' or 'all-timber', and (b) composite or hybrid materials, which entail arrangements, timber combined with concrete, timber combined with steel, or timber combined with both concrete and steel. Our study particularly focuses on main structural elements such as columns and beams, while disregarding the examination of slabs. It is crucial to highlight that the material makeup of structural members on the basement floor does not change the categorization of the load-bearing system.

Figure 2. Case studies from various locations on the world map (image created by authors).

According to the categorization of structural materials, a primary criterion for a structure to be classed as 'timber' is that both its main vertical and horizontal load-bearing components are completely crafted from timber [50]. It is important to note that a 'timber' building can combine non-timber connections in certain areas among timber members, as in the case of Lighthouse Joensuu [75].

On the contrary, in composite or hybrid types encompassing timber, an extensive share of the structural framework includes materials other than timber, notably steel, concrete, or a blend of both. In buildings amalgamating timber and concrete, it is frequent to observe a concrete core supporting a timber structure, as demonstrated by HAUT [76,77]. Alternatively, in buildings merging timber and steel, an extensive share of the structure relies on steel, as evidenced by Tallwood 1 at District 56 [78]. Analogously, a hybrid category integrating timber, concrete, and steel utilizes an amalgamation of all three materials to bear main forces. A common pattern encompasses a concrete service core working alongside steel beams and columns, whereas timber is employed in separation walls as in the case of De Karel Doorman in Netherlands [79].

In the realm of enhancing lateral stiffness in high-rise buildings, especially concerning challenges like wind and earthquake forces, numerous load-bearing systems and categorizations have been applied in real-world situations and scrutinized within the existing literature [80]. In this paper, the authors used the structural system categorization proposed by [33] owing to its complete coverage: (i) rigid-frame system; (ii) shear-frame system including shear-trussed frame and shear-walled frame systems; and (iii) shear-wall system, as seen in Figure 3.

Rigid frame system (Composed of beams and columns)	**Shear-frame system** (Composed of shear wall/truss and frame) with the subgroups 'shear trussed frame' and 'shear walled frame'	**Shear wall system** (Composed of perforated or solid shear walls)
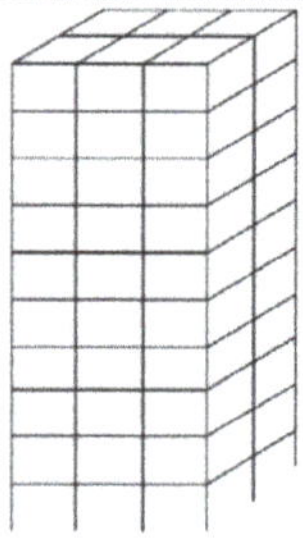	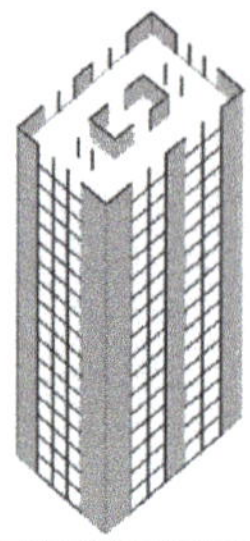	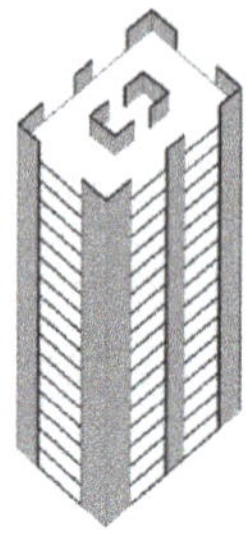

Figure 3. High-rise timber tower load-bearing systems (image created by authors).

Moreover, it is noteworthy to highlight that, in the realm of timber construction, configurations pertaining to structural integrity such as outriggered frames [79,80], assorted tube systems (comprising framed, diagrid-framed, trussed, and bundled tubes) [81,82], as well as buttressed cores [83,84] predominantly find application in buildings of extraordinary height exceeding 300 m. These systems are favored due to their efficacy and economic viability. Therefore, these load-bearing systems were omitted from consideration in our study, as its focus lies on tall buildings rather than exceptionally tall ones. Nevertheless, as illustrated by the case of Treet (Figure 4) in Norway [85,86], there exists a singular instance of a timber-based residential high-rise structure incorporating a tube system.

Figure 4. Treet (photo courtesy of ARTEC).

Space efficiency concerns the relationship between net floor area (NFA) and GFA. Spatial effectiveness has considerable importance, notably for stakeholders, as it involves the efficient utilization of floor plan regions to realize maximal returns on financial commitments. The level of spatial effectiveness predominantly depends on diverse factors, such as the selection of structural frameworks and architectural design [25].

In the methodology of this manuscript, the evaluation of spatial effectiveness was performed by computing the ratio of NFA to GFA, furnishing a quantitative gauge of the efficiency of space utilization. Concurrently, the computation of core relative to GFA entailed assessing the proportion of the service core to GFA, providing perspectives into the proportional distribution of space allocated to fundamental structural and service elements within the comprehensive building framework.

The computation of spatial effectiveness encompasses two primary ratios [26,27]: NFA to GFA and core to GFA. NFA is derived by deducting the service core area from GFA. This deduction aids in isolating the functional zones where activities occur, excluding areas designated for infrastructure and auxiliary services. The ratio of NFA to GFA offers a quantitative assessment of the efficiency with which the accessible floor area is utilized for practical purposes, indicating the effectiveness of space distribution.

For exhaustive particulars and precise results of these computations, readers are urged to consult Appendix C. This part furnishes a more profound comprehension of the utilized methodology, guaranteeing transparency and enabling stakeholders to evaluate the authenticity and dependability of the spatial effectiveness measurements presented.

4. Results

In this part, key architectural design parameters, like core planning and form, key structural design parameters, including structural material and structural system, and space efficiency, with its interrelationship to several design considerations, are provided.

4.1. Main Architectural Design Parameters: Core Planning and Building Form

Through the analysis of Figure 5, it becomes apparent that the central core predominates as the frequently employed core configuration, comprising 55% of occurrences. Subsequently, peripheral arrangement follows, representing 45% of occurrences.

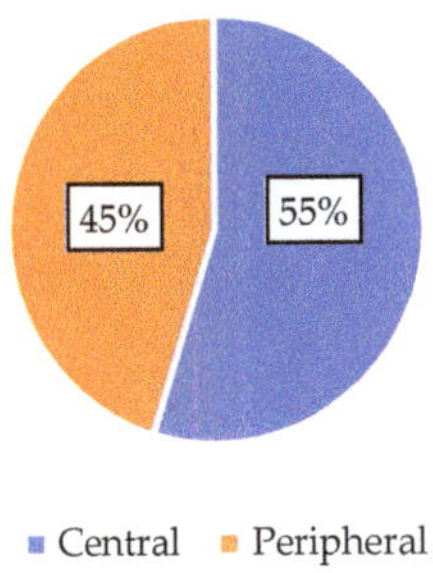

Figure 5. Case studies categorized by core planning.

In high-rise residential wooden projects, a central core configuration offers numerous benefits [87]: 1. Structural stability against lateral forces, reducing reinforcement needs. 2. Maximized usable space by centralizing essential services. 3. Enhanced efficiency with shorter travel distances and streamlined access. 4. Improved natural light and views due to larger windows. 5. Flexibility for future adaptations with minimal structural changes. 6. Cost savings through optimized structural efficiency. 7. Facade design freedom with core bearing most of the load. 8. Enhanced fire safety with fire-resistant zones in the core.

In accordance with the morphological categorization for high-rise residential timber structures, the prismatic configuration, constituting 78% of instances, emerges as the prevailing architectural inclination, whereas unbounded configurations comprise 22% of the aggregate, as shown in Figure 6.

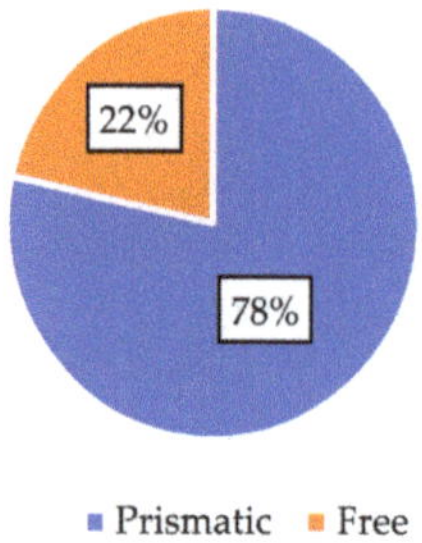

Figure 6. Case studies categorized by building form.

The prevalent application of prismatic forms in high-rise timber residential buildings can be attributed to several factors [33]: 1. Structural efficiency, distributing loads uniformly. 2. Material optimization, aligning with standardized wooden construction materials. 3. Construction simplicity, with regular geometry reducing time and labor costs. 4. Space utilization, maximizing interior space with rectangular or square floor plans. 5. Facade design flexibility, allowing for various architectural features and energy efficiency. 6. Modular construction potential, enhancing efficiency and quality control. 7. Fire safety considerations, with efficient compartmentalization and modern techniques mitigating risks.

Table 1 shows the comparison between our findings on core planning and building form and findings on non-timber skyscrapers [25–27].

Table 1. Comparison of core planning and building form.

	Our Findings	Mixed-Use Skyscrapers [25] (64 Cases)	Residential Skyscrapers [26] (27 Cases)	Office Skyscrapers [27] (44 Cases)
Core planning	Central (55%)	Central (100%)	Central (93%)	Central (95%)
	Peripheral (45%)		Peripheral (7%)	External (5%)
Building form	Prismatic (78%) Free (22%)	Prismatic (17%) Setback (17%) Tapered (38%) Twisted (1%) Free (27%)	Prismatic (56%) Setback (11%) Tapered (11%) Twisted (7%) Free (15%)	Prismatic (21%) Setback (18%) Tapered (27%) Twisted (2%) Free (32%)

4.2. Main Structural Design Parameters: Structural Material and Structural System

Figure 7 highlights a significant instance of pure timber assembly, constituting 53% of the data set, whereas composite structures follow closely, representing 47% within a collection encompassing 51 high-rise residential timber structures.

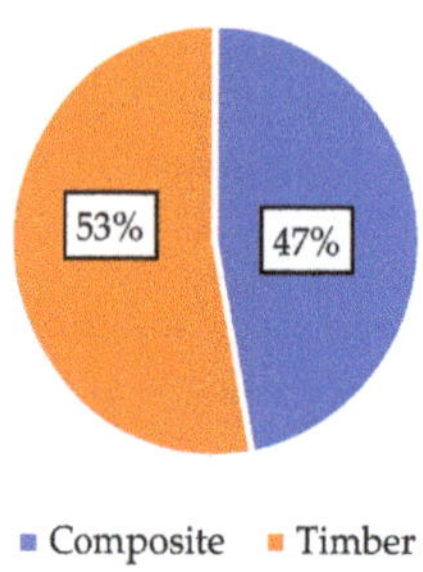

Figure 7. Case studies categorized by structural material.

In European high-rise residential wooden constructions, wood is preferred due to proximity to woodlands, sustainability goals, and cost savings [88]. Availability of indigenous wood supports sustainability. Wood construction helps reduce carbon emissions and aligns with Europe's sustainability objectives. Its suitability for tall buildings, ease of construction, and environmental benefits make it a favored choice for many European projects.

Figure 8 depicts composite configurations characterized by the amalgamation of load-bearing components. Timber paired with concrete emerged as the prevailing preference, comprising 67% of occurrences. Subsequently, timber used with both concrete and steel is noted in 21% of instances. Conversely, timber used with steel is the least common, manifesting in merely three occurrences.

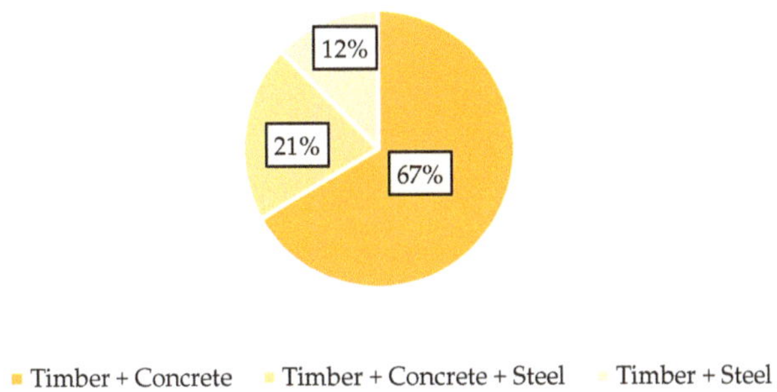

Figure 8. Composite case studies categorized by structural material combinations.

In hybrid configurations, incorporating concrete in the core element offers several advantages. Firstly, it boosts the overall transverse stiffness and durability, addressing structural needs [89]. Secondly, it enhances fire resilience due to concrete's properties. Thirdly, concrete helps dampen vibrations caused by wind forces, common in tall buildings [80]. Notably, timber and concrete hybrid techniques benefit from concrete cores, significantly improving transverse rigidity [90]. Additionally, including concrete cores in architectural plans, exemplified by the Brock Commons Tallwood House project, simplifies authorization processes [91]. This ease can be attributed to the widespread use of concrete cores in traditional skyscrapers. Crucially, in this context, fire escape staircases are housed within concrete cores, ensuring their non-flammable construction.

Illustrated in Figure 9, shear-frame systems are favored, utilized at a frequency of 57%, followed by the shear-wall mechanism, constituting 41% of cases. Concerning shear-frame systems, shear-walled frames are primarily used, totaling 26 instances, with a significant lead.

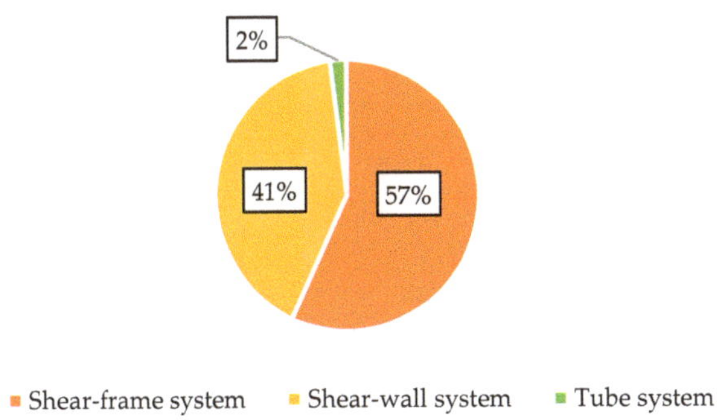

Figure 9. Case studies categorized by structural system.

In shear-frame systems, which include shear-trussed frame and shear-walled frame setups, the drawbacks of rigid frames compared to shear trusses or walls, and the limitations of shear trusses or walls compared to rigid frames, are balanced out when used together [92]. For instance, Ascent, the tallest timber building, demonstrates this synergy. This explains why shear-frame systems are commonly used. The popularity of shear-wall systems in 21 towers can be attributed to several factors [93,94]: faster construction due to their simple design, compatibility with prefabrication, effective distribution of lateral forces promoting sustainability, and providing sufficient rigidity for structures up to around 35 stories high, ensuring safety against wind and earthquakes.

Table 2 demonstrates the comparison between our findings on structural material and structural system and findings on non-timber skyscrapers [25–27].

Table 2. Comparison of structural material and system.

	Our Findings	Mixed-Use Skyscrapers [25] (64 Cases)	Residential Skyscrapers [26] (27 Cases)	Office Skyscrapers [27] (44 Cases)
Structural material	Timber (53%) Composite (47%)	Concrete (27%) Composite (72%) Steel (1%)	Concrete (96%) Composite (4%)	Concrete (14%) Composite (80%) Steel (6%)
Structural system	Shear frame (57%) Shear wall (41%) Tube (2%)	Outriggered frame (73%) Tube (17%) Buttressed core (3%) Mega column & core (6%) Shear frame (1%)	Outriggered frame (74%) Tube (22%) Buttressed core (4%)	Outriggered frame (59%) Tube (23%) Mega column & core (7%) Shear frame (11%)

4.3. Space Efficiency in High-Rise Timber Residential Buildings

In our paper, by examining 51 cases, the average spatial effectiveness and ratio of core to GFA were determined to be approximately 83% and 10%, respectively. The spectrum of measurements varied from a minimum of 70% and 4% to a maximum of 93% and 21%, respectively. Table 3 indicates the comparison between our findings on space efficiency and ratio of core to GFA and findings on non-timber skyscrapers [25–27].

Table 3. Comparison of average space efficiency and ratio of core to GFA.

	Our Findings	Mixed-Use Skyscrapers [25] (64 Cases)	Residential Skyscrapers [26] (27 Cases)	Office Skyscrapers [27] (44 Cases)
Average space efficiency	83% (max. 93%, min. 70%)	71% (max. 84%, min. 55%)	76% (max. 84%, min. 56%)	71% (max. 82%, min. 63%)
Average ratio of core to GFA	10% (max. 21%, min. 4%)	26% (max. 38%, min. 16%)	19% (max. 36%, min. 11%)	26% (max. 36%, min. 15%)

The INTRO Residential Tower, Aveo Bella Vista, Arbora Condominium, and Ascent exemplify remarkable advancements in space utilization and structural design, boasting unparalleled spatial efficiencies ranging from 91% to 93% and setting a benchmark with the smallest core-to-GFA ratio within the study sample, as illustrated in Appendix C. This exceptional performance is underpinned by meticulously engineered cores that prioritize compactness through an optimized service area and shaft organization, resulting in maximized usable floor space. Moreover, these towers employ a shear-walled frame system, enhancing their resilience against both vertical and horizontal loads by leveraging the compact cross sections of their structural members to efficiently distribute and withstand forces. This combination of innovative design principles not only ensures superior space efficiency but also reinforces the structural integrity of these buildings, setting a new standard for sustainable urban development.

4.3.1. Interrelationship of Space Efficiency and Core Planning

Figure 10 offers a graphical representation summarizing data related to selected towers, with a focus on their core planning strategies. The graph illustrates bars on the right side, indicating the total number of such structures, classified by their core types. Furthermore, blue dots are utilized to indicate the space efficiency of these towers for each respective core type. Additionally, a gray bar is included in the graph to visually represent the frequency of structures within the case study set that possess the same core type.

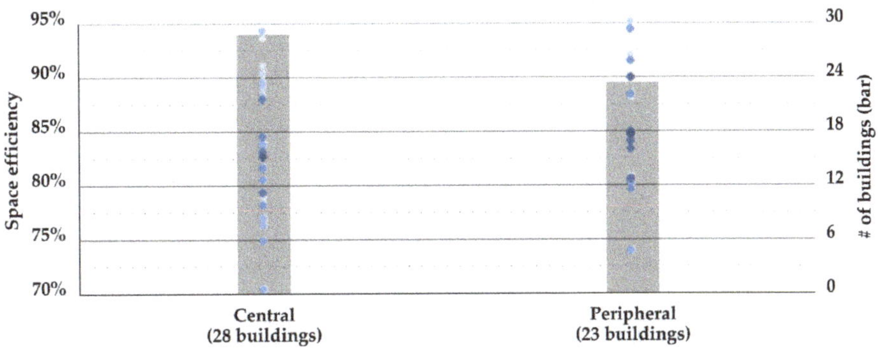

Figure 10. Interrelationship between space efficiency and core planning (figure by authors).

In the realm of the core arrangement utilized in high-rise timber residential towers, the central type emerges as the predominant option, encompassing 28 instances. This type showcases spatial efficiency levels ranging from 71% to 92%, averaging 82%. Conversely, case studies with the peripheral core type, totaling 23, exhib spatial efficiency ranging from 70% to 93%, averaging 83%. Thus, the average spatial efficiency demonstrates little variance between different core types in this study.

4.3.2. Interrelationship of Space Efficiency and Building Form

Figure 11 shows the interrelationship of space efficiency and building form. The graph demonstrates bars on the right side, representing the total count of these structures classified by their forms. Additionally, green dots are used to denote the spatial efficiency of these towers for each specific form. Furthermore, a gray bar is incorporated in the graph to visually depict the frequency of towers within the case study set sharing the same form.

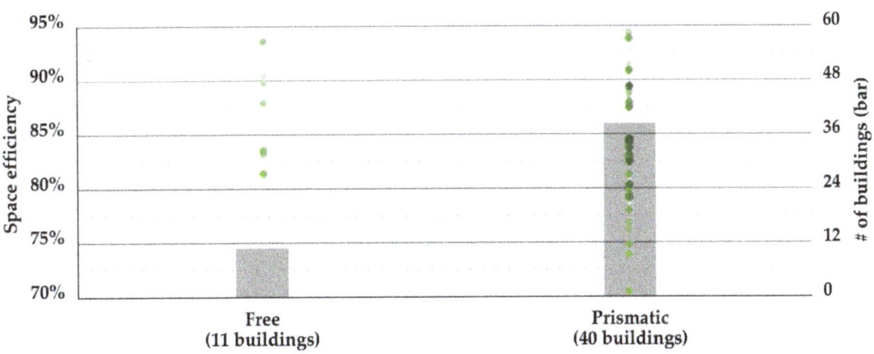

Figure 11. Interrelationship between space efficiency and building form (figure by authors).

In the realm of building forms of case study towers, prismatic forms emerge as the prevailing choice, observed in 40 cases. These configurations demonstrate spatial efficiency levels spanning from 70% to 93%, with an average of 82%. Conversely, examples featuring the free type, totaling 11 cases, display spatial efficiency ranging from 81% to 92%, with an average of 85%. Hence, the average spatial efficiency shows no major variation between building forms.

4.3.3. Interrelationship of Space Efficiency and Structural System

Figure 12 gives a graphical overview of the information about the analyzed buildings, specifically emphasizing their load-bearing systems. The chart presents bars on the right-hand side, indicating the total number of cases grouped by their structural systems. Moreover, orange dots are employed to indicate the spatial efficiency of these buildings for each respective structural system. Furthermore, a gray bar is integrated into the graph to visually represent the occurrence frequency of buildings sharing identical structural systems within the case study set.

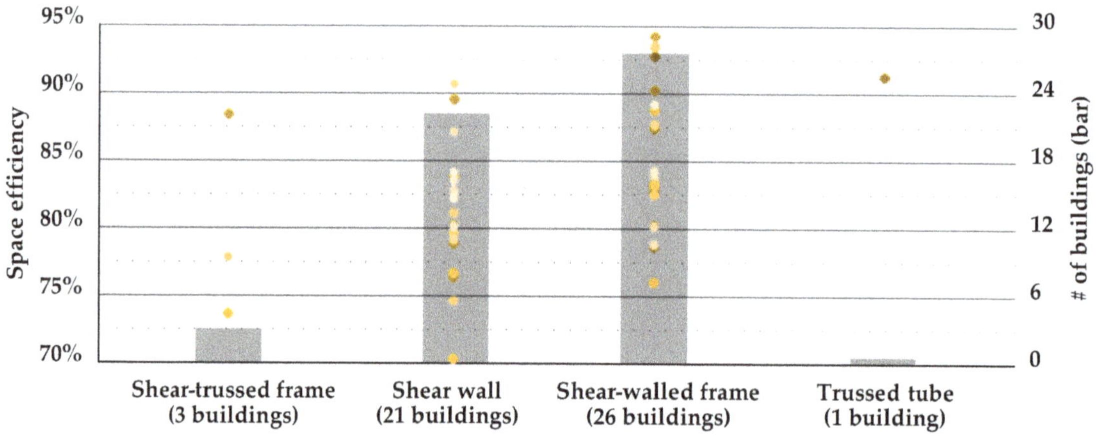

Figure 12. Interrelationship between space efficiency and structural system (figure by authors).

In the realm of structural systems used in our case study towers, shear-walled frames have become the dominant preference, noted in 26 instances. These designs show spatial efficiency levels ranging from 76% to 93%, averaging 85%. Conversely, instances utilizing the second most favored structural system, namely shear wall, totaling 21 cases, demonstrate spatial efficiency varying from 70% to 90%, with an average of 80%. Hence, there exists a 5% variance in average spatial efficiency between the predominantly employed load-bearing systems.

4.3.4. Interrelationship of Space Efficiency and Structural Material

Figure 13 offers an image representation of the data concerning the examined buildings, with a particular focus on their structural materials. The graph displays bars on the right side, showing the total count of buildings categorized by their structural materials. Additionally, orange dots are used to indicate the space efficiency of these towers for each specific structural material. Furthermore, a gray bar is incorporated in the graph to visually depict the frequency of cases within the case study set sharing the same structural material.

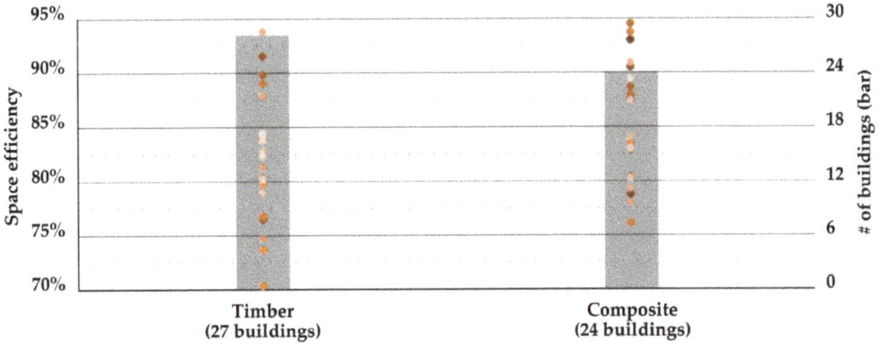

Figure 13. Interrelationship between space efficiency and structural material (figure by authors).

In high-rise timber residential towers, (pure) timber and composite materials are predominantly utilized for structural purposes. Timber and composite materials exhibit spatial efficiency levels ranging from 70% to 92% and 76% to 93%, with an average of 81% and 84%, respectively. Thus, there is only a 3% difference in average spatial efficiency between the primarily utilized structural materials in our case studies.

4.3.5. Interrelationship of Space Efficiency and Composite Structural Material Combinations

Figure 14 presents a graphical overview of the data pertaining to the examined towers, specifically highlighting their composite structural material combinations. The chart features bars on the right-hand side, illustrating the total count of composite structures categorized by material combinations. Additionally, orange dots are employed to signify the spatial efficiency of these towers for each respective composite material combination. Furthermore, a gray bar is integrated into the graph to visually represent the occurrence frequency of buildings within the case study set sharing identical combinations.

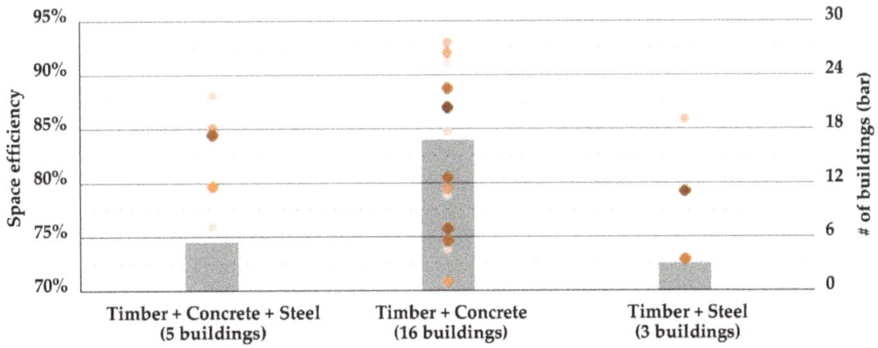

Figure 14. Interrelationship between space efficiency and composite structural material combinations (figure by authors).

In composite structural material combinations employed in the case study sample, the amalgamation of timber and concrete occurs as the most adopted option, accounting for 16 cases out of 24, followed by the combination of timber, concrete, and steel composite. The timber and concrete hybrid composition demonstrates spatial efficiency levels ranging from 76% to 93%, with an average of 84%, while the combination of timber, concrete, and steel has a spatial efficiency ranging from 80% to 89%, averaging 85%. Consequently, there is a minimal difference in average spatial efficiency between these combinations.

5. Discussion

This article aims to systematically collect and amalgamate exhaustive information pertaining to 51 modern high-rise residential timber edifices. The primary emphasis of this inquiry is on spatial effectiveness concerning architectural and structural aspects of these green towers, aiming to augment our understanding of the intricacies embedded within the design and erection of high-rise structures. Eventually, this study endeavors to contribute significantly to the extant repository of knowledge in the realm of design and construction of these green structures.

The outcomes delineated in this manuscript exhibit resemblances and distinctions in contrast to prior research endeavors, exemplified by the study conducted by, e.g., [33]. Succinctly summarized, the principal discoveries from this investigation are as follows:

(1) Central cores were the predominant core arrangements, while peripheral configurations ranked as the other preferred alternative. Prismatic forms emerged as the most preferred selections.
(2) The widespread utilization of timber as a building material was predominant, with amalgamations of timber and concrete following closely as the subsequent choice. Load-bearing systems were primarily defined by the application of shear-walled frames and shear-wall configurations.
(3) The average space utilization within the examined cases was recorded at 83%, exhibiting variances ranging from 70% to 93% across different cases.
(4) The average proportion of core area-to-GFA was calculated at 10%, demonstrating fluctuations within the range of 4% to 21% across diverse scenarios.
(5) No notable distinction was observed in the effect of various core planning strategies on spatial efficiency when properly designed, and similar conclusions were drawn regarding building forms and structural materials.

In modern high-rise timber residential buildings, the central core configuration prevails, supported by extensive research [25–27,33], reflecting a broader architectural and construction preference. This configuration, housing essential vertical circulation elements, offers numerous advantages. It bolsters structural stability, maximizes floor area efficiency by consolidating circulation and service functions, and facilitates smooth vertical transportation, reducing congestion. Additionally, it allows for architectural flexibility, particularly beneficial in mixed-use developments. In this study, the majority of structures exhibit a prismatic configuration, favored for its simplicity compared to more complex geometries [25,33]. Prismatic structures align well with conventional building techniques, aiding in project execution. Their rectangular floor plans optimize spatial utilization and functional efficiency, simplifying room layouts for occupants. Moreover, their modular characteristics contribute to sustainable and cost-effective construction processes by optimizing material and resource utilization.

A notable trend in high-rise residential construction is the increasing use of timber and concrete composite materials, with this blend becoming the preferred choice due to its flexibility and numerous advantages [95]. This combination capitalizes on timber's warmth and esthetic appeal along with concrete's strength and versatility [96,97]. In the structural systems of timber buildings, a clear hierarchy has emerged, with tall timber projects favoring shear-walled frames and shear-wall systems for their effective handling of lateral loads and provision of stability, particularly against wind and seismic forces.

A proposed standard for evaluating the spatial efficiency of tall buildings aims for a benchmark of 75% utilization [98]. Investigations into high-rise edifices designed for office, residential, and mixed-use purposes, employing non-timber materials, have unveiled a spectrum of space optimization levels, typically clustering within the range of 71% to 76% [25–27]. These analyses delineate variances in the allocation of core area relative to GFA, spanning from 19% to 26%. Critically, these inquiries capture the extremities of space efficiency and the core area-to-GFA ratio.

Within the realm of office skyscrapers, the least efficient space utilization recorded was 63%, while the pinnacle reached 82%. Similarly, the core area-to-GFA ratio oscillated between a minimum of 15% and a maximum of 36%. Analogous fluctuations were observed in residential and mixed-use high-rise developments. Furthermore, a study focused on over fifty mid-rise timber residential units in Finland depicted a space utilization range between approximately 78% and 88%, with an average efficiency of 83% [33].

In our research, drawing upon an analysis of 51 case studies, the average space utilization and core area-to-GFA ratio were calculated at approximately 83% and 10%, respectively. The spectrum of measurements portrayed a wide gamut, with the lowest recorded space efficiency at 70% and the highest at 93%, while the core area-to-GFA ratio ranged from 4% to 21%.

This study did not identify notable distinctions in the impact of different core planning strategies on space efficiency, nor were distinct conclusions drawn regarding building forms and structural materials in modern high-rise residential timber structures. Choosing an optimal core type could have potentially mitigated notable ratio variations among these clusters. Likewise, concerning spatial efficiency, implementing custom-tailored building forms and structural materials could have prevented significant disparities among the various groups of buildings. This highlights the importance of thoughtful selection and customization in architectural and structural design to maintain consistency and effectiveness across different building typologies. On the other hand, within the Finnish mid-rise timber apartment cases [33], no discernible scientific link was found between building height and space efficiency. Nevertheless, according to [25–27], there exists a trend wherein, as the building height escalates, space efficiency tends to decline. This phenomenon is attributed to the expansion of load-bearing components accompanying taller structures. However, noteworthy empirical observations from timber constructions across Europe suggest a potential remedy. Specifically, they indicate that, as the building ascends in height, it becomes feasible to decrease cross-sectional area of load-bearing walls on upper stories [33]. This implies that, despite the conventional notion of decreasing space efficiency with building height, strategic design considerations, especially in timber constructions, can potentially mitigate this effect.

Prospective research avenues aimed at advancing the understanding of space efficiency in high-rise timber residential towers could include the following initiatives [99–105]: 1. Comparative analysis: conduct comparative assessments of space efficiency metrics between high-rise timber structures and conventional materials like concrete and steel. 2. Spatial optimization: utilize computational methodologies to maximize spatial efficiency in timber buildings through the application of optimization algorithms. 3. User-centered design: investigate residents' spatial preferences through surveys and interviews to inform design improvements focused on enhancing usability. 4. Parametric design: employ parametric tools to model and optimize design parameters with the aim of improving space efficiency. 5. Life cycle assessment: perform comprehensive assessments of the environmental impact associated with space-efficient designs in timber constructions. 6. Flexible space design: explore adaptable design strategies to accommodate changing occupancy requirements over time. 7. Smart technology integration: evaluate the potential of Internet of Things (IoT) and automation systems in optimizing spatial utilization within timber structures. 8. Cross-disciplinary collaboration: foster collaborative efforts among architects, engineers, urban planners, and sociologists to develop holistic solutions addressing spatial efficiency challenges. 9. Policy implications: examine existing policies and regulations governing timber construction to identify opportunities for promoting and enhancing space-efficient design practices, while also proposing recommendations for policy improvements.

6. Conclusions

Our study focuses on space efficiency in contemporary high-rise timber residential constructions, addressing a gap in the existing literature. Through a systematic examination of data from 51 buildings via case studies, this article addresses the critical factors that have the most impact on space efficiency in wood-frame dwellings, including core planning, but also the size of the service core, building form, and selection of structural materials and load-bearing systems, taking into account the dimensions of structural members. The practical application of the findings from this study in timber office constructions encompasses several pivotal components. Firstly, central core layouts emerged as the predominant choice, providing guidance to architects in crafting designs that optimize space utilization and accessibility while simplifying construction processes. Prismatic building forms with shear-frame and shear-wall systems were endorsed for their contributions to both structural stability and spatial efficiency. The prevalent use of pure timber, occasionally complemented by concrete, indicated a material strategy that harmonizes sustainability with structural robustness. With an average space utilization rate of 83% and a core area-to-GFA ratio of 10%, architects and designers are provided with benchmarks for maximizing leasable space and attaining efficient residential layouts. By adhering to these insights, architectural practitioners can devise timber residential constructions that are not only more efficient and sustainable but also financially lucrative. By striking this intricate balance, they can conceive distinctive residential spaces that embody contemporary design principles and ecological awareness.

Author Contributions: Conceptualization, H.E.I.; methodology, H.E.I. and Ö.N.A.; software, H.E.I.; formal analysis, H.E.I. and Ö.N.A.; investigation, H.E.I. and Ö.N.A.; writing—original draft preparation, H.E.I.; writing—review and editing, H.E.I. and Ö.N.A.; supervision, H.E.I. All authors have read and agreed to the published version of the manuscript.

Funding: This research received no external funding.

Institutional Review Board Statement: Not applicable.

Informed Consent Statement: Not applicable.

Data Availability Statement: The data presented in this study are available in this article.

Conflicts of Interest: The authors declare no conflicts of interest.

Appendix A

Table A1. High-rise residential timber buildings.

#	Building Name	Country	City	Height (Meters)	# of Stories	Completion Date
1	Ascent	United States	Milwaukee	87	25	2022
2	HAUT	Netherlands	Amsterdam	73	22	2022
3	De Karel Doorman	Netherlands	Rotterdam	71	22	2012
4	Roots Tower	Germany	Hamburg	65	19	UC
5	Brock Commons Tallwood House	Canada	Vancouver	58	18	2017
6	Hyperion	France	Bordeaux	55	16	2021
7	Rundeskogen Hus B	Norway	Sandnes	55	16	2013
8	Treet	Norway	Bergen	49	14	2015
9	Lighthouse Joensuu	Finland	Joensuu	48	14	2019
10	Cederhusen	Sweden	Stockholm	44	13	UC
11	Hoas Tuuliniitty	Finland	Espoo	44	13	2021
12	Tallwood 1 at District 56	Canada	Victoria	42	12	UC

Table A1. *Cont.*

#	Building Name	Country	City	Height (Meters)	# of Stories	Completion Date
13	Origine	Canada	Quebec	41	13	2017
14	INTRO Residential Tower	United States	Cleveland	40	9	2022
15	Sensations	France	Strasbourg	38	11	2019
16	Rundeskogen Hus C	Norway	Sandnes	38	11	2013
17	Monterey	Australia	Brisbane	37	12	2021
18	Trafalgar Place	UK	London	36	10	2015
19	Aveo Bella Vista	Australia	Sydney	36	11	2018
20	Kringsja Studentby	Norway	Oslo	34	10	2018
21	Rundeskogen Hus A	Norway	Sandnes	34	10	2012
22	SKAIO	Germany	Heilbronn	34	10	2019
23	Dalston Works	UK	London	34	10	2017
24	The Cube Building	UK	London	33	10	2015
25	Forte	Australia	Melbourne	32	10	2012
26	Botanikern	Sweden	Uppsala	31	9	2019
27	Cenni di Cambiamento	Italy	Milan	31	9	2013
28	Kajstaden	Sweden	Vasteras	31	9	2019
29	Press House	UK	London	31	9	2017
30	Vallen	Sweden	Vaxjo	31	9	2015
31	Stadthaus	UK	London	29	9	2009
32	Carbon12	United States	Portland	29	8	2018
33	Moholt 50/50	Norway	Trondheim	28	9	2016
34	Arbora Condominium	Canada	Montreal	27	8	2019
35	DAS Kelo	Finland	Rovaniemi	27	8	2019
36	Docenten	Sweden	Vaxjo	27	8	2018
37	Dramsvegen	Norway	Tromso	27	8	2017
38	Frostaliden	Sweden	Skövde	27	8	2018
39	Highpoint Terrace	UK	London	27	8	2017
40	Jo & Joe	France	Gentilly	27	8	2019
41	Limnologen	Sweden	Vaxjo	27	8	2014
42	Maskinparken TRE	Norway	Trondheim	27	8	2018
43	Puukuokka Housing Block	Finland	Jyvaskyla	27	8	2018
44	Residences J. Ferry	France	Saint-Dié-des-Vosges	27	8	2014
45	Strandparken	Sweden	Stockholm	27	8	2014

Table A1. *Cont.*

#	Building Name	Country	City	Height (Meters)	# of Stories	Completion Date
46	The Gardens Macarthur	Australia	Sydney	27	8	2018
47	Trummens Strand	Sweden	Vaxjo	27	8	2019
48	Wood City Residential Buildings	Finland	Helsinki	27	8	2018
49	Lucien Cornil Student Residence	France	Marseille	27	8	2017
50	Pentagon I	Norway	As	27	8	2013
51	Bridport House	UK	London	27	8	2010

Note on abbreviation: 'UC' indicates under construction.

Appendix B

Table A2. High-rise residential timber buildings by building form, core type, structural system, and structural material.

	Building Name	Building Form	Core Type	Structural System	Structural Material
1	Ascent	Prismatic	Central	Shear-walled frame	Composite (T + C)
2	HAUT	Free	Peripheral	Shear-walled frame	Composite (T + C)
3	De Karel Doorman	Prismatic	Peripheral	Shear-walled frame	Composite (T + C + S)
4	Roots Tower	Prismatic	Central	Shear-walled frame	Composite (T + C)
5	Brock Commons Tallwood House	Prismatic	Peripheral	Shear-walled frame	Composite (T + C)
6	Hyperion	Free	Central	Shear-walled frame	Composite (T + C + S)
7	Rundeskogen Hus B	Free	Central	Shear-walled frame	Composite (T + C)
8	Treet	Prismatic	Peripheral	Trussed tube	Timber
9	Lighthouse Joensuu	Prismatic	Central	Shear wall	Timber
10	Cederhusen	Prismatic	Central	Shear wall	Timber
11	Hoas Tuuliniitty	Prismatic	Peripheral	Shear wall	Timber
12	Tallwood 1 at District 56	Prismatic	Central	Shear-trussed frame	Composite (T + S)
13	Origine	Free	Central	Shear wall	Timber
14	INTRO Residential Tower	Prismatic	Peripheral	Shear-walled frame	Composite (T + C)
15	Sensations	Free	Central	Shear-walled frame	Timber
16	Rundeskogen Hus C	Free	Central	Shear-walled frame	Composite (T + C)
17	Monterey	Free	Peripheral	Shear-walled frame	Composite (T + C + S)
18	Trafalgar Place	Prismatic	Peripheral	Shear wall	Timber
19	Aveo Bella Vista	Free	Central	Shear-walled frame	Composite (T + C)
20	Kringsja Studentby	Prismatic	Central	Shear-walled frame	Timber
21	Rundeskogen Hus A	Free	Central	Shear-walled frame	Composite (T + C)
22	SKAIO	Prismatic	Central	Shear-walled frame	Composite (T + C)

Table A2. *Cont.*

	Building Name	Building Form	Core Type	Structural System	Structural Material
23	Dalston Works	Prismatic	Central	Shear wall	Timber
24	The Cube Building	Free	Central	Shear-walled frame	Composite (T + C + S)
25	Forte	Prismatic	Central	Shear wall	Timber
26	Botanikern	Prismatic	Peripheral	Shear-trussed frame	Timber
27	Cenni di Cambiamento	Free	Central	Shear wall	Timber
28	Kajstaden	Prismatic	Peripheral	Shear wall	Timber
29	Press House	Prismatic	Central	Shear-walled frame	Timber
30	Vallen	Prismatic	Central	Shear-walled frame	Composite (T + C)
31	Stadthaus	Prismatic	Central	Shear wall	Timber
32	Carbon12	Prismatic	Central	Shear-trussed frame	Composite (T + S)
33	Moholt 50/50	Prismatic	Central	Shear wall	Timber
34	Arbora Condominium	Prismatic	Peripheral	Shear-walled frame	Timber
35	DAS Kelo	Prismatic	Peripheral	Shear-walled frame	Timber
36	Docenten	Prismatic	Peripheral	Shear wall	Composite (T + C)
37	Dramsvegen	Prismatic	Peripheral	Shear wall	Composite (T + C)
38	Frostaliden	Prismatic	Central	Shear-walled frame	Composite (T + C)
39	Highpoint Terrace	Prismatic	Peripheral	Shear wall	Timber
40	Jo & Joe	Prismatic	Peripheral	Shear-walled frame	Composite (T + C)
41	Limnologen	Prismatic	Central	Shear wall	Composite (T + C + S)
42	Maskinparken TRE	Prismatic	Central	Shear wall	Timber
43	Puukuokka Housing Block	Prismatic	Central	Shear-walled frame	Timber
44	Residences J. Ferry	Prismatic	Peripheral	Shear-walled frame	Timber
45	Strandparken	Prismatic	Peripheral	Shear wall	Composite (T + S)
46	The Gardens Macarthur	Prismatic	Peripheral	Shear wall	Timber
47	Trummens Strand	Prismatic	Peripheral	Shear wall	Timber
48	Wood City Residential Buildings	Prismatic	Peripheral	Shear-walled frame	Timber
49	Lucien Cornil Student Residence	Prismatic	Peripheral	Shear-walled frame	Composite (T + C)
50	Pentagon I	Prismatic	Central	Shear wall	Timber
51	Bridport House	Prismatic	Peripheral	Shear wall	Timber

Note on abbreviations: '(T + C + S)' indicates composite/hybrid structures combining timber, concrete, and steel; '(T + C)' indicates composite/hybrid structures combining timber and concrete; '(T + S)' indicates composite/hybrid structures combining timber and steel.

Appendix C

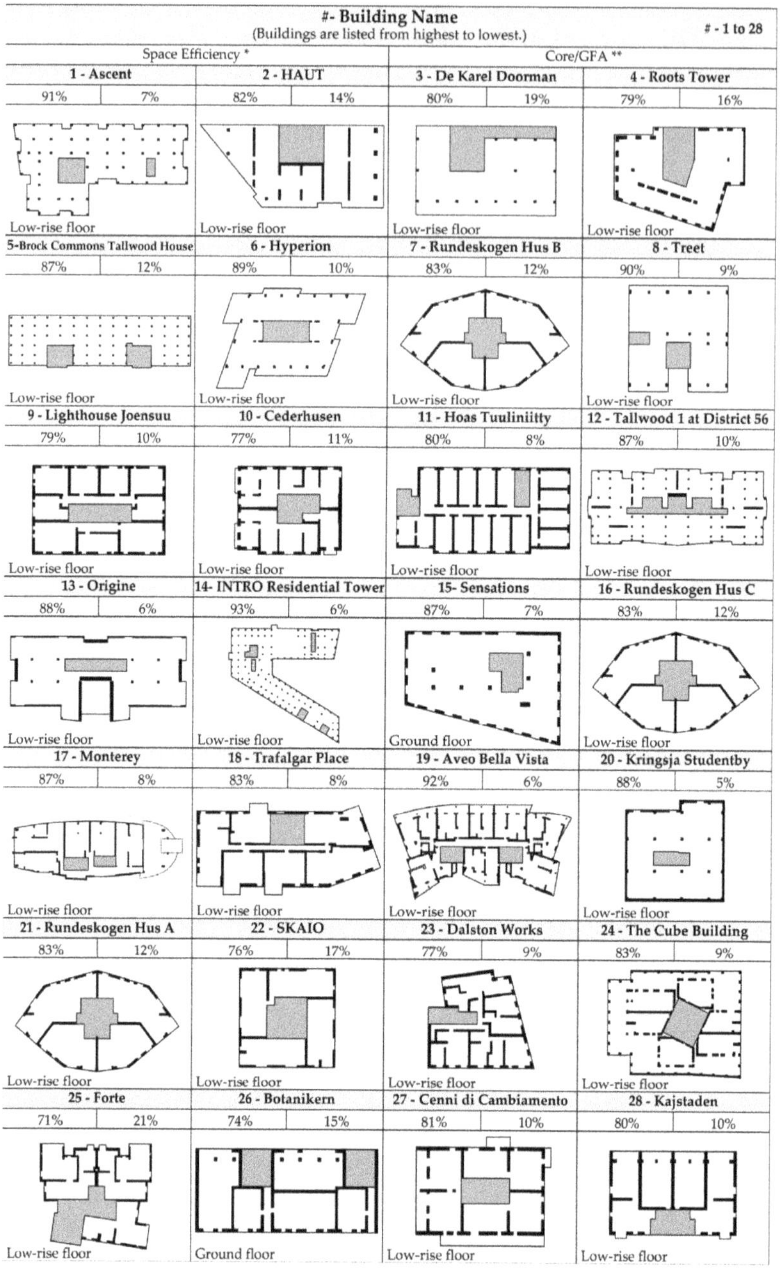

Figure A1. *Cont.*

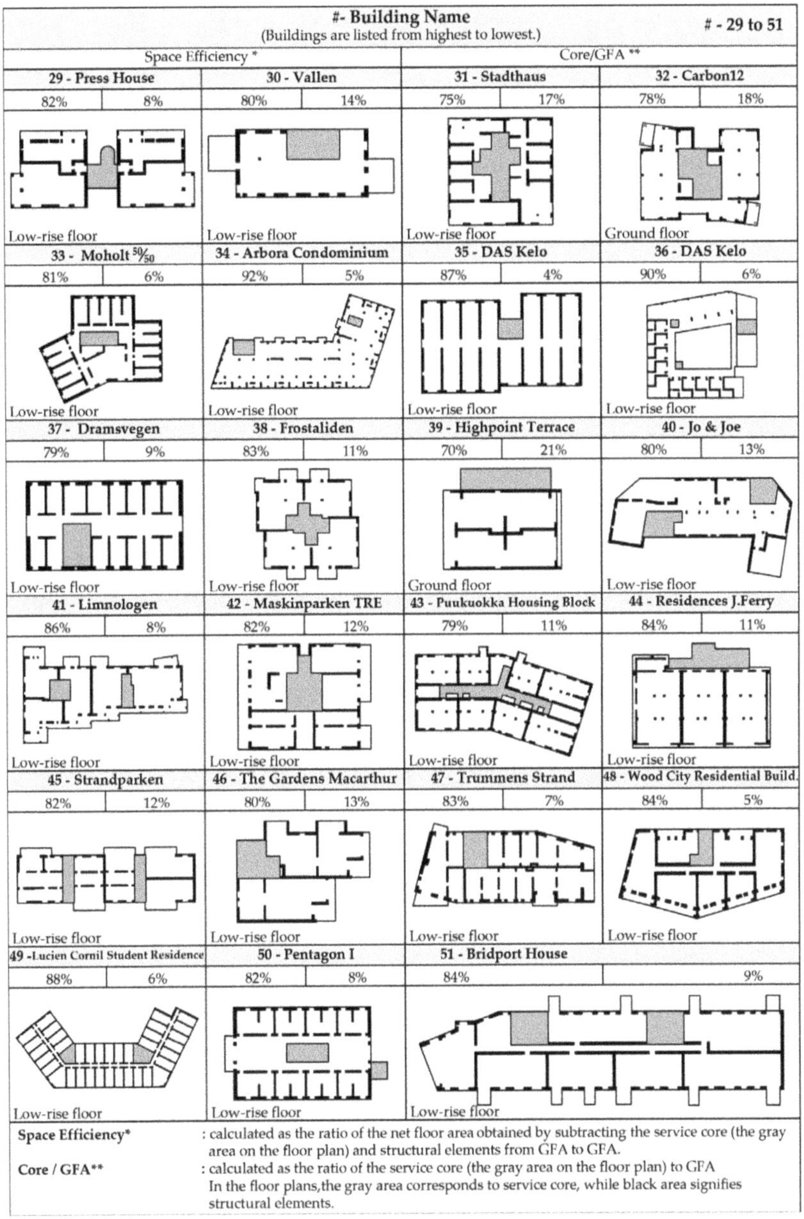

Figure A1. Space efficiency and Core/GFA of high-rise residential timber buildings.

References

1. Baiz, W.H.; Hoskara, E. Developing a measurement scale for sustainable high-rise building in city of Erbil. *J. Asian Archit. Build. Eng.* **2022**, *21*, 717–734. [CrossRef]
2. United Nations. *World Population Prospects—The 2017 Revision: Key Findings and Advance Tables*; Department of Economic and Social Affairs Population Division: New York, NY, USA, 2017; pp. 1–46.
3. Švajlenka, J.; Pošiváková, T. Innovation potential of wood constructions in the context of sustainability and efficiency of the construction industry. *J. Clean. Prod.* **2023**, *411*, 137209. [CrossRef]

4. Schubert, M.; Panzarasa, G.; Burgert, I. Sustainability in wood products: A new perspective for handling natural diversity. *Chem. Rev.* **2022**, *123*, 1889–1924. [CrossRef] [PubMed]
5. Aleksandr, C.; Tatiana, B.; Viktor, T.; Anton, K. On The Possibility of Using Timber Structures in the Construction of High-Rise Buildings in Seismic Areas. *Archit. Eng.* **2023**, *8*, 60–70.
6. van Veelen, B.; Knuth, S. An urban 'age of timber'? Tensions and contradictions in the low-carbon imaginary of the bioeconomic city. *Environ. Plan. E Nat. Space* **2023**, 25148486231179815. [CrossRef]
7. Zahiri, N. Timber High-Rises in Nordic Countries: Current Trends. *CTBUH J.* **2023**, 44–50. Available online: https://www.proquest.com/openview/18bd24c28e7b2ef07815ea4e8423a3b6/1?pq-origsite=gscholar&cbl=6578554 (accessed on 12 May 2024).
8. Kazulis, V.; Muizniece, I.; Zihare, L.; Blumberga, D. Carbon storage in wood products. *Energy Procedia* **2017**, *128*, 558–563. [CrossRef]
9. Abed, J.; Rayburg, S.; Rodwell, J.; Neave, M. A Review of the Performance and Benefits of Mass Timber as an Alternative to Concrete and Steel for Improving the Sustainability of Structures. *Sustainability* **2022**, *14*, 5570. [CrossRef]
10. Petrović, B.; Eriksson, O.; Zhang, X. Carbon assessment of a wooden single-family building—A novel deep green design and elaborating on assessment parameters. *Build. Environ.* **2023**, *233*, 110093. [CrossRef]
11. Arehart, J.H.; Hart, J.; Pomponi, F.; D'Amico, B. Carbon sequestration and storage in the built environment. *Sustain. Prod. Consum.* **2021**, *27*, 1047–1063. [CrossRef]
12. Sakuragawa, S.; Miyazaki, Y.; Kaneko, T.; Makita, T. Influence of wood wall panels on physiological and psychological responses. *J. Wood Sci.* **2005**, *51*, 136–140. [CrossRef]
13. Gold, S.; Rubik, F. Consumer attitudes towards timber as a construction material and towards timber frame houses—Selected findings of a representative survey among the German population. *J. Clean. Prod.* **2009**, *17*, 303–309. [CrossRef]
14. Kim, S.; Yun, B.Y.; Choi, J.Y.; Kim, Y.U.; Kim, S. Quantification of visual thermal perception changes in a wooden interior environment using physiological responses and immersive virtual environment. *Build. Environ.* **2023**, *240*, 110420. [CrossRef]
15. Rhee, P. Beyond green: Environmental building technologies for social and economic equity. *Arch. Des.* **2018**, *88*, 94–101. [CrossRef]
16. Öberg, J.; Wiege, E. Moisture Risks with CLT-Panels Subjected to Outdoor Climate during Construction: Focus on Mould and Wetting Processes. 2018. Available online: http://kth.diva-portal.org/smash/get/diva2:1222631/FULLTEXT01.pdf (accessed on 12 May 2024).
17. Strang, M.; Leardini, P.; Brambilla, A.; Gasparri, E. Mass Timber Envelopes in Passivhaus Buildings: Designing for Moisture Safety in Hot and Humid Australian Climates. *Buildings* **2021**, *11*, 478. [CrossRef]
18. Harada, H.; Fukushima, T.; Hatori, T.; Aoyagi, H. W350-The Roadmap of Super High-Rise Timber Building. *Int. J. High-Rise Build.* **2020**, *9*, 255–260.
19. Fernandez, A.; Komp, J.; Peronto, J. Ascent-challenges and advances of tall mass timber construction. *Int. J. High-Rise Build.* **2020**, *9*, 235–244.
20. Lehmann, S.; Kremer, P.D. Filling the Knowledge Gaps in Mass Timber Construction. *Mass Timber Constr. J.* **2023**, *6*, 1–10.
21. Yu, C.; Moslehpour, M.; Tran, T.K.; Trung, L.M.; Ou, J.P.; Tien, N.H. Impact of non-renewable energy and natural resources on economic recovery: Empirical evidence from selected developing economies. *Resour. Policy* **2023**, *80*, 103221. [CrossRef]
22. Yesügev, C.; Yılmaz Kahraman, Ö.; Güzel, N. *Ahşap Malzemeli Konut Teknolojisi*; Yalın Yayıncılık: Istanbul, Turkey, 2014; ISBN 978-605-4539-67-3. (In Turkish)
23. Amin, N.; Song, H. The role of renewable, non-renewable energy consumption, trade, economic growth, and urbanization in achieving carbon neutrality: A comparative study for South and East Asian countries. *Environ. Sci. Pollut. Res.* **2023**, *30*, 12798–12812. [CrossRef]
24. Szewczyk, J. Timber in contemporary architecture Part 3. "Plyscrapers". *Builder* **2019**, *268*, 40–44.
25. Ilgın, H.E. A study on space efficiency in contemporary supertall mixed-use buildings. *J. Build. Eng.* **2023**, *69*, 106223. [CrossRef]
26. Ilgın, H.E. Space Efficiency in Contemporary Supertall Residential Buildings. *Architecture* **2021**, *1*, 25–37. [CrossRef]
27. Ilgın, H.E. Space Efficiency in Contemporary Supertall Office Buildings. *J. Archit. Eng.* **2021**, *27*, 04021024. [CrossRef]
28. Lin, C.L.; Chiang, W.H.; Weng, Y.S.; Wu, H.P. Assessing the anthropogenic carbon emission of wooden construction: An LCA study. *Build. Res. Inf.* **2023**, *51*, 138–157. [CrossRef]
29. Karjalainen, M.; Ilgın, H.E.; Metsäranta, L.; Norvasuo, M. *Wooden Facade Renovation and Additional Floor Construction for Suburban Development in Finland*; IntechOpen: London, UK, 2022.
30. Ilgın, H.E.; Karjalainen, M. *Perceptions, Attitudes, and Interests of Architects in the Use of Engineered Wood Products for Construction: A Review*; IntechOpen: London, UK, 2021.
31. Jayawardana, J.; Kulatunga, A.K.; Jayasinghe, J.A.S.C.; Sandanayake, M.; Zhang, G. Environmental Sustainability of Off-Site Construction in Developed and Developing Regions: A Systematic Review. *J. Archit. Eng.* **2023**, *29*, 04023008. [CrossRef]
32. Bahrami, A.; Jakobsson, J.; Söderroos, T. Factors Influencing Choice of Wooden Frames for Construction of Multi-Story Buildings in Sweden. *Buildings* **2023**, *13*, 217. [CrossRef]
33. Tuure, A.; Ilgın, H.E. Space Efficiency in Finnish Mid-Rise Timber Apartment Buildings. *Buildings* **2023**, *13*, 2094. [CrossRef]
34. Ilgın, H.E.; Aslantamer, Ö.N. Investigating Space Utilization in Skyscrapers Designed with Prismatic Form. *Buildings* **2024**, *14*, 1295. [CrossRef]
35. Ilgın, H.E. Space Efficiency in Tapered Super-Tall Towers. *Buildings* **2023**, *13*, 2819. [CrossRef]

36. Ilgın, H.E. An analysis of space efficiency in Asian supertall towers. *Int. J. Build. Pathol. Adapt.* **2023**, *41*, 237–253. [CrossRef]
37. Okbaz, F.T.; Sev, A. A model for determining the space efficiency in non-orthogonal high rise office buildings. *J. Fac. Eng. Archit. Gazi Univ.* **2023**, *38*, 113–125.
38. Ibrahimy, R.; Mohmmand, M.A.; Elham, F.A. An Evaluation of Space Use Efficiency in Residential Houses, Kabul City. *J. Res. Appl. Sci. Biotechnol.* **2023**, *2*, 1–6. [CrossRef]
39. Goessler, T.; Kaluarachchi, Y. Smart Adaptive Homes and Their Potential to Improve Space Efficiency and Personalisation. *Buildings* **2023**, *13*, 1132. [CrossRef]
40. Hamid, G.M.; Elsawi, M.; Yusra, O. The Impacts of Spatial Parameters on Space Efficiency in Hybrid Villa-Apartments in Greater Khartoum, Sudan. *J. Archit. Plan.* **2022**, *34*, 425–440.
41. Suga, R. Space Efficiency in Hotel Development. Master's Thesis, MODUL University Vienna, Vienna, Austria, 2021.
42. Arslan Kılınç, G. Improving a Model for Determining Space Efficiency of Tall Office Buildings. Ph.D. Thesis, Department of Architecture, Mimar Sinan Fine Art University, Istanbul, Turkey, 2019. (In Turkish).
43. Von Both, P. A stakeholder-and function-based planning method for space-efficient buildings. *IOP Conf. Ser. Earth Environ. Sci.* **2019**, *323*, 012040. [CrossRef]
44. Höjer, M.; Mjörnell, K. Measures and steps for more efficient use of buildings. *Sustainability* **2018**, *10*, 1949. [CrossRef]
45. Nam, H.J.; Shim, J.H. An analysis of the change in space efficiency based on various tall building corner shapes and lease spans. *J. Archit. Inst. Korea Plan. Des.* **2016**, *32*, 13–20.
46. Zhang, L.; Zhang, L.; Wang, Y. Shape optimization of free-form buildings based on solar radiation gain and space efficiency using a multi-objective genetic algorithm in the severe cold zones of China. *Sol. Energy* **2016**, *132*, 38–50. [CrossRef]
47. Sev, A.; Özgen, A. Space efficiency in high-rise office buildings. *METU J. Fac. Archit.* **2009**, *26*, 69–89. [CrossRef]
48. Saari, A.; Tissari, T.; Valkama, E.; Seppänen, O. The effect of a Redesigned Floor Plan, Occupant Density and the Quality of Indoor Climate on the Cost of Space, Productivity and Sick Leave in an Office Building—A Case Study. *Build. Environ.* **2006**, *41*, 1961–1972. [CrossRef]
49. Kim, H.; Elnimeiri, M. Space efficiency in multi-use tall building. In Proceedings of the Tall Buildings in Historical Cities-Culture and Technology for Sustainable Cities, Seoul, Republic of Korea, 10–13 October 2004.
50. CTBUH Council on Tall Buildings Urban Habitat Illinois Institute of Technology, S.R. Crown Hall, 3360 South State Street, Chicago, IL, USA. Available online: https://www.ctbuh.org (accessed on 12 May 2024).
51. Hill, C.; Altgen, M.; Rautkari, L. Thermal modification of wood—A review: Chemical changes and hygroscopicity. *J. Mater. Sci.* **2021**, *56*, 6581–6614. [CrossRef]
52. Song, X.; Han, G.; Jiang, K.; Chi, X.; Liu, D.; Zhang, H.; Cheng, W.; Shi, S.Q. Effect of wood microstructure and hygroscopicity on the drying characteristics of waterborne wood coating. *Wood Sci. Technol.* **2022**, *56*, 743–758. [CrossRef]
53. Xu, K.; Gao, Y.; Zhang, X.; Li, Z.; Song, S.; Wu, Y.; Li, X.; Lu, J. Effects of drying temperature on hygroscopicity and mechanical performance of resin-impregnated wood. *Dry. Technol.* **2022**, *40*, 2414–2426. [CrossRef]
54. Tanthanawiwat, K.; Gheewala, S.H.; Nilsalab, P.; Schoch, M.; Silalertruksa, T. Environmental sustainability and cost performances of construction and demolition waste management scenarios: A case study of timber and concrete houses in Thailand. *J. Clean. Prod.* **2024**, *436*, 140652. [CrossRef]
55. Holtström, J.; Kenlind, S.S.; Nord, T. Project-based business models in the construction industry–key success factors for sustainable timber extension projects. *Constr. Manag. Econ.* **2024**, 1–14. [CrossRef]
56. Balasbaneh, A.T.; Sher, W. Economic and environmental life cycle assessment of alternative mass timber walls to evaluate circular economy in building: MCDM method. *Environ. Dev. Sustain.* **2024**, *26*, 239–268. [CrossRef]
57. Ascione, F.; Esposito, F.; Iovane, G.; Faiella, D.; Faggiano, B.; Mele, E. Sustainable and Efficient Structural Systems for Tall Buildings: Exploring Timber and Steel–Timber Hybrids through a Case Study. *Buildings* **2024**, *14*, 524. [CrossRef]
58. Tupėnaitė, L.; Žilėnaitė, V.; Kanapeckienė, L.; Sajjadian, S.M.; Gečys, T.; Sakalauskienė, L.; Naimavičienė, J. Multiple criteria assessment of high-rise timber buildings. *Eng. Struct. Technol.* **2019**, *11*, 87–94. [CrossRef]
59. Basterra, L.A.; Baño, V.; López, G.; Cabrera, G.; Vallelado-Cordobés, P. Identification and Trend Analysis of Multistorey Timber Buildings in the SUDOE Region. *Buildings* **2023**, *13*, 1501. [CrossRef]
60. Ghobadi, M.; Sepasgozar, S.M. Circular economy strategies in modern timber construction as a potential response to climate change. *J. Build. Eng.* **2023**, *77*, 107229. [CrossRef]
61. Litleskare, S.; Wuyts, W. Planning Reclamation, Diagnosis and Reuse in Norwegian Timber Construction with Circular Economy Investment and Operating Costs for Information. *Sustainability* **2023**, *15*, 10225. [CrossRef]
62. Aghamohammadi, N.; Shahmohammadi, M. Towards sustainable development goals and role of bio-based building materials. In *Bio-Based Materials and Waste for Energy Generation and Resource Management*; Elsevier: Amsterdam, The Netherlands, 2023; pp. 243–279.
63. Mamić, D.; Domljan, D. Positive Aspects of Using Solid Wood in Interiors on Human Wellbeing: A Review. *Drv. Ind.* **2023**, *74*, 379–391. [CrossRef]
64. Frunzio, G.; De Simone, M.; Loreto, G.; Di Gennaro, L.; Massaro, L. *The Use of Wood Betters the Relationship between People and Places*; Building Technology Educators' Society: Beijing, China, 2023. [CrossRef]

65. Ojala, A.; Kostensalo, J.; Viik, J.; Matilainen, H.; Wik, I.; Virtanen, L.; Muilu-Mäkelä, R. Psychological and physiological effects of a wooden office room on human well-being: Results from a randomized controlled trial. *J. Environ. Psychol.* **2023**, *89*, 102059. [CrossRef]
66. Cao, A.S.; Esser, L.; Glarner, B.; Frangi, A. A nonlinear dynamic model for collapse investigations in tall timber buildings-Preliminary results. In Proceedings of the 3th World Conference on Timber Engineering (WCTE 2023), Oslo, Norway, 19–22 June 2023; pp. 2268–2277.
67. Ehtisham, R.; Qayyum, W.; Camp, C.V.; Plevris, V.; Mir, J.; Khan, Q.U.Z.; Ahmad, A. Computing the characteristics of defects in wooden structures using image processing, C.N.N. *Autom. Constr.* **2024**, *158*, 105211. [CrossRef]
68. Loh, T.W.; Barnett, J.R.; Zhang, G.K.; Nguyen, K.T. Fire-Induced Damage and Postfire Analysis of the Compression Properties of Cross-Laminated Timber. *J. Mater. Civ. Eng.* **2024**, *36*, 04023595. [CrossRef]
69. Bahrami, A.; Rashid, S.P. Sustainable Development of Recent High-Rise Timber Buildings. In *Sustainable Structures and Buildings*; Springer International Publishing: Cham, Switzerland, 2024; pp. 1–16.
70. Thinley, J.; Hengrasmee, S. Innovating Bhutan's residential construction with mass timber for economic and environmental sustainability. *J. Build. Eng.* **2023**, *78*, 107763. [CrossRef]
71. Ahmad, F.; Allan, K.; Phillips, A.R. Multicriteria decision analysis of steel and mass timber prototype buildings in the Pacific Northwest. *J. Archit. Eng.* **2023**, *29*, 04023001. [CrossRef]
72. Wang, P.; Yang, Y.; Ji, C.; Huang, L. Influence of built environment on building energy consumption: A case study in Nanjing, China. *Environ. Dev. Sustain.* **2024**, *26*, 5199–5222. [CrossRef]
73. Miao, J.T.; Aritenang, A.F.; Gissma, N. Smart city in the creativity-built environment nexus: A case study of Bandung. In *Routledge Companion to Creativity and the Built Environment*; Routledge: London, UK, 2024; pp. 435–447.
74. Carapellucci, F.; Conti, V.; Lelli, M.; Liberto, C.; Orchi, S.; Valenti, G.; Valentini, M.P. Tools and Methodologies for the Analysis of Home-to-Work Shuttle Service Impacts: The ENEA "Casaccia" Case Study. *Future Transp.* **2023**, *3*, 901–917. [CrossRef]
75. Svatoš-Ražnjević, H.; Orozco, L.; Achim, M. Advanced Timber Construction Industry: A Review of 350 Multi-Story Timber Projects from 2000–2021. *Buildings* **2022**, *12*, 404. [CrossRef]
76. Kuzmanovska, I.; Gasparri, E.; Monne, D.T.; Aitchison, M. Tall timber buildings: Emerging trends and typologies. In Proceedings of the 2018 World Conference on Timber Engineering (WCTE 2018), Seoul, Republic of Korea, 20–23 August 2018.
77. Safarik, D.; Elbrecht, J.; Miranda, W. State of tall timber 2022. *CTBUH J.* **2022**, 22–31. Available online: https://s3.eu-west-2.amazonaws.com/construo-storage/attachments/9a13793fe644586846b1b92c4901ae721325e3e88e4eb40b8e83a354c75d0d97/CTBUH%20-%20State%20of%20Tall%20Timber%202022,%20Issue%201.pdf (accessed on 12 May 2024).
78. Verhaegh, R.; Vola, M.; de Jong, J. Haut-a 21-storey tall timber residential building. *Int. J. High-Rise Build.* **2020**, *9*, 213–220.
79. Xue, H.; Zhao, Z.; Li, P.; Shu, W.; Guan, Y.; He, D.; Wei, D.; Song, J. Application of amplified damped outrigger in seismic design of high-rise buildings. *Struct. Des. Tall Spec. Build.* **2024**, *33*, e2077. [CrossRef]
80. Abdulhadi, M.; Xun'an, Z.; Atroshchenko, E.; Rungamornrat, J. Vibration control of inerter-enhanced mega sub-controlled structure system (MSCSS) and the reliability analysis of the structure under seismic action. *Eng. Struct.* **2024**, *304*, 117508. [CrossRef]
81. Zain, M.; Keawsawasvong, S.; Thongchom, C.; Sereewatthanawut, I.; Usman, M.; Prasittisopin, L. Establishing efficacy of machine learning techniques for vulnerability information of tubular buildings. *Eng. Sci.* **2024**, *27*, 1008. [CrossRef]
82. Ghongade, D.H.P.; Bhadre, D.A.A. Dynamic Analysis of Tall Buildings in Various Seismic Zones with Central Shear Walls And Diagonal Bracings Using E-Tabs Software. *Eur. Chem. Bull.* **2023**, *12*, 3117–3131.
83. Yusuf, D.A.; Ahmed, A.; Sagir, A.; Yusuf, A.A.; Yakubu, A.; Zakari, A.T.; Usman, A.M.; Nashe, A.S.; Hamma, A.S. A Review of Conceptual Design and Self Health Monitoring Program in a Vertical City: A Case of Burj Khalifa, U.A.E. *Buildings* **2023**, *13*, 1049. [CrossRef]
84. Jiang, B.; Wang, J.; Li, Q.; Zhu, Y.; Sang, S.; Xu, J. The Typical Case Analysis of Special Ultra-high-rise Building. In *IOP Conference Series: Materials Science and Engineering*; IOP Publishing: Bristol, UK, 2020; Volume 780, p. 032014.
85. Ussher, E.; Aloisio, A.; Rathy, S. Effect of lateral resisting systems on the wind-induced serviceability response of tall timber buildings. *Case Stud. Constr. Mater.* **2023**, *19*, e02540. [CrossRef]
86. Hayajneh, S.M.; Naser, J. Fire Spread in Multi-Storey Timber Building, a CFD Study. *Fluids* **2023**, *8*, 140. [CrossRef]
87. Oldfield, P.; Doherty, B. Offset Cores: Trends, Drivers and Frequency in Tall Buildings. *CTBUH J.* **2019**, 40–45. Available online: https://global.ctbuh.org/resources/papers/download/4186-offset-cores-trends-drivers-and-frequency-in-tall-buildings.pdf (accessed on 12 May 2024).
88. Sikkema, R.; Styles, D.; Jonsson, R.; Tobin, B.; Byrne, K.A. A market inventory of construction wood for residential building in Europe–in the light of the Green Deal and new circular economy ambitions. *Sustain. Cities Soc.* **2023**, *90*, 104370. [CrossRef]
89. Angelucci, G.; Mollaioli, F.; Molle, M.; Paris, S. Performance assessment of Timber High-rise Buildings: Structural and Technological Considerations. *Open Constr. Build. Technol. J.* **2022**, *16*, e187483682206270. [CrossRef]
90. Binck, C.; Cao, A.S.; Frangi, A. Lateral stiffening systems for tall timber buildings–tube-in-tube systems. *Wood Mater. Sci. Eng.* **2022**, *17*, 309–316. [CrossRef]
91. Stulen, J. The use of mass timber—An update. *N. Z. J. For.* **2019**, *64*, 26–31.
92. Wang, W. Research on Seismic Design of High-Rise Buildings Based on Framed-Shear Structural System. *Front. Res. Archit. Eng.* **2020**, *3*, 87–90. [CrossRef]

93. Li, Z.; Wang, X.; He, M. Experimental and analytical investigations into lateral performance of cross-laminated timber (CLT) shear walls with different construction methods. *J. Earthq. Eng.* **2022**, *26*, 3724–3746. [CrossRef]
94. Gunel, M.H.; Ilgın, H.E. *Tall Buildings: Structural Systems and Aerodynamic Form*; Routledge: London, UK; New York, NY, USA, 2014.
95. Dias, A.M.P.G.; Skinner, J.; Crews, K.; Tannert, T. Timber-concrete-composites increasing the use of timber in construction. *Eur. J. Wood Wood Prod.* **2016**, *74*, 443–451. [CrossRef]
96. Hanták, J.; Končeková, D. Positive effects of wood in Vorarlberg's (Austria) timber kindergartens. *Archit. Pap. Fac. Archit. Des. STU* **2023**, *28*, 36–49. [CrossRef]
97. Dharanidharan, K.; Selvanayaki, S.; Divya, K.; Ravikumar, R.; Raj, S.V. Exploring Household Wood Preferences among Consumers in Coimbatore. *Asian J. Agric. Ext. Econ. Sociol.* **2023**, *41*, 196–203. [CrossRef]
98. Yeang, K. *Service Cores: Detail in Building*; Wiley-Academy: London, UK, 2000.
99. Ilgın, H.E.; Karjalainen, M.; Koponen, O. *Various Geometric Configuration Proposals for Dovetail Wooden Horizontal Structural Members in Multi-Story Building Construction, Engineered Wood Products for Construction*; IntechOpen: London, UK, 2022.
100. Lucchi, E.; Delera, A.C. Enhancing the historic public social housing through a user-centered design-driven approach. *Buildings* **2020**, *10*, 159. [CrossRef]
101. Domljan, D.; Janković, L. Design of sustainable modular wooden booths inspired by revitalization of Croatian traditional construction and new user needs due to COVID-19 pandemic. *Sustainability* **2022**, *14*, 720. [CrossRef]
102. Wagner, H.J.; Alvarez, M.; Groenewolt, A.; Menges, A. Towards digital automation flexibility in large-scale timber construction: Integrative robotic prefabrication and co-design of the BUGA Wood Pavilion. *Constr. Robot.* **2020**, *4*, 187–204. [CrossRef]
103. Wiegand, E.; Ramage, M. The impact of policy instruments on the first generation of Tall Wood Buildings. *Build. Res. Inf.* **2022**, *50*, 255–275. [CrossRef]
104. Maniak-Huesser, M.; Tellnes, L.G.; Zea Escamilla, E. Mind the gap: A policy gap analysis of programmes promoting timber construction in nordic countries. *Sustainability* **2021**, *13*, 11876. [CrossRef]
105. Ilgın, H.E.; Karjalainen, M. *Tallest Timber Buildings: Main Architectural and Structural Design Considerations, Wood Industry—Past, Present and Future Outlook*; IntechOpen: London, UK, 2022.

applied sciences

MDPI

Article

Space Efficiency in European High-Rise Timber Buildings

Hüseyin Emre Ilgın [1,*] and Özlem Nur Aslantamer [2]

1 School of Architecture, Faculty of Built Environment, Tampere University, P.O. Box 600, FI-33014 Tampere, Finland

2 Department of Interior Architecture and Environmental Design, Faculty of Art, Design and Architecture, Atılım University, Ankara 06830, Turkey; ozlem.aslantamer@atilim.edu.tr

* Correspondence: emre.ilgin@tuni.fi

Abstract: As towering wooden edifices (≥8 stories) become a rapidly expanding and promising field, they provide substantial environmental and economic advantages throughout their entire lifespans, leading to their increasing popularity, especially in the European context. Similar to various other construction forms, spatial efficiency is a vital design criterion in timber buildings to guarantee the viability of a project. Currently, there is no thorough study on spatial efficiency in these towers in Europe, which is home to the majority of the world's timber towers. This paper examined data from 56 cases to improve comprehension of the planning factors affecting space efficiency in these buildings. The main findings showed that the average space efficiency across the analyzed examples was documented at 82%, with deviations spanning from 70% to 90%, the average core area to gross floor area (GFA) ratio was determined to be 11%, ranging from 4% to 21%, and no substantial difference was noted in the impact of core arrangements on space efficiency, and parallel findings were observed for forms and construction materials. This article aspires to provide architectural designers with essential perspectives, assisting and directing them in the conception and realization of upcoming ventures both across Europe and internationally in this domain.

Keywords: space efficiency; high rise; timber; wood; Europe

1. Introduction

In the past few decades, the global building sector has encountered the difficulty of harmonizing urban development with the necessity for ecological sustainability amidst worries about climate change [1]. As urban areas become densified, there is an increased need for creative and environmentally-friendly construction methods that surpass simple structural performance, integrating design and long-term ecological factors [2].

High-rise timber structures have surfaced as an innovative answer to urban development issues, showcasing a dedication to environmental stewardship [3]. Timber, a renewable and low-carbon resource, differs from conventional materials such as concrete, helping to reduce the construction industry's carbon emissions [4]. Besides their ecological benefits, high-rise timber towers offer a distinctive aesthetic warmth, contrasting with the uniform façades of contemporary buildings [5]. This blend of sustainability and aesthetic charm aids in crafting metropolitan areas that emphasize well-being and a relation to natural environment.

It is worth noting that historical timber structures serve as foundational predecessors to contemporary high-rise timber buildings, exemplifying enduring principles of architectural innovation and structural engineering [6–8]. The Great Pagoda of Nara in Japan, constructed in the 8th century, stands as one of the earliest timber high-rises, showcasing a nine-story wooden tower that attests to the skillful craftsmanship and advanced timber joinery techniques of its time [9]. Similarly, medieval European timber-framed buildings such as the 14th-century Tallinn Town Hall in Estonia [10] and the Old St. Peter's Basilica in Rome [11], dating back to the 4th century, demonstrated the feasibility of constructing tall and expansive spaces with timber while adhering to stringent safety and durability requirements. These historical structures not only utilized timber's strength and flexibility but also pioneered construction methods that balanced architectural grandeur with structural integrity. By examining their enduring legacy, contemporary architects and engineers draw inspiration for modern timber high-rises, leveraging technological advancements in materials and construction techniques to redefine sustainable urban architecture in the 21st century.

High-rise timber buildings are most viable in urban contexts that prioritize sustainability, innovation in construction, and environmental stewardship [12]. These contexts typically include cities aiming to reduce carbon footprints and promote renewable resources. Additionally, urban areas undergoing redevelopment or seeking distinctive architectural solutions find high-rise wood buildings appealing for their aesthetic and environmental benefits. Mixed-use developments that integrate residential, commercial, and public spaces also benefit from the flexibility and sustainability of timber construction. Furthermore, regions with seismic activity may favor wooden buildings for their resilience when engineered appropriately. Overall, high-rise wooden buildings thrive in urban settings where there is a convergence of sustainable development goals, architectural innovation, and a supportive regulatory environment.

The construction of tall wooden towers corresponds with the increasing movement towards eco-friendly urban development, fostering green areas and sustainable architecture [13]. As these sustainable buildings gain traction, they represent a notable shift in the construction industry, harmonizing the practical requirements of urban living with a conscientious focus on environmental well-being. In Europe, particularly in the Nordic region, a longstanding affinity for wooden construction highlights a rich cultural legacy, underscoring artisanal quality, sustainability, and a strong sense of locality [14]. The revival of timber as a primary material in Europe marks a crucial transition towards more eco-conscious options, recognizing the ecological effect of traditional materials.

It was worth mentioning that in Europe, the demographics of investors involved in the construction of high-rise timber buildings exhibit a diverse profile [15], influenced by multifaceted considerations. Typically, these investors encompass a spectrum ranging from visionary institutional entities, such as pension funds and real estate investment trusts, driven by long-term sustainability goals and the potential for robust returns on investment, to innovative private equity firms and venture capitalists seeking early-mover advantages in burgeoning markets. Moreover, government-backed entities and development agencies play pivotal roles, leveraging policy frameworks incentivizing timber construction to foster urban regeneration and environmental stewardship. Collaboratively, these stakeholders navigate intricate financial landscapes, integrating risk assessment models tailored to novel materials and construction methodologies while championing sustainable urban development paradigms across the continent.

As a pioneer in mass timber skills, Europe has numerous advantages that make it a leading international hub for building tall wooden constructions. This dominance stems from several key factors. First, Europe has well-managed forests subjected to rigorous oversight, ensuring a consistent and sustainable timber supply essential for mass timber projects [16]. Additionally, the area possesses a strong structure of stringent ecological rules, highlighting a dedication to sustainable construction methods and conscientious resource utilization [17]. Given these favorable conditions, it is unsurprising that Europe holds a remarkable two-thirds share of the high-rise timber residential market worldwide.

North America, including the United States, has witnessed a burgeoning interest and substantial investment in the development of tall wooden structures, mirroring a broader global change towards sustainable building practices [18–20]. This region is well-known for its long-established tradition of wooden construction [21]. The momentum behind the growth of high-rise wooden towers is driven by a confluence of environmental, economic, and technological factors. From an environmental perspective, these structures present a substantial reduction in carbon footprint when compared to traditional concrete and steel constructions, as wood acts as a carbon sink, sequestering carbon dioxide during its growth. This sustainability aspect aligns with the increasing global efforts to combat climate change. Economically, the advancements in engineered wood products, such as cross-laminated timber (CLT) and glued laminated timber (glulam), have significantly enhanced the feasibility and cost-effectiveness of constructing high-rise timber buildings. These materials offer superior strength and versatility, making them viable alternatives to traditional building materials. Technologically, innovations in fire-resistant treatments and structural engineering have successfully addressed historical concerns regarding the safety and durability of wooden high-rises. These advancements have paved the way for regulatory changes and the adaptation of building codes to accommodate and promote timber construction. Notable projects exemplify this trend, including Europe's tallest timber tower, the 18-story Mjøstårnet in Norway, which has served as a source of inspiration for similarly ambitious projects in the United States. Among these is the 25-story Ascent in Milwaukee, which stands as a testament to the potential of high-rise timber construction. These projects not only highlight the technical and economic viability of timber structures but also underscore a growing commitment to sustainable building practices in the architectural and construction industries.

The choice of timber structures for residential purposes significantly enhances positive environmental impacts because of numerous pivotal considerations deeply grounded in sustainability and human well-being. Residential buildings constructed from timber not only leverage its renewable nature and carbon sequestration capabilities but also promote a healthier indoor environment for occupants. Wood's natural thermal insulation properties contribute to reduced energy consumption for heating and cooling, thereby lowering carbon emissions associated with building operation. Moreover, the aesthetic and psychological benefits of wood in residential settings foster a connection with nature, promoting mental well-being and sustainable living practices among residents. By choosing timber for residential construction, stakeholders not only support sustainable forestry practices but also contribute to mitigating climate change, improving indoor comfort, and enhancing overall quality of life in urban environments. This holistic approach underscores the pivotal role of residential wood construction in advancing both environmental sustainability and human-centric design principles.

In timber construction, architectural and structural considerations are crucial in the development of high-rise edifices. Among these factors, space efficiency is particularly crucial for optimizing the usable area to boost their economic attractiveness [22,23]. Despite its significance, there is limited academic research delving into the spatial efficiency of such edifices [22]. Furthermore, despite thorough study being carried out on the technical, ecological, and economic dimensions of wooden structures (e.g., [24–29]), there is a significant lack of focus on spatial efficiency within the European context. This discrepancy underscores the need for additional studies focused on improving space utilization in tall timber buildings while balancing design, financial, and ecological considerations. These studies would enhance understanding in the area and offer crucial visions for especially architects and contractors aiming to create more effective wooden buildings in Europe [30–35]. Additionally, this research can inform building regulations and standards, ensuring that timber buildings meet safety, durability, and performance requirements while fostering innovation in European architectural design and construction practices.

In research on spatial efficiency in buildings, Ilgın and Aslantamer [36] and Aslantamer and Ilgın [37] focused on wooden residential and office buildings, finding average efficiencies 83% and 88%, respectively. Tuure and Ilgın [38] analyzed the spatial effectiveness of 55 mid-rise timber buildings in Finland, discovering efficiencies between 78% and 88%. Ilgın [22–24] studied skyscrapers of various functions—office, residential, and mixed-use—and forms and found a common utilization of outriggered frame systems and central cores, with an inverse correlation between tower height and spatial efficiency. Ilgın and Aslantamer [39] and Ilgın [40] examined the space efficiency in prismatic and tapered tall buildings and it was revealed that the average space efficiency in both forms was 72%. Ilgın [41,42] focused on skyscrapers in different contexts including Asia and the Middle East in terms of space efficiency and found that average space efficiencies were 67.5% and 75.5%, respectively. Okbaz and Sev [43] performed a simulation for spatial effectiveness in 11 freeform high-rise office buildings, identifying structural design as a significant factor, with conical shapes achieving the highest efficiency. Ibrahimy et al. [44] assessed spatial usage in Kabul's residential structures, spotting deviations from norms due to substandard interior planning and insufficient regulatory supervision. Goessler and Kaluarachchi [45] examined the impact of cutting-edge technologies on high-density urban areas, demonstrating a two- to threefold enhancement in spatial efficiency over traditional configurations. Hamid et al. [46] discovered that positioning homes at corners enhances land efficiency in detached residences in Sudan. Suga [47] investigated space optimization in hotels, highlighting the advantages of focusing on spatial efficiency. Arslan Kılınç [48] explored the structural and load-bearing aspects of extruded-form skyscrapers, noting that the service core space expands with tower height and finding no relationship between materials and space efficiency. Von Both [49] conducted a stakeholder assessment to improve regional and spatial effectiveness in city planning. Höjer and Mjörnell [50] proposed a four-stage construction model that considers the effects of digitalization on interior space demands and availability. Nam and Shim [51] explored lease span and corner patterns in tall towers, concluding that corner cuts have minimal effect on spatial efficiency. Zhang et al. [52] developed a freeform design for cold climates, increasing solar gains with minimal impact on spatial efficiency. Sev and Özgen [53] examined space utilization in high-rise office construction, focusing on the effects of load-bearing and core configurations. Saari et al. [54] studied the connection between space optimization and overall costs in tall office structures, demonstrating that higher spatial efficiency significantly aids in meeting indoor environmental comfort goals. Kim and Elnimeiri [55] assessed space efficiency in multi-use tall edifices, emphasizing the importance of ideal structural systems and designs in improving space use.

The literature review indicates that the existing academic research inadequately addresses spatial efficiency in high-rise timber towers in Europe, which hosts most of the timber towers globally. To bridge this knowledge gap and better understand European trends, based on insights from 56 towers, this paper evaluates space efficiency by examining essential architectural and structural components in modern high-rise timber towers in Europe.

The description of tall or high-rise towers remains unclear, as there is no global criterion for their height or number of floors. The classification of a structure as 'tall' or 'high-rise', particularly in the context of wooden projects, is still a matter of debate. However, within this discourse, a 'high-rise timber building' is typically described as a structure with eight or more stories [56]. Due to the focus of the study, proportional evaluations of key building zones like the service core were used, without considering the floor area dimensions and building size of the case study towers and structural elements such as columns.

This study aims to understand the influence of various planning factors on space efficiency, having four main aspects. These aspects are critical architectural design factors including function, core arrangement, and form; primary structural design considerations including material and system; and spatial efficiency, and the correlation between this efficiency and the aforementioned design factors. This comprehensive analysis, reflecting the latest developments in these construction applications, aims to provide insights for more informed decision-making in the planning of future wooden construction in Europe.

2. Methods

The utilization of the case study technique was employed for the purpose of classifying and consolidating information concerning 56 modern high-rise timber projects in the European context (Figure 1). This approach is often used in academic research to structure and unite information about current timber-based office projects. It integrates both qualitative and quantitative data, in addition to a comprehensive review of the existing literature [57,58], allowing for a systematic investigation of their architectural and structural aspects. This method facilitates a comprehensive assessment of particular instances in practical environments, fostering a deep comprehension of the distinct attributes and trends in contemporary office designs that utilize timber as a primary material.

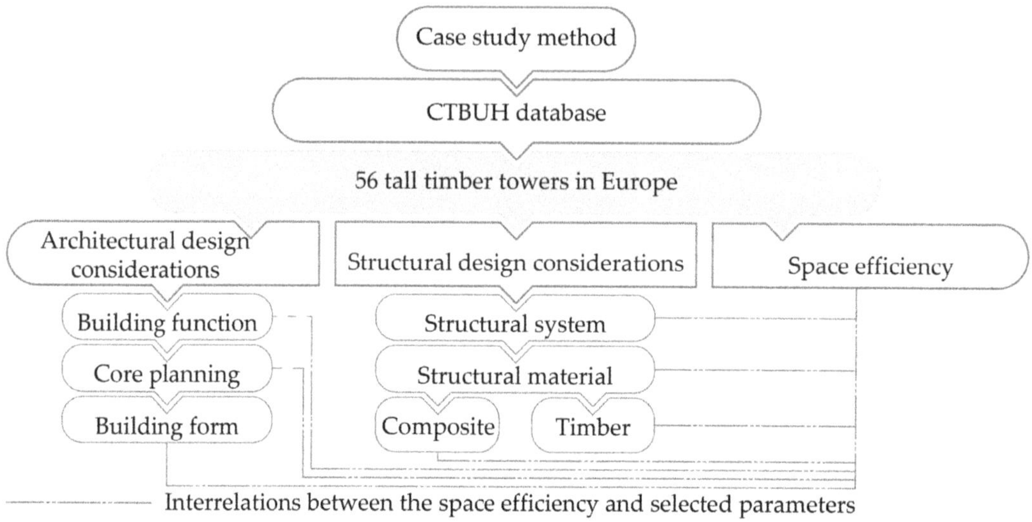

Figure 1. Research method (created by the authors).

The selection of 56 cases in this study ensures robust and representative results, particularly given the inherent scarcity of accessible contemporary European high-rise timber buildings. This sample size allows for the identification of significant trends and patterns in spatial utilization and architectural features. The diversity among the selected cases, spanning various geographical locations in Europe, building heights, and design typologies, enhances the generalizability of the findings. Notably, the study examines timber towers across Europe from a wide array of countries including eleven in Norway and Sweden, nine in France, seven in the United Kingdom, five in Finland, four in Germany, three in the Netherlands, two in Switzerland, and one in Denmark, Austria, Italy, and Spain (as depicted in Figure 2 and detailed in Sections A and B). By encompassing such diversity, the research offers a comprehensive view of current practices in high-rise timber construction. Methodologically, a sample size of approximately 30 cases is adequate for deriving meaningful conclusions in the existing literature, balancing depth and breadth of analysis as evidenced by prior research [23,39,42]. Thus, the examination of 56 cases forms a robust foundation for understanding spatial utilization in European high-rise timber buildings, ensuring the study's insights are both reliable and widely applicable, facilitating the identification of recurring themes such as function, core typology, form, structural system, and different construction materials.

In an extensive undertaking, this paper thoroughly assessed the floor plans of a variety of timber edifices, including typical stories. This thorough method promised the gathering of accurate and dependable information, establishing a robust foundation for assessing spatial effectiveness in the sample cases. In general, when classifying high-rise structures according to their anticipated use, they are divided into those designed for a single function or for multiple functions. This article bases its evaluation of utility on the following categories: residential, office, hotel, and mixed-use. Moreover, following the established academic literature [59–61], we utilized the thorough classification system [36,37] for the key components in architectural and structural planning. This choice was motivated by the specific features of these classes, as illustrated in Figure 3.

Figure 2. High-rise timber buildings located throughout Europe (created by the authors).

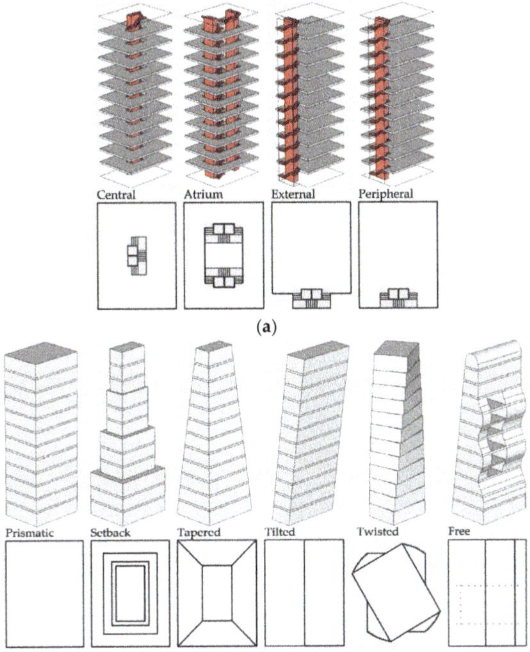

Figure 3. *Cont.*

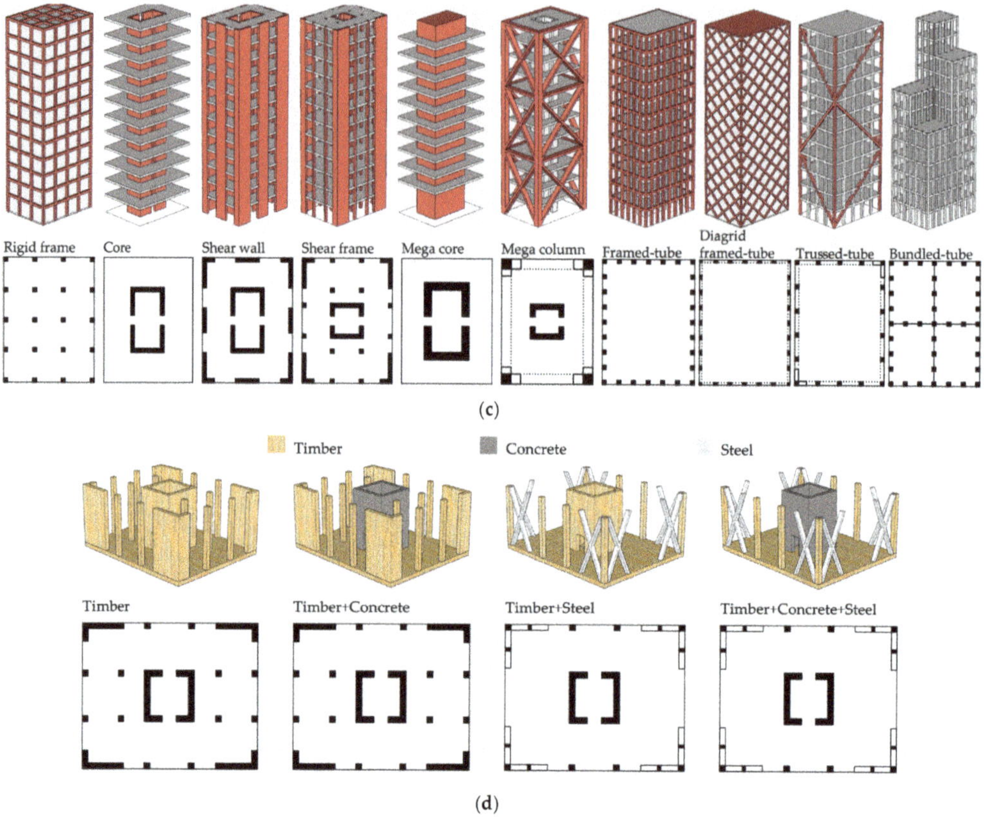

Figure 3. (**a**) Core, (**b**) building form, (**c**) structural system, and (**d**) material classifications.

In this study's methodology, space efficiency was evaluated by determining the ratio of NFA to GFA, thereby providing a quantitative metric for the effective utilization of the available space. The service core to GFA ratio was calculated, offering insights into the allocation of space dedicated to the main structural and service components within the building.

Assessment of space efficiency includes two critical ratios: NFA to GFA and service core to GFA. The NFA is derived by subtracting the service core area from the GFA, isolating functional areas where activities occur and excluding spaces allocated to infrastructure and supplementary facilities. The NFA to GFA ratio quantitatively measures how efficiently the floor area is employed for actual use, reflecting space allocation efficiency.

The service core to GFA ratio is another vital consideration in space efficiency assessments. It measures the proportion of the GFA occupied by the service core, which includes the main amenities like elevators, stairs, and utility rooms. This ratio highlights the extent to which the building's infrastructure occupies valuable floor space, affecting whole efficiency.

Also, architectural voids, like atriums, exert a substantial influence on space computations. When determining the NFA, these voids are deducted from the GFA. This deduction accommodates regions within the structure that may not actively support functional spaces but enhance the aesthetic or environmental quality. The incorporation of voids acknowledges both practical and aesthetic considerations when assessing spatial efficiency. For comprehensive details and precise results of these computations, readers are advised to consult Appendix C.

3. Results

Figures 4a–c and 5a,b illustrate the key architectural and structural considerations related to space efficiency.

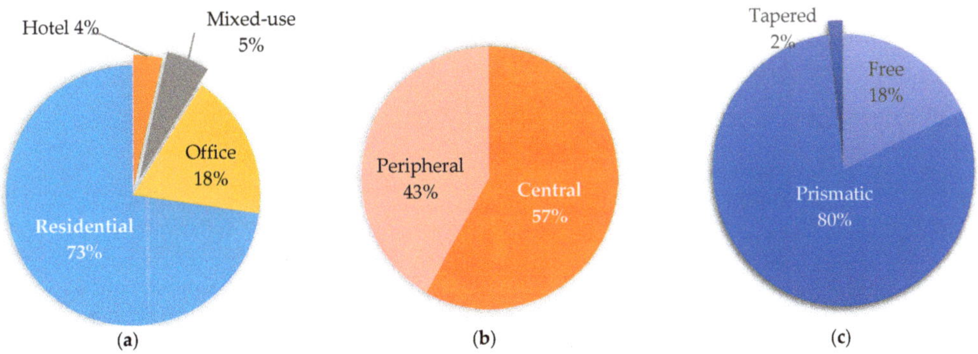

Figure 4. Examined cases by (**a**) function, (**b**) core type, and (**c**) form.

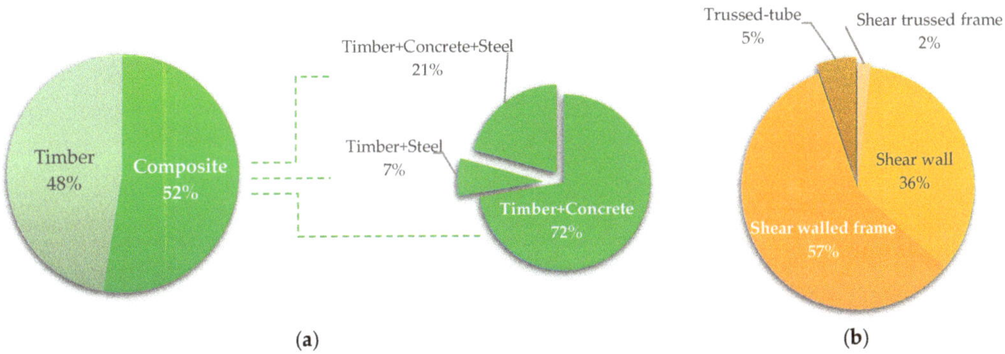

Figure 5. Examined cases by (**a**) material and (**b**) system.

3.1. Key Architectural Considerations: Function, Core Planning, and Building Form

Figure 4a shows a predominant preference for residential use, comprising over 70%. This trend is driven by several factors [62–64], including the environmental benefits of timber. Additionally, timber buildings offer psychological and health benefits, improving mental and physical health. High-rise timber towers meet the market demand, are economically viable, and address housing needs in crowded cities in Europe.

Analyzing Figure 4b reveals that the central core is the most used configuration, accounting for 57%. The peripheral arrangement comes next, making up 43%. Within high-rise timber developments, a focal core arrangement presents myriad advantages [65–67]:

- Stability against sideways pressures, lessening requirements for reinforcement.
- Augmented available area by concentrating vital amenities.
- Increased effectiveness via reduced circulation lengths and simplified entryways.
- Elevated daylight and vistas owing to expanded window areas.
- Versatility for forthcoming adjustments requiring minimal structural alterations.
- Economies achieved via enhanced structural effectiveness.
- Liberty in façade design as the core structure bears the primary load.
- Reinforced fire safety with fire-retardant zones.

Following the morphological classification of high-rise timber constructions, prismatic arrangement, representing 82%, occurred as the leading architectural choice, while free configurations make up 16%, as depicted in Figure 4c. The widespread adoption of prismatic designs in the case study towers might be ascribed to various considerations [37,68,69]: 1. Structural efficacy, evenly dispersing loads. 2. Optimization of materials, in line with standard wooden construction supplies. 3. Simplified construction, as regular shapes reduce time and labor expenses. 4. Space use maximization, utilizing rectangular or square floor plans to their fullest extent. 5. Flexibility in façade design, accommodating various architectural elements and promoting energy efficiency. 6. Potential for modular construction, bolstering efficiency and quality management. 7. Considerations pertaining to fire safety involve effective compartmentalization and contemporary methodologies aimed at mitigating risks.

3.2. Key Structural Considerations: Material and System

Figure 5a shows that composites dominate, representing 52% of cases, with timber at 48%. Composites are advantageous for high-rise timber towers [70–73]. Timber's lightweight strength suits these towers and combining it with concrete enhances the load-bearing capacity and stability, ensuring safety during environmental events. Non-combustible materials improve fire resistance, meeting building codes. Hybrid designs offer flexibility, meeting aesthetic and structural needs while reducing environmental impact. Timber's natural insulation improves energy efficiency, and prefabricating components speeds up construction and reduces costs.

As depicted in Figure 5b, shear-frame structures, notably including 'shear walled frame' and 'shear trussed frame' variants, predominate, constituting about 60%. Following them is the shear wall system, accounting for 35%. Notably, among shear-frame structures, 'shear walled frame' systems are the predominant choice in the analyzed cases, representing 97% of instances. In contrast, the tube system exhibits a lower adoption rate, appearing in only three instances among load-bearing systems for European cases. Within the realm of shear-frame systems, which incorporate both shear trussed/walled frames, a nuanced and comprehensive understanding emerges when evaluating the respective limitations and drawbacks of each system. Inherent weaknesses of a rigid frame, in comparison to the advantages offered by shear truss or wall systems, and vice versa, can effectively be diminished through the strategic integration of these components [74].

3.3. Space Efficiency

In this research, following an analysis of 56 cases, we observed that the average spatial efficiency and core-to-GFA ratio were 82% and 11%, correspondingly. The spectrum of these metrics ranged from a smallest of 70% and 4% to a highest of 90% and 21%, respectively.

Table 1 presents a comparative analysis of the findings above on average space efficiency and the ratio of core to GFA with those documented for Middle Eastern non-timber skyscrapers as referenced in Asia [41], and Middle Eastern skyscrapers as detailed in [42].

Table 1. Comparison of average space efficiency and ratio of core to GFA.

	The Findings	Asian Skyscrapers (75 Cases)	Middle Eastern Skyscrapers (27 Cases)
Average space efficiency	82% (max. 90%, min. 70%)	67.5% (max. 82%, min. 56%)	75.5% (max. 84%, min. 63%)
Average ratio of core to GFA	11% (max. 21%, min. 4%)	29.5% (max. 38%, min. 14%)	21.3% (max. 36%, min. 11%)

Treet, Timber Pioneer, and Docenten buildings demonstrate outstanding progress in space usage and architectural design, achieving impressive spatial efficiencies of up to 90%. These structures have set a new benchmark within the study group by boasting the smallest GFA ratios. This exceptional performance is attributable to meticulously designed cores that concentrate on compactness through the optimization of service areas and shaft arrangements, thereby maximizing usable floor space. Furthermore, these towers employ innovative structural systems—Treet with a tube system, Timber Pioneer with a shear-walled frame, and Docenten with a shear wall system—each enhancing their resilience against both vertical and horizontal forces. By utilizing compact cross-sections of structural elements, these buildings competently distribute and resist loads, ensuring both stability and durability. The integration of these cutting-edge design norms not only guarantees superior space efficiency but also significantly bolsters the structural robustness of these towers, establishing a new model for sustainable urban design. This holistic approach to architectural and structural innovation underscores the potential for combining advanced engineering techniques with optimized space utilization strategies to create resilient, efficient, and sustainable urban environments.

3.3.1. Inter-Relationship of Space Efficiency and Function

In Figure 6, observational data clarify the connection between spatial effectiveness and functional use. Bars on the right depict the total number of towers sorted by function type. Colored dots denote the spatial efficiency of high-rise cases according to their functions. Also, the gray bar implies the frequency of structures in the sampled group that have the same functions.

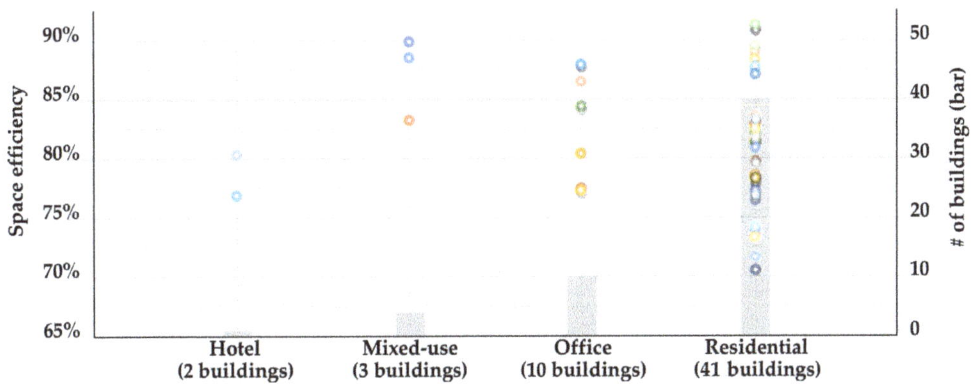

Figure 6. Inter-relationship between space efficiency and function.

The analysis of spatial effectiveness across 56 buildings reveals notable distinctions: residential timber buildings demonstrate an average effectiveness of 82%, spanning a range of 70% to 90% in efficiency. In contrast, office-functioned timber structures exhibit a slightly higher average effectiveness of 84%, with efficiency levels ranging from 79% to 90%. These findings underscore the variability and context-specific nature of timber structures' performance in different building types, emphasizing the nuanced considerations required for optimizing spatial efficiency in architectural design.

3.3.2. Inter-Relationship of Space Efficiency and Core Planning

In Figure 7, observed measurements point out the relationship between space utilization and core typology. Bars on the right side symbolize the number of instances grouped by core type. Each orange marker signifies the spatial efficiency of selected cases associated with specific core types. Additionally, a gray bar highlights the frequency of towers in the sample population that have the same core types.

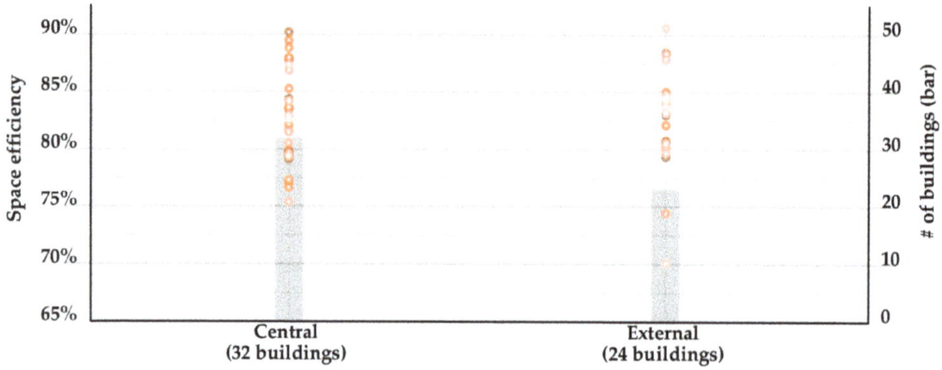

Figure 7. Inter-relationship between space efficiency and core planning.

Within the context of core layouts adopted in high-rise wooden buildings across Europe, the central configuration occurs as the leading choice, observed in 32 instances. This layout demonstrates a range of spatial efficiency levels from 75% to 89%, with an average efficiency of 82%. Conversely, buildings employing a peripheral core, totaling 24 cases, exhibit spatial efficiency levels spanning from 70% to 90%, averaging at 83%. The marginal difference in average spatial efficiency between central and peripheral core typologies, with central cores averaging 82% and peripheral cores at 83%, underscores the subtle interplay of architectural and engineering factors. These findings emphasize that while core layout is a significant determinant of spatial efficiency, the performance outcomes are closely aligned, reflecting the optimized design strategies employed across both configurations.

3.3.3. Inter-relationship of Space Efficiency and Form

Figure 8 illustrates the relationship between spatial efficiency and different architectural configurations. Bars on the right each indicate the total number of towers grouped by their forms. Each blue dot corresponds to the spatial efficiency of these cases according to specific forms. Also, the gray bar denotes the frequency of towers in the sample that have similar form configurations.

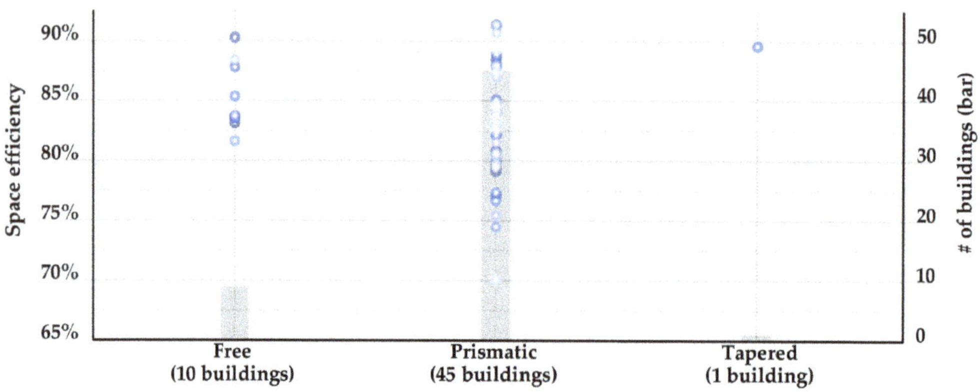

Figure 8. Inter-relationship between space efficiency and form.

Prismatic configurations, characterized by their geometric simplicity and structural efficiency, are favored due to their ease of construction and predictable load-bearing capacities, as evidenced by their prevalence in 45 instances. The spatial effectiveness in these designs ranges broadly from 70% to 90%, averaging 82%, indicating robust performance in spatial utilization. In contrast, the 10 instances of free-form configurations, although fewer, reveal spatial effectiveness levels between 81% and 89%, with an average of 84%. This slight edge in average effectiveness highlights the potential for innovative design approaches to achieve high spatial performance. The minimal variation in average spatial effectiveness between prismatic and free-form designs (82% vs. 84%) underscores the versatility and adaptability of timber as a building material, maintaining consistent spatial efficiency regardless of geometric complexity.

3.3.4. Inter-Relationship of Space Efficiency and Structural Material

Figure 9 depicts the relationship between spatial effectiveness and materials. The diagram features vertical columns to the right, showing the overall occurrence rate of instances classified by their individual building materials. Each green point represents the spatial effectiveness of structures linked to particular materials. Furthermore, a grey column is included to display the frequency of instances that utilize consistent building materials.

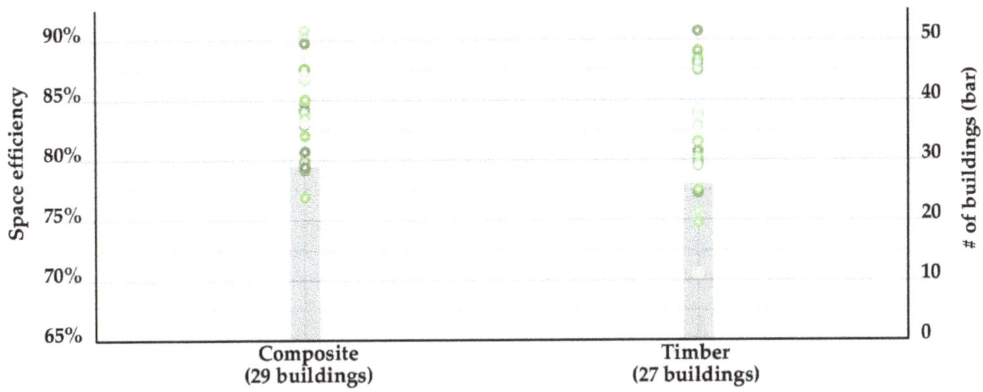

Figure 9. Inter-relationship between space efficiency and structural material.

Empirical data from multiple high-rise timber structures reveal that composite timber materials exhibit spatial efficiency levels ranging from 76% to 90%, while pure timber materials range from 70% to 90%. These figures translate to average spatial efficiencies of approximately 83% and 82%, respectively. The narrow variance of merely 1% between the average efficiencies of composite and pure timber materials signifies a substantial overlap in their performance metrics. This negligible difference in spatial efficiency highlights the comparable suitability of both material types for structural applications in high-rise constructions. The consistency in spatial efficiency metrics across diverse instances underscores the effectiveness of both composite and pure timber materials in meeting the structural demands of contemporary European architectural designs. Such findings are indicative of the robust versatility and efficacy of timber as a fundamental structural component, whether used in its pure form or in composite configurations.

3.3.5. Inter-Relationship of Space Efficiency and Structural System

In Figure 10, the observed information regarding the connection between spatial effectiveness and load-carrying frameworks is displayed. The vertical column to the right illustrates the overall occurrence of instances by their individual structural frameworks. Each yellow spot corresponds to the spatial effectiveness of constructions within the examination, linked with particular structural frameworks. Additionally, a grey pole demonstrates the predominance of towers sharing identical structural frameworks.

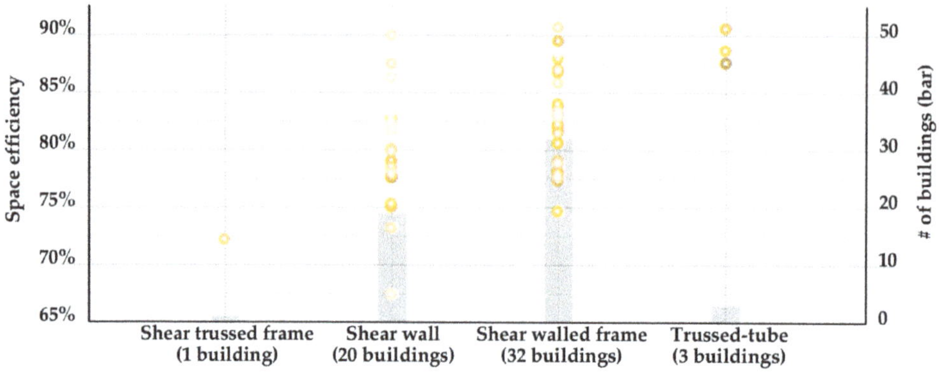

Figure 10. Inter-relationship between space efficiency and structural system.

In the investigation of structural frameworks utilized in European towers, shear walled frame systems have emerged as the dominant choice, evident in 32 analyzed instances. These systems exhibit substantial spatial efficiency, with recorded levels ranging from 76% to 90%, and averaging at 83%. In contrast, shear walls represent the second most preferred framework, employed in 20 cases, demonstrating spatial efficiencies spanning from 70% to 90%, and averaging 81%. This 2% disparity in average spatial efficiency between the two prevalent structural systems highlights nuanced differences crucial in high-rise timber tower construction. The prominence of shear walled frame systems underscores their efficacy in optimizing spatial utilization, reflecting advancements in design and construction methodologies tailored to modern architectural demands and environmental sustainability imperatives.

4. Discussion

This article aims to thoroughly compile and integrate extensive data from 56 tall timber buildings across Europe, centering on spatial efficiency, specifically examining the architectural and structural aspects of these ecologically sustainable towers. By doing so, we aim to deepen the understanding of the intricate processes involved in the design and development of high-rise timber edifices. Overall, this study seeks to significantly contribute to the current body of literature in the global and European context of designing and constructing these sustainable buildings. The results have both parallels and differences when compared to earlier research endeavors, as exemplified by the research [36]. Prominent studies presented highlight distinct approaches and findings in high-rise construction, focusing on timber and composite materials. Tall timber buildings, as supported by study [37], emphasize a centralized core design that enhances structural robustness, integrates circulation efficiently, and fosters architectural adaptability. These buildings typically employ straightforward rectangular layouts that streamline construction and optimize space utilization, reflecting traditional practices and modular construction principles. In contrast, the trend in Europe towards composite materials (notably discussed in [75]) combines timber's aesthetic appeal with concrete's strength, particularly favoring shear-walled frames for stability against horizontal forces. Both studies stress the importance of space efficiency, with non-timber high-rises averaging 71% to 76% efficiency and varying core area-to-GFA ratios, whereas timber constructions, especially in Finland and other studies, achieve higher efficiencies ranging from 70% to 95%.

In tall timber buildings, the predominant arrangement often features a centralized core design, supported by various studies such as [33]. This layout, accommodating main vertical circulation elements, provides several benefits. It enhances structural robustness, expands floor area by integrating circulation and service purposes, and promotes efficient vertical movement, mitigating congestion. Moreover, it fosters architectural adaptability, specifically useful in multi-use projects. The majority of structures examined in this research exhibit a straightforward rectangular design, favored for its simplicity compared to more intricate shapes. Rectangular layouts align well with traditional building practices, streamlining the building process. Their floor plans efficiently utilize space and simplify room configurations. Additionally, their modular nature advances sustainable and cost-efficient construction by optimizing materials and resources.

An observable shift in high-rise construction across Europe involves the increasing adoption of composite materials combining timber and concrete, evolving as the favored option due to their adaptability and myriad advantages [75]. This blend harnesses the cozy charm of timber with the robustness and adaptability of concrete [76,77]. Within the structural frameworks of timber constructions, a distinct ranking has surfaced, wherein lofty timber endeavors lean towards shear-walled frames and shear-wall arrangements for their adept management of horizontal forces and assurance of steadiness, principally in confronting wind and earthquake impacts.

A suggested criterion to assess the space efficiency of high-rise structures is a target level of 75% [78]. Studies on non-timber high-rise buildings intended for office, residential, and mixed-use functions revealed various levels of space optimization, generally ranging from 71% to 76% [22–24]. These evaluations highlight differences in the proportion of core area to GFA, ranging from 19% to 26%. Regarding office towers, the lowest recorded space utilization was 63%, whereas the highest was 82%. Likewise, the core area-to-GFA ratio varied from the lowest of 15% to the highest of 36%. Similar variations were noted in residential and mixed-use towers. Additionally, research on over fifty Finnish mid-rise timber apartments showed space use ranging from 78% to 88%, with an average efficiency of 83% [38]. Moreover, studies on high-rise wooden residential [34] and office towers [37] showed space efficiencies ranging from 70% to 93% and 75% to 95%, with averages of 83% and 88%, respectively. In this paper, the average space utilization and core area-to-GFA percentage were 82% and 11%, respectively. The range displayed a broad spectrum, with the lowest space efficiency at 70% and the highest at 90%, whereas core area-to-GFA percentage varied from 4% to 21%.

The marked disparity in space efficiency between timber and non-timber structural systems can be attributed to fundamental differences in their material characteristics and construction methodologies [79–81]. Timber structures capitalize on lightweight frames and panels that enable the use of slender structural elements, contrasting with the thicker components necessary in conventional concrete or steel buildings. This inherent advantage allows timber constructions to achieve reduced wall thicknesses and shallower floor depths, effectively maximizing the usable floor area. Moreover, timber's excellent strength-to-weight ratio facilitates longer spans and fewer internal supports, enhancing interior layout flexibility and spatial optimization. In contrast, non-timber systems such as concrete and steel require bulkier structural members to withstand comparable loads, resulting in thicker walls and deeper floor sections that inherently diminish overall usable space efficiency.

To further improve space efficiency, this study suggests several strategies: optimizing core designs to minimize space consumption, employing advanced construction technologies to reduce material footprint, and exploring innovative architectural forms that enhance usable floor area without compromising structural integrity. Additionally, designers should integrate multifunctional spaces, implement flexible layouts that can adapt to changing needs, and utilize modular construction techniques to streamline the building process and enhance spatial utilization. By adopting these approaches, timber towers can achieve higher space efficiency, thereby contributing to their sustainability and economic viability while setting new benchmarks in urban design.

The research presented valuable insights into the space efficiency of towering timber edifices in Europe, a rapidly expanding domain with significant environmental and economic advantages. However, several limitations need to be acknowledged. Firstly, the study primarily focused on European cases, potentially limiting the generalizability of its findings to other regions with distinct regulatory, cultural, and environmental contexts. Secondly, the analysis relied on a relatively modest sample size of 56 case studies, which might not fully capture the diversity and complexity of timber tower projects across Europe. Despite these limitations, the study provided crucial baseline data and findings that can inform architectural design practices, but further research with larger and more diverse datasets is warranted to comprehensively understand the factors influencing space efficiency in towering timber structures.

5. Conclusions

This research centers on the spatial efficiency of high-rise wooden buildings throughout Europe, bridging a void in prior research. By methodically examining information from 56 towers through individual case analyses, we have identified critical parameters affecting spatial efficiency. These factors include the layout and size of the central and service cores, functionality, design, and choice of structural materials and systems, considering the size of load-bearing elements. The main conclusions from this study can be summarized as follows:

- Residential use is the top choice in Europe. Central core designs are the most common, showing a liking for structures with a strong and stable center. Prismatic shapes are favored, indicating a preference for smooth and simple architectural styles.
- A significant trend is the extensive use of composite construction, primarily combining timber and concrete. Structural systems used often favor shear-frame designs, particularly shear-walled frames, underscoring a dedication to strong structural solutions among European high-rise towers.
- The average space efficiency observed in the analyzed towers was noted at 82%, with fluctuations spanning from 70% to 90% among various cases.
- The average core area to GFA ratio was 11%, indicating differences between 4% and 21%.

The practical application of these results in European tall timber buildings involves several key aspects. Firstly, central core arrangements were found to be the most effective, providing architects with a framework for designing spaces that optimize utility and accessibility, thereby streamlining the construction process. Prismatic configurations with shear-frame and shear wall systems are recommended for structural integrity and space efficacy. The widespread utilization of timber, supplemented with concrete, suggests a material approach that balances sustainability and structural resilience.

With an average space efficiency of 82% and a core area to gross floor area ratio of 10%, architectural designers now have standards for maximizing rentable space and crafting efficient architectural plan layouts. Utilizing these guidelines, architectural professionals can develop timber structures that are more efficient, sustainable, and economically feasible. This balance fosters the development of distinctive high-rise spaces that symbolize the modern design philosophy and environmental consciousness. Moreover, although there are no significant differences between the building groups and an insufficient number of towers in certain groups such as tapered forms to make definitive scientific inferences, it can be suggested that composite buildings featuring peripheral cores, free-form designs, and tube systems were more efficient in terms of space utilization among the case study sample.

Future research on spatial efficiency holds immense potential for advancing sustainable construction practices and urban development. By delving into the development of novel timber materials with enhanced structural properties and exploring innovative construction techniques such as modular assembly and digital fabrication, researchers can push the boundaries of what is achievable in tall timber buildings. The integration of optimization algorithms into the design process can optimize layouts to maximize usable space while ensuring structural integrity and meeting occupants' needs. Furthermore, investigations into biophilic design principles and flexible space planning strategies can enhance occupants' well-being and productivity while accommodating diverse uses within a limited footprint. Assessments of energy efficiency and sustainability, coupled with empirical studies on occupant behavior, can provide valuable insights into optimizing building performance and user satisfaction. Moreover, the incorporation of advanced building technologies, such as IoT sensors and data analytics, allows for real-time monitoring and management of space utilization and environmental conditions. This integration enhances efficiency and promotes the conservation of resources. Through these interdisciplinary efforts, researchers can contribute to the advancement of space-efficient, environmentally sustainable high-rise timber towers that serve as beacons of innovation and resilience in the urban landscape.

Author Contributions: Conceptualization, H.E.I.; methodology, H.E.I. and Ö.N.A.; software, H.E.I.; formal analysis, H.E.I. and Ö.N.A.; investigation, H.E.I. and Ö.N.A.; writing—original draft preparation, H.E.I.; writing—review and editing, H.E.I. and Ö.N.A.; supervision, H.E.I. All authors have read and agreed to the published version of the manuscript.

Funding: This research received no external funding.

Institutional Review Board Statement: Not applicable.

Informed Consent Statement: Not applicable.

Data Availability Statement: The data presented in this study are available in this article.

Conflicts of Interest: The authors declare no conflicts of interest.

Appendix A

Table A1. European High-Rise Timber Buildings.

#	Building Name	Country	City	Height (m)	# of Stories	Completion Date	Function
1	Mjøstårnet	Norway	Brumunddal	85	18	2019	M(R + O + H)
2	HoHo Wien	Austria	Vienna	84	23	2020	M(R + O + H)
3	TRAE	Denmark	Aarhus	82	20	UC	Office
4	HAUT	Netherlands	Amsterdam	73	22	2022	Residential
5	Sara Kulturhus	Sweden	Skellefteå	73	20	2021	Hotel
6	De Karel Doorman	Netherlands	Rotterdam	71	22	2012	Residential
7	Roots Tower	Germany	Hamburg	65	19	UC	Residential
8	Abro (Baufeld 1 Suurstoffi Campus)	Switzerland	Risch-Rotkreuz	60	15	2019	Office
9	Hyperion	France	Bordeaux	55	16	2021	Residential
10	Rundeskogen Hus B	Norway	Sandnes	55	16	2013	Residential
11	Treet	Norway	Bergen	49	14	2015	Residential
12	Lighthouse Joensuu	Finland	Joensuu	48	14	2019	Residential
13	Cederhusen	Sweden	Stockholm	44	13	UC	Residential
14	Hoas Tuuliniitty	Finland	Espoo	44	13	2021	Residential
15	Palazzo Nice Meridia	France	Nice	44	10	2019	Office
16	Spor X	Norway	Drammen	41	10	2021	Office
17	Sensations	France	Strasbourg	38	11	2019	Residential
18	Rundeskogen Hus C	Norway	Sandnes	38	11	2013	Residential
19	Fyrtornet	Sweden	Malmö	38	11	UC	M(R+O)
20	Trafalgar Place	UK	London	36	10	2015	Residential
21	Suurstoffi 22	Switzerland	Risch-Rotkreuz	36	10	2018	Office

Table A1. *Cont.*

#	Building Name	Country	City	Height (m)	# of Stories	Completion Date	Function
22	Opalia	France	Saint-Ouen-sur-Seine	35	8	2017	Office
23	Green Office ENJOY	France	Paris	35	8	2018	Office
24	Pont de Flandres Batiment 007	France	Paris	35	8	2019	Office
25	Hotel Jakarta	Netherlands	Amsterdam	34	9	2018	Hotel
26	Kringsja Studentby	Norway	Oslo	34	10	2018	Residential
27	Rundeskogen Hus A	Norway	Sandnes	34	10	2012	Residential
28	SKAIO	Germany	Heilbronn	34	10	2019	Residential
29	Dalston Works	UK	London	34	10	2017	Residential
30	The Cube Building	UK	London	33	10	2015	Residential
31	Botanikern	Sweden	Uppsala	31	9	2019	Residential
32	Cenni di Cambiamento	Italy	Milan	31	9	2013	Residential
33	Kajstaden	Sweden	Vasteras	31	9	2019	Residential
34	Press House	UK	London	31	9	2017	Residential
35	Vallen	Sweden	Vaxjo	31	9	2015	Residential
36	Timber Pioneer	Germany	Frankfurt am Main	30	8	UC	Office
37	Stadthaus	UK	London	29	9	2009	Residential
38	EDGE Suedkreuz	Germany	Berlin	29	8	2022	Office
39	Moholt 50/50	Norway	Trondheim	28	9	2016	Residential
40	Cirerers	Spain	Barcelona	27	8	2022	Residential
41	DAS Kelo	Finland	Rovaniemi	27	8	2019	Residential
42	Docenten	Sweden	Vaxjo	27	8	2018	Residential
43	Dramsvegen	Norway	Tromso	27	8	2017	Residential
44	Frostaliden	Sweden	Skövde	27	8	2018	Residential
45	Highpoint Terrace	UK	London	27	8	2017	Residential
46	Jo & Joe	France	Gentilly	27	8	2019	Residential
47	Limnologen	Sweden	Vaxjo	27	8	2014	Residential
48	Maskinparken TRE	Norway	Trondheim	27	8	2018	Residential
49	Puukuokka Housing Block	Finland	Jyvaskyla	27	8	2018	Residential
50	Residences J.Ferry	France	Saint-Dié-des Vosges	27	8	2014	Residential
51	Strandparken	Sweden	Stockholm	27	8	2014	Residential
52	Trummens Strand	Sweden	Vaxjo	27	8	2019	Residential
53	Wood City Residential Buildings	Finland	Helsinki	27	8	2018	Residential
54	Lucien Cornil Student Residence	France	Marseille	27	8	2017	Residential
55	Pentagon I	Norway	As	27	8	2013	Residential
56	Bridport House	UK	London	27	8	2010	Residential

Note on abbreviations: 'M' indicates Mixed-use; 'R' indicates Residential; 'O' indicates Office; 'H' indicates Hotel; 'UC' indicates under construction.

Appendix B

Table A2. European High-Rise Timber Buildings by Form, Core Type, Structural System, and Structural Material.

#	Building Name	Building Form	Core Type	Structural System	Structural Material
1	Mjøstårnet	Prismatic	Peripheral	Trussed-tube	Timber
2	HoHo Wien	Prismatic	Central	Shear walled frame	Composite (T + C)
3	TRAE	Prismatic	Central	Shear walled frame	Composite (T + C)
4	HAUT	Free	Peripheral	Shear walled frame	Composite (T + C)
5	Sara Kulturhus	Prismatic	Peripheral	Shear wall	Composite (T + S)
6	De Karel Doorman	Prismatic	Peripheral	Shear walled frame	Composite (T + C+S)
7	Roots Tower	Prismatic	Central	Shear walled frame	Composite (T + C)
8	Abro (Baufeld 1 Suurstoffi Campus)	Prismatic	Central	Shear walled frame	Composite (T + C)
9	Hyperion	Free	Central	Shear walled frame	Composite (T + C + S)
10	Rundeskogen Hus B	Free	Central	Shear walled frame	Composite (T + C)
11	Treet	Prismatic	Peripheral	Trussed-tube	Timber
12	Lighthouse Joensuu	Prismatic	Central	Shear wall	Timber
13	Cederhusen	Prismatic	Central	Shear wall	Timber
14	Hoas Tuuliniitty	Prismatic	Peripheral	Shear wall	Timber
15	Palazzo Nice Meridia	Prismatic	Peripheral	Shear walled frame	Composite (T + C)
16	Spor X	Prismatic	Central	Shear walled frame	Timber
17	Sensations	Free	Central	Shear walled frame	Timber
18	Rundeskogen Hus C	Free	Central	Shear walled frame	Composite (T + C)
19	Fyrtornet	Tapered	Central	Trussed-tube	Timber
20	Trafalgar Place	Prismatic	Peripheral	Shear wall	Timber
21	Suurstoffi 22	Prismatic	Central	Shear walled frame	Composite (T + C)
22	Opalia	Prismatic	Peripheral	Shear walled frame	Composite (T + C + S)
23	Green Office ENJOY	Prismatic	Central	Shear walled frame	Composite (T + C)
24	Pont de Flandres Batiment 007	Free	Central	Shear walled frame	Composite (T + C + S)
25	Hotel Jakarta	Prismatic	Central	Shear walled frame	Composite (T + C)
26	Kringsja Studentby	Prismatic	Central	Shear walled frame	Timber
27	Rundeskogen Hus A	Free	Central	Shear walled frame	Composite (T + C)
28	SKAIO	Prismatic	Central	Shear walled frame	Composite (T + C)
29	Dalston Works	Prismatic	Central	Shear wall	Timber
30	The Cube Building	Free	Central	Shear walled frame	Composite (T + C + S)
31	Botanikern	Prismatic	Peripheral	Shear trussed frame	Timber
32	Cenni di Cambiamento	Free	Central	Shear wall	Timber
33	Kajstaden	Prismatic	Peripheral	Shear wall	Timber
34	Press House	Prismatic	Central	Shear walled frame	Timber
35	Vallen	Prismatic	Central	Shear walled frame	Composite (T + C)
36	Timber Pioneer	Prismatic	Peripheral	Shear walled frame	Composite (T + C)
37	Stadthaus	Prismatic	Central	Shear wall	Timber
38	EDGE Suedkreuz	Prismatic	Central	Shear walled frame	Composite (T + C)
39	Moholt 50/50	Prismatic	Central	Shear wall	Timber
40	Cirerers	Free	Peripheral	Shear wall	Timber
41	DAS Kelo	Prismatic	Peripheral	Shear walled frame	Timber
42	Docenten	Prismatic	Peripheral	Shear wall	Composite (T + C)
43	Dramsvegen	Prismatic	Peripheral	Shear wall	Composite (T + C)
44	Frostaliden	Prismatic	Central	Shear walled frame	Composite (T + C)
45	Highpoint Terrace	Prismatic	Peripheral	Shear wall	Timber
46	Jo & Joe	Prismatic	Peripheral	Shear walled frame	Composite (T + C)
47	Limnologen	Prismatic	Central	Shear wall	Composite (T + C + S)
48	Maskinparken TRE	Prismatic	Central	Shear wall	Timber
49	Puukuokka Housing Block	Prismatic	Central	Shear walled frame	Timber
50	Residences J.Ferry	Prismatic	Peripheral	Shear walled frame	Timber
51	Strandparken	Prismatic	Peripheral	Shear wall	Composite (T + S)
52	Trummens Strand	Prismatic	Peripheral	Shear wall	Timber
53	Wood City Residential Buildings	Prismatic	Peripheral	Shear walled frame	Timber
54	Lucien Cornil Student Residence	Prismatic	Peripheral	Shear walled frame	Composite (T + C)
55	Pentagon I	Prismatic	Central	Shear wall	Timber
56	Bridport House	Prismatic	Peripheral	Shear wall	Timber

Note on abbreviations: '(T + C + S)' indicates composite/hybrid structures combining timber and concrete and steel; '(T + C)' indicates composite/hybrid structures combining timber and concrete; '(T + S)' indicates composite/hybrid structures combining timber and steel.

Appendix C

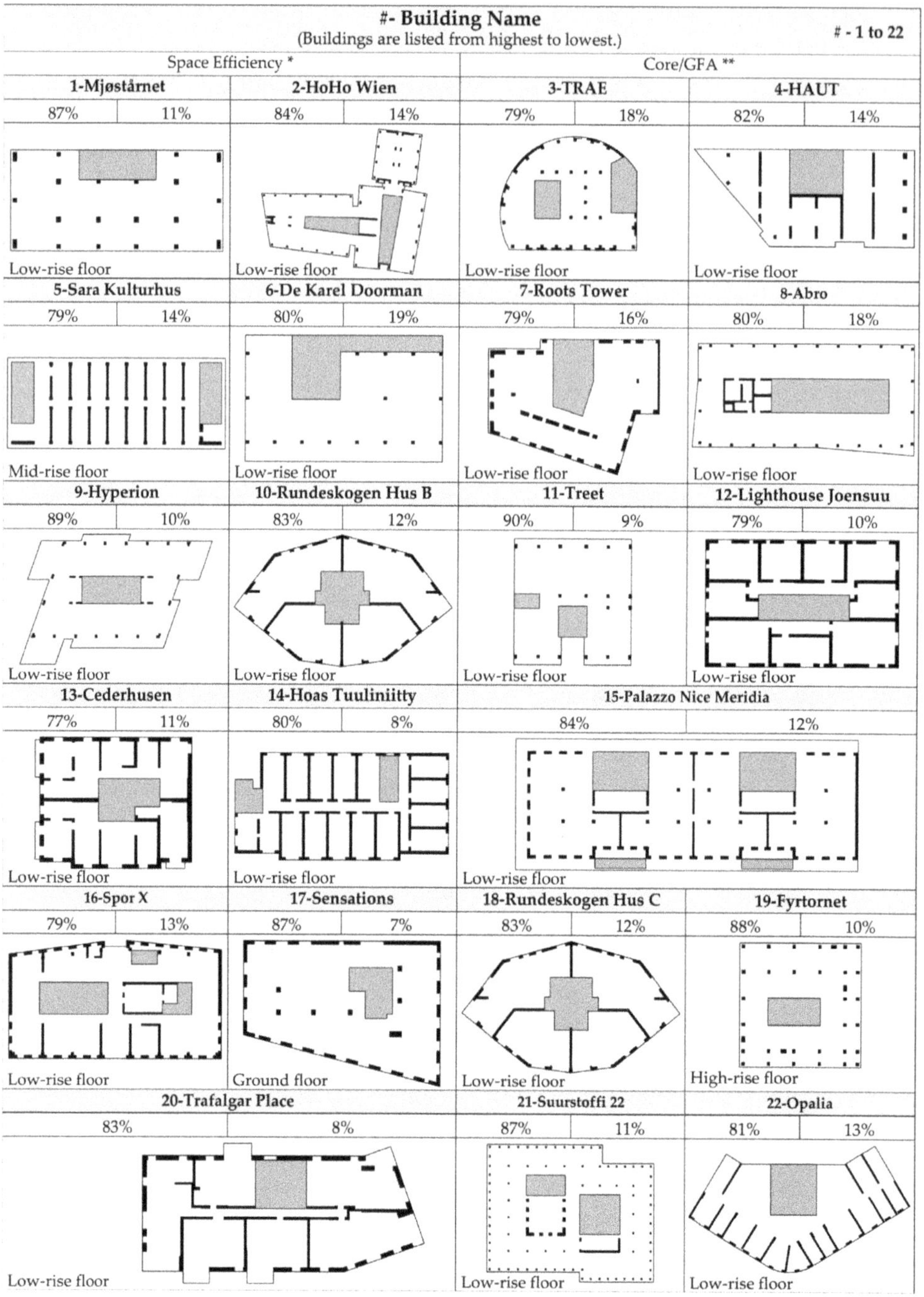

Figure A1. *Cont.*

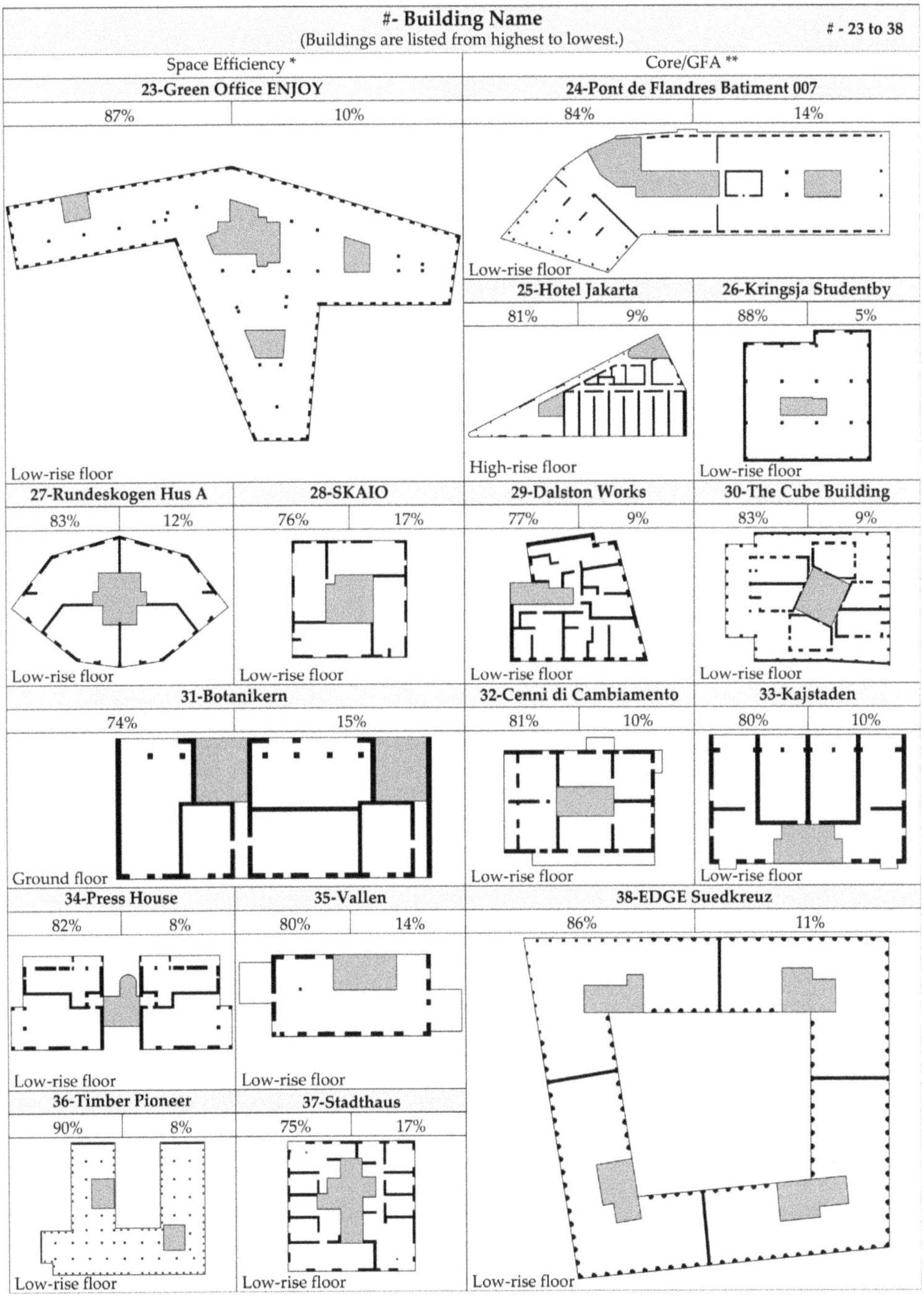

Figure A1. *Cont.*

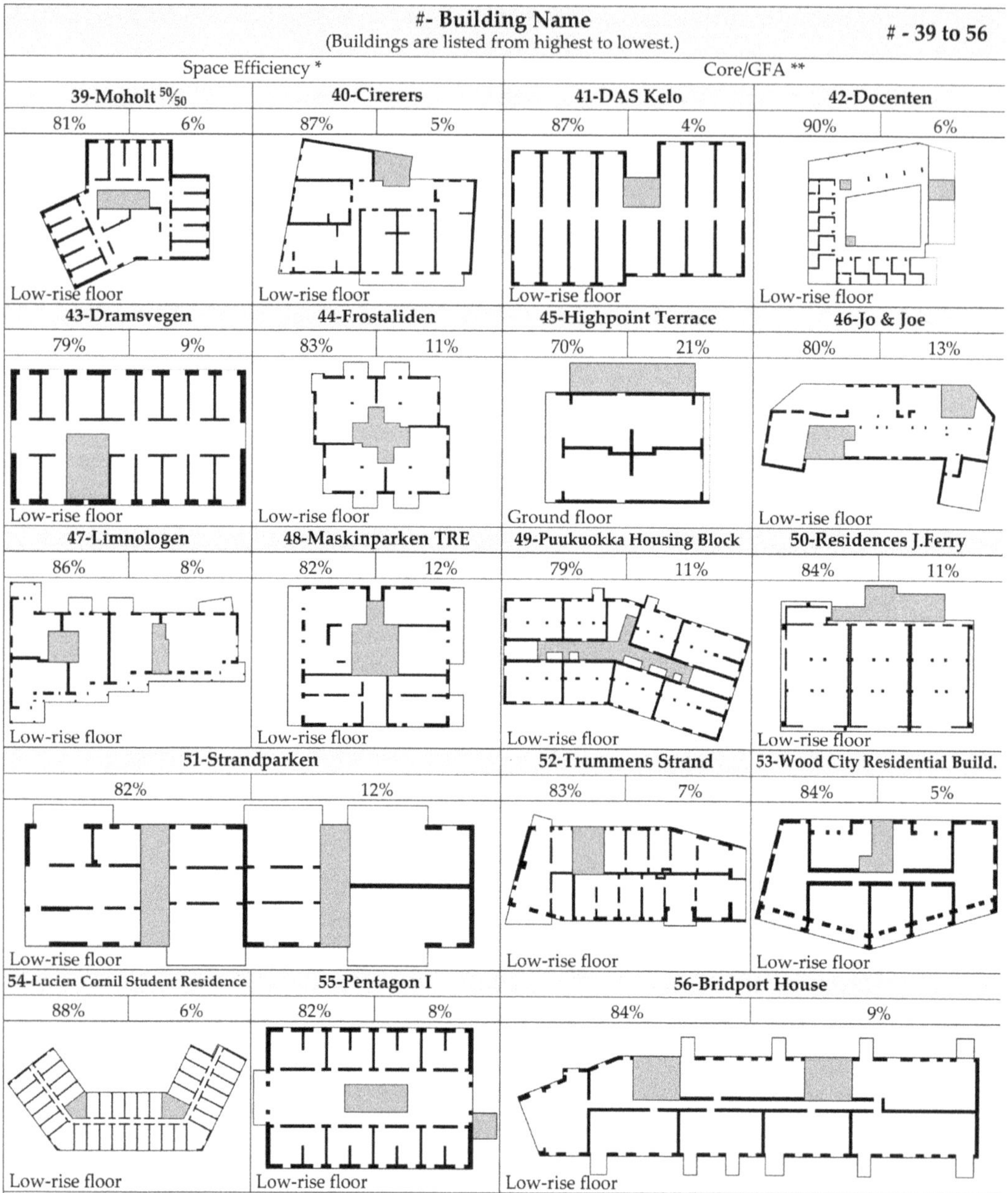

Space Efficiency* : calculated as the ratio of the net floor area obtained by subtracting the service core (the gray area on the floor plan) and structural elements from NFA to GFA.

Core / GFA** : calculated as the ratio of the service core (the gray area on the floor plan) to GFA.
In the floor plans the gray area corresponds to service core, while black area signifies structural elements.

Figure A1. Space Efficiency and Core/GFA of European High-Rise Timber Buildings (Created by the Authors).

References

1. Rahko, J.; Alola, A.A. Examining green productivity amidst climate change technological development and spillovers in the Nordic economies. *J. Clean. Prod.* **2024**, *434*, 140028. [CrossRef]
2. Meena, C.S.; Kumar, A.; Jain, S.; Rehman, A.U.; Mishra, S.; Sharma, N.K.; Bajaj, M.; Shafiq, M.; Eldin, E.T. Innovation in green building sector for sustainable future. *Energies* **2022**, *15*, 6631. [CrossRef]
3. Safarik, D.; Elbrecht, J.; Miranda, W. State of Tall Timber 2022. *CTBUH J.* **2022**, 22–31. Available online: https://s3.eu-west-2.amazonaws.com/construo-storage/attachments/9a13793fe644586846b1b92c4901ae721325e3e88e4eb40b8e83a354c75d0d97/CTBUH%20-%20State%20of%20Tall%20Timber%202022,%20Issue%201.pdf (accessed on 29 June 2024).
4. Morales-Beltran, M.; Engur, P.; Sisman, Ö.A.; Aykar, G.N. Redesigning for disassembly and carbon footprint reduction: Shifting from reinforced concrete to hybrid timber–steel multi-story building. *Sustainability* **2023**, *15*, 7273. [CrossRef]
5. Laboy, M.M. Reimagining low-carbon futures: Architectural and ecological tradeoffs of mass timber for durable buildings. *Archit. Struct. Constr.* **2022**, *2*, 723–741. [CrossRef]
6. Cavalli, A.; Togni, M. Monitoring of historical timber structures: State of the art and prospective. *J. Civ. Struct. Health Monit.* **2015**, *5*, 107–113. [CrossRef]
7. Holzer, S.M. Analysis of historical timber structures. In *Structural Analysis of Historical Constructions: Anamnesis, Diagnosis, Therapy, Controls*; CRC Press: Boca Raton, FL, USA, 2016; pp. 1203–1210.
8. Cruz, H.; Yeomans, D.; Tsakanika, E.; Macchioni, N.; Jorissen, A.; Touza, M.; Mannucci, M.; Lourenço, P.B. Guidelines for on-site assessment of historic timber structures. *Int. J. Archit. Herit.* **2015**, *9*, 277–289. [CrossRef]
9. Le, V.A.; Nguyen, T.K.N. Discussion of So-Called "Architectural Heritage DNA" via a Case Study of the Conservation of the Nara Palace Site, Japan. *Buildings* **2024**, *14*, 132. [CrossRef]
10. Cristea, I. Architecture of Estonia–old and new together. *Urbanism. Arhitectură. Construcţii* **2014**, *5*, 73–86.
11. McKitterick, R.; Osborne, J.; Richardson, C.M.; Story, J. *Old Saint Peter's, Rome*; Cambridge University Press: Cambridge, UK, 2013.
12. Goubran, S.; Masson, T.; Walker, T. Diagnosing the local suitability of high-rise timber construction. *Build. Res. Inf.* **2020**, *48*, 101–123. [CrossRef]
13. Ahn, N.; Dodoo, A.; Riggio, M.; Muszynski, L.; Schimleck, L.; Puettmann, M. Circular economy in mass timber construction: State-of-the-art, gaps and pressing research needs. *J. Build. Eng.* **2022**, *53*, 104562. [CrossRef]
14. Zahiri, N. Timber High-Rises in Nordic Countries: Current Trends. *CTBUH J.* **2023**, 44–50. Available online: https://www.proquest.com/openview/18bd24c28e7b2ef07815ea4e8423a3b6/1?pq-origsite=gscholar&cbl=6578554 (accessed on 29 June 2024).
15. Hurmekoski, E.; Jonsson, R.; Nord, T. Context, drivers, and future potential for wood-frame multi-story construction in Europe. *Technol. Forecast. Soc. Change* **2015**, *99*, 181–196. [CrossRef]
16. Vacek, Z.; Vacek, S.; Cukor, J. European forests under global climate change: Review of tree growth processes, crises and management strategies. *J. Environ. Manag.* **2023**, *332*, 117353. [CrossRef]
17. Matiiuk, Y.; Krikštolaitis, R.; Liobikienė, G. The COVID-19 pandemic in context of climate change perception and resource-saving behavior in the European Union countries. *J. Clean. Prod.* **2023**, *395*, 136433. [CrossRef] [PubMed]
18. Scouse, A.; Kelley, S.S.; Liang, S.; Bergman, R. Regional and net economic impacts of high-rise mass timber construction in Oregon. *Sustain. Cities Soc.* **2020**, *61*, 102154. [CrossRef]
19. Liang, S.; Gu, H.; Bergman, R. Environmental life-cycle assessment and life-cycle cost analysis of a high-rise mass timber building: A case study in Pacific Northwestern United States. *Sustainability* **2021**, *13*, 7831. [CrossRef]
20. Nepal, P.; Prestemon, J.P.; Ganguly, I.; Kumar, V.; Bergman, R.D.; Poudyal, N.C. The potential use of mass timber in mid-to high-rise construction and the associated carbon benefits in the United States. *PLoS ONE* **2024**, *19*, e0298379. [CrossRef] [PubMed]
21. Arriaga, F.; Wang, X.; Íñiguez-González, G.; Llana, D.F.; Esteban, M.; Niemz, P. Mechanical Properties of Wood: A Review. *Forests* **2023**, *14*, 1202. [CrossRef]
22. Ilgın, H.E. A study on space efficiency in contemporary supertall mixed-use buildings. *J. Build. Eng.* **2023**, *69*, 106223. [CrossRef]
23. Ilgın, H.E. Space Efficiency in Contemporary Supertall Residential Buildings. *Architecture* **2021**, *1*, 25–37. [CrossRef]
24. Ilgın, H.E. Space Efficiency in Contemporary Supertall Office Buildings. *J. Archit. Eng.* **2021**, *27*, 04021024. [CrossRef]
25. Lin, C.L.; Chiang, W.H.; Weng, Y.S.; Wu, H.P. Assessing the anthropogenic carbon emission of wooden construction: An LCA study. *Build. Res. Inf.* **2023**, *51*, 138–157. [CrossRef]
26. Karjalainen, M.; Ilgın, H.E.; Metsäranta, L.; Norvasuo, M. *Wooden Facade Renovation and Additional Floor Construction for Suburban Development in Finland*; IntechOpen: London, UK, 2022.
27. Ilgın, H.E.; Karjalainen, M. *Perceptions, Attitudes, and Interests of Architects in the Use of Engineered Wood Products for Construction: A Review*; IntechOpen: London, UK, 2021.
28. Jayawardana, J.; Kulatunga, A.K.; Jayasinghe, J.A.S.C.; Sandanayake, M.; Zhang, G. Environmental Sustainability of Off-Site Construction in Developed and Developing Regions: A Systematic Review. *J. Archit. Eng.* **2023**, *29*, 04023008. [CrossRef]
29. Bahrami, A.; Jakobsson, J.; Söderroos, T. Factors Influencing Choice of Wooden Frames for Construction of Multi-Story Buildings in Sweden. *Buildings* **2023**, *13*, 217. [CrossRef]
30. Balasbaneh, A.T.; Sher, W. Economic and environmental life cycle assessment of alternative mass timber walls to evaluate circular economy in building: MCDM method. *Environ. Dev. Sustain.* **2024**, *26*, 239–268. [CrossRef]
31. Ascione, F.; Esposito, F.; Iovane, G.; Faiella, D.; Faggiano, B.; Mele, E. Sustainable and Efficient Structural Systems for Tall Buildings: Exploring Timber and Steel–Timber Hybrids through a Case Study. *Buildings* **2024**, *14*, 524. [CrossRef]

32. Ghobadi, M.; Sepasgozar, S.M. Circular economy strategies in modern timber construction as a potential response to climate change. *J. Build. Eng.* **2023**, *77*, 107229. [CrossRef]
33. Mamić, D.; Domljan, D. Positive Aspects of Using Solid Wood in Interiors on Human Wellbeing: A Review. *Drv. Ind.* **2023**, *74*, 379–391. [CrossRef]
34. Loh, T.W.; Barnett, J.R.; Zhang, G.K.; Nguyen, K.T. Fire-Induced Damage and Postfire Analysis of the Compression Properties of Cross-Laminated Timber. *J. Mater. Civ. Eng.* **2024**, *36*, 04023595. [CrossRef]
35. Ahmad, F.; Allan, K.; Phillips, A.R. Multicriteria decision analysis of steel and mass timber prototype buildings in the Pacific Northwest. *J. Archit. Eng.* **2023**, *29*, 04023001. [CrossRef]
36. Ilgın, H.E.; Aslantamer, Ö.N. Analysis of Space Efficiency in High-Rise Timber Residential Towers. *Appl. Sci.* **2024**, *14*, 4337. [CrossRef]
37. Aslantamer, Ö.N.; Ilgın, H.E. Space efficiency in timber office buildings. *J. Build. Eng.* **2024**, *91*, 109618. [CrossRef]
38. Tuure, A.; Ilgın, H.E. Space Efficiency in Finnish Mid-Rise Timber Apartment Buildings. *Buildings* **2023**, *13*, 2094. [CrossRef]
39. Ilgın, H.E.; Aslantamer, Ö.N. Investigating Space Utilization in Skyscrapers Designed with Prismatic Form. *Buildings* **2024**, *14*, 1295. [CrossRef]
40. Ilgın, H.E. Space Efficiency in Tapered Super-Tall Towers. *Buildings* **2023**, *13*, 2819. [CrossRef]
41. Ilgın, H.E. An analysis of space efficiency in Asian supertall towers. *Int. J. Build. Pathol. Adapt.* **2023**, *41*, 237–253. [CrossRef]
42. Ilgın, H.E. Examination of spatial efficiency in super-tall towers within the Middle Eastern context. *Open House Int.* 2024; *ahead-of-print*. [CrossRef]
43. Okbaz, F.T.; Sev, A. A model for determining the space efficiency in non-orthogonal high rise office buildings. *J. Fac. Eng. Archit. Gazi Univ.* **2023**, *38*, 113–125.
44. Ibrahimy, R.; Mohmmand, M.A.; Elham, F.A. An Evaluation of Space Use Efficiency in Residential Houses, Kabul City. *J. Res. Appl. Sci. Biotechnol.* **2023**, *2*, 1–6. [CrossRef]
45. Goessler, T.; Kaluarachchi, Y. Smart Adaptive Homes and Their Potential to Improve Space Efficiency and Personalisation. *Buildings* **2023**, *13*, 1132. [CrossRef]
46. Hamid, G.M.; Elsawi, M.; Yusra, O. The Impacts of Spatial Parameters on Space Efficiency in Hybrid Villa-Apartments in Greater Khartoum, Sudan. *J. Archit. Plan.* **2022**, *34*, 425–440.
47. Suga, R. Space Efficiency in Hotel Development. Master's Thesis, MODUL University Vienna, Vienna, Austria, 2021.
48. Arslan Kılınç, G. Improving a Model for Determining Space Efficiency of Tall Office Buildings. Ph.D. Thesis, Department of Architecture, Mimar Sinan Fine Art University, Istanbul, Turkey, 2019. (In Turkish).
49. Von Both, P. A stakeholder-and function-based planning method for space-efficient buildings. *IOP Conf. Ser. Earth Environ. Sci.* **2019**, *323*, 012040. [CrossRef]
50. Höjer, M.; Mjörnell, K. Measures and steps for more efficient use of buildings. *Sustainability* **2018**, *10*, 1949. [CrossRef]
51. Nam, H.J.; Shim, J.H. An analysis of the change in space efficiency based on various tall building corner shapes and lease spans. *J. Archit. Inst. Korea Plan. Des.* **2016**, *32*, 13–20.
52. Zhang, L.; Zhang, L.; Wang, Y. Shape optimization of free-form buildings based on solar radiation gain and space efficiency using a multi-objective genetic algorithm in the severe cold zones of China. *Sol. Energy* **2016**, *132*, 38–50. [CrossRef]
53. Sev, A.; Özgen, A. Space efficiency in high-rise office buildings. *METU J. Fac. Archit.* **2009**, *26*, 69–89. [CrossRef]
54. Saari, A.; Tissari, T.; Valkama, E.; Seppänen, O. The effect of a Redesigned Floor Plan, Occupant Density and the Quality of Indoor Climate on the Cost of Space, Productivity and Sick Leave in an Office Building—A Case Study. *Build. Environ.* **2006**, *41*, 1961–1972. [CrossRef]
55. Kim, H.; Elnimeiri, M. Space efficiency in multi-use tall building. In Proceedings of the Tall Buildings in Historical Cities-Culture and Technology for Sustainable Cities, Seoul, Republic of Korea, 10–13 October 2004.
56. CTBUH Council on Tall Buildings Urban Habitat Illinois Institute of Technology, S.R. Crown Hall, 3360 South State Street, Chicago, IL, USA. Available online: https://www.ctbuh.org (accessed on 14 June 2024).
57. Opoku, A.; Ahmed, V.; Akotia, J. Choosing an Appropriate Research Method. In *Research Methodology in the Built Environment*; Routledge: London, UK, 2016; pp. 32–49.
58. Noor, K.B.M. Case study: A strategic research methodology. *Am. J. Appl. Sci.* **2008**, *5*, 1602–1604. [CrossRef]
59. Ali, M.M.; Moon, K.S. Advances in Structural Systems for Tall Buildings: Emerging Developments for Contemporary Urban Giants. *Buildings* **2018**, *8*, 104. [CrossRef]
60. Taranath, B.S. *Structural Analysis and Design of Tall Buildings: Steel and Composite Construction*; CRC Press: Boca Raton, FL, USA, 2016.
61. Ali, M.M.; Al-Kodmany, K. Structural Systems for Tall Buildings. *Encyclopedia* **2022**, *2*, 1260–1286. [CrossRef]
62. Li, S.; Shao, Y.; Hong, M.; Zhu, C.; Dong, B.; Li, Y.; Lin, Y.; Wang, K.; Gan, M.; Zhu, J.; et al. Impact mechanisms of urbanization processes on supply-demand matches of cultivated land multifunction in rapid urbanization areas. *Habitat Int.* **2023**, *131*, 102726. [CrossRef]
63. Tupenaite, L.; Kanapeckiene, L.; Naimaviciene, J.; Kaklauskas, A.; Gecys, T. Timber construction as a solution to climate change: A systematic literature review. *Buildings* **2023**, *13*, 976. [CrossRef]

64. Ojala, A.; Kostensalo, J.; Viik, J.; Matilainen, H.; Wik, I.; Virtanen, L.; Muilu-Mäkelä, R. Psychological and physiological effects of a wooden office room on human well-being: Results from a randomized controlled trial. *J. Environ. Psychol.* **2023**, *89*, 102059. [CrossRef]
65. Oldfield, P.; Doherty, B. Offset Cores: Trends, Drivers and Frequency in Tall Buildings. *CTBUH J.* **2019**, 40–45. Available online: https://global.ctbuh.org/resources/papers/download/4186-offset-cores-trends-drivers-and-frequency-in-tall-buildings.pdf (accessed on 14 June 2024).
66. Nehdi, M.L. Only tall things cast shadows: Opportunities, challenges and research needs of self-consolidating concrete in super-tall buildings. *Constr. Build. Mater.* **2013**, *48*, 80–90. [CrossRef]
67. Szolomicki, J.; Golasz-Szolomicka, H. Technological advances and trends in modern high-rise buildings. *Buildings* **2019**, *9*, 193. [CrossRef]
68. James, M. Twisted, Tapered, Tilted (Leaning) Towers: Structural Complexities & Advantages. *Int. J. Innov. Res. Adv. Stud. (IJIRAS)* **2017**, *4*, 78–83.
69. Demirel, G.N.; Ay, B.Ö. A comparative study on structural design alternatives for twisted tall buildings with outriggers. *Archit. Sci. Rev.* **2022**, *65*, 436–448. [CrossRef]
70. Pastori, S.; Mazzucchelli, E.S.; Wallhagen, M. Hybrid timber-based structures: A state of the art review. *Constr. Build. Mater.* **2022**, *359*, 129505. [CrossRef]
71. Shin, B.; Kim, S. Advancing the circular economy and environmental sustainability with timber hybrid construction in South Korean public building. *Build. Environ.* **2024**, *257*, 111543. [CrossRef]
72. Zhang, X.; Huang, W.; Khajehpour, M.; Asgari, M.; Tannert, T. Seismic Performance and LCA Comparison between Concrete and Timber–Concrete Hybrid Buildings. *Buildings* **2023**, *13*, 1714. [CrossRef]
73. Altaher Omer Ahmed, A.; Garab, J.; Horváth-Szováti, E.; Kozelka, J.; Bejó, L. The Bending Properties of Hybrid Cross-Laminated Timber (CLT) Using Various Species Combinations. *Materials* **2023**, *16*, 7153. [CrossRef] [PubMed]
74. Wang, W. Research on seismic design of high-rise buildings based on framed-shear structural system. *Front. Res. Archit. Eng.* **2020**, *3*, 87–90. [CrossRef]
75. Dias, A.M.P.G.; Skinner, J.; Crews, K.; Tannert, T. Timber-concrete-composites increasing the use of timber in construction. *Eur. J. Wood Wood Prod.* **2016**, *74*, 443–451. [CrossRef]
76. Hanták, J.; Končeková, D. Positive effects of wood in Vorarlberg's (Austria) timber kindergartens. *Archit. Pap. Fac. Archit. Des. STU* **2023**, *28*, 36–49. [CrossRef]
77. Dharanidharan, K.; Selvanayaki, S.; Divya, K.; Ravikumar, R.; Raj, S.V. Exploring Household Wood Preferences among Consumers in Coimbatore. *Asian J. Agric. Ext. Econ. Sociol.* **2023**, *41*, 196–203. [CrossRef]
78. Yeang, K. *Service Cores: Detail in Building*; Wiley-Academy: London, UK, 2000.
79. Abed, J.; Rayburg, S.; Rodwell, J.; Neave, M. A Review of the Performance and Benefits of Mass Timber as an Alternative to Concrete and Steel for Improving the Sustainability of Structures. *Sustainability* **2022**, *14*, 5570. [CrossRef]
80. Caniato, M.; Marzi, A.; da Silva, S.M.; Gasparella, A. A review of the thermal and acoustic properties of materials for timber building construction. *J. Build. Eng.* **2021**, *43*, 103066. [CrossRef]
81. Ramage, M.H.; Burridge, H.; Busse-Wicher, M.; Fereday, G.; Reynolds, T.; Shah, D.U.; Wu, G.; Yu, L.; Fleming, P.; Densley-Tinglet, D.; et al. The wood from the trees: The use of timber in construction. *Renew. Sustain. Energy Rev.* **2017**, *68*, 333–359. [CrossRef]

applied sciences

MDPI

Article

Analysis of Planar Double-Layer Timber Spatial Frames by Using Parametric Tools

Maddi Manterola-Ubillos [1], Francisco Gonzalez-Quintial [2], Jose Miguel Rico-Martinez [2,*], Josu Benito Ayucar [3] and Jon Andoni Begiristain-Mitxelena [2]

[1] Independent Researcher, 20018 Donostia-San Sebastián, Spain; maddimanterola@gmail.com
[2] Architecture Department, School of Architecture, University of the Basque Country (UPV/EHU), Oñati Plaza 2, 20018 Donostia-San Sebastián, Spain; francisco.gonzalez@ehu.eus (F.G.-Q.); jon.begiristain@ehu.eus (J.A.B.-M.)
[3] TECNALIA, Basque Research and Technology Alliance (BRTA), Construction Materials—Lab Services, Polígono Lasao—Área Anardi, 5, 20730 Azpeitia, Spain; josu.benito@tecnalia.com
* Correspondence: j.rico@ehu.eus; Tel.: +34-657712374 or +34-943015896

Abstract: It is in the preliminary design phase of a project that the designer makes decisions concerning the global geometry of the structure. When working with space frames, the choice of the frame topology is key for the structural behavior. It is difficult to find manuals that provide guidance on which of the most common topologies is the right one for the project, let alone in wood construction. In response to this shortcoming, the use of parametric software is proposed (Grasshopper build1.0.0007 and Karamba 3D 2.2.0.16-220828). The aim is to create a dynamic catalog that responds instantaneously to changes in the parameters to provide information on structural behavior, predimensioning and metrics. With the display of all this information, the architect will have enough technical argumentation to choose or reject options. The proposal is developed through a case study: the early design and analysis stages of flat double-layer timber spatial frames as for rectangular medium-span roofs.

Keywords: computational design; parametric tools; Grasshopper; Karamba 3D; timber construction; flat double-layer space frames

1. Introduction

1.1. Space Frames

A space frame is often defined as a structural system designed from linear elements arranged in such a way that the transmission of forces is three-dimensional. This leads to an idea of opposition between spatial structures and so-called "flat" ones. However, the division between "flat" and "spatial" structures refers not to the structure itself, which is always spatial, but to the methods of analysis [1].

In planar systems, stresses are transmitted from the less rigid elements to the more rigid ones up to the foundations. On the way to the foundation, the forces become, as well as the section, larger and larger. On the contrary, in a space frame structure, all the elements contribute according to the three-dimensional geometry given to them, with no obvious force transfer sequence. In other words, the transmission of forces to the support is bifurcated in several elements so as not to have a concentration of forces in just a few elements.

The objective of this type of construction is to use the minimum amount of material to obtain lightweight structures without losing stiffness, stability and resistance. These are appealing structures not only from the engineering point of view, but also from the architectural one. These systems have led to the understanding of the building as a living being, as an organic structure, capable of adapting to the uncertainty of the future. There are approaches such as those of Louis Kahn, who plays with the natural growth of the spatial frames from the multiplication of the base spatial module. Another well-known project is the Fun Palace by Cedric Price and Joan Littlewood (Figure 1), a dismountable and reconfigurable artefact capable of housing any program that could respond to social needs. This logic is even extrapolated to the city and territory, giving rise to projects such as, among others, the megastructures of Archigram (Figure 1) or the three-dimensional urbanism of Yona Friedman.

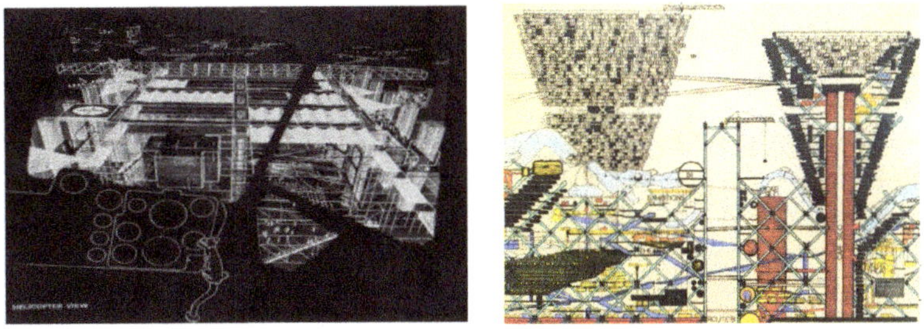

(a) (b)

Figure 1. (**a**) Cedric Price, Fun Palace. (Source: toran, iv. https://www.flickr.com/photos/136374633@N04/23341131945/, accessed on 6 April 2023); (**b**) Archigram, Plug in City. (Source: Wyliepoon https://www.flickr.com/photos/wyliepoon/49224543288/, accessed on 7 April 2023).

In recent years, research has been carried out in several related areas. On the one hand, calculation methods have been proposed to measure the carbon emissions for the lifetime of the frames [2]. On the other hand, robotization processes are being developed for assembly and disassembly [3,4]. In this regard, a very promising line of research has been opened with respect to the possibilities of disassembly and reuse of the frames [5–8]. Given the current emerging context, this is a path that undoubtedly deserves to be investigated.

Back to the strictly structural-engineering aspect, spatial structures can be grouped into three main categories [9,10] (Figure 2):

1. lattice discrete structures: they are based on discretization to create a more or less regular frame (grid, barrel vaults, domes, towers and free forms).
2. continuous structures: a single surface acting as a membrane (slabs, shell, fabric).
3. biform structures: combination of both discrete and continuous parts.

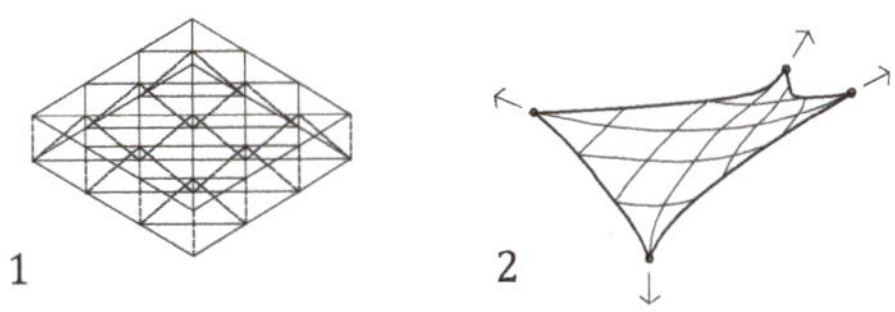

Figure 2. Space structure categories. (**1**) lattice discrete structure; (**2**) continuous structure.

This paper will be centered on the first group for two main reasons:

- Both planar and curved surfaces can be used. This research is focused on flat frames with the awareness that processes designed for a simple shape can be extrapolated to more complex ones. Parametrically, it is possible to apply any algorithmically designed process to any type of geometry.

That said, in shell structures, stiffness is achieved through shape. Their geometry directly derives from their flow of forces and defines their load-bearing behavior and lightness [11]. On the other hand, lattice spatial structures achieve resistance through cross-section. Being multiple-layer structures, the very composition of the sheet has a lever arm that stiffens them. By not being so form-dependent, the range of application of lattices is considerably wider, and that makes them appealing as the scope of this work.

- A wide range of patented systems is available on the market. Even though most of them are aimed at metal construction, development in timber construction is also being accomplished. The main focus of the research is the node solution. It is at the joint where the stress transmission is concentrated and where the constructive requirements tend to reduce the cross-section of the timber bar. This contradiction poses a challenge in the development of the connection. It is, however, a difficulty that can be overcome, since timber construction is an option that looks very promising in the current climate crisis [3,6,7].

1.2. Construction Systems

Historically, space frames have belonged to the field of steel construction. Steel allows for thin sections without losing structural efficiency. Timber, while requiring care in the design of the joints, is also a suitable material for this type of system.

The present paper will focus on timber construction, and a brief state of the art is presented about the application of timber in space frames. The elements that form the spatial system will be developed individually as follows:

1.2.1. Bar Elements

This is an element in which the longitudinal dimension is of higher order than the other two. In the built examples, three types of wood bars have generally been used (Figure 3): bamboo bars (a), solid wood logs (b) and laminated timber sections (c1, c2).

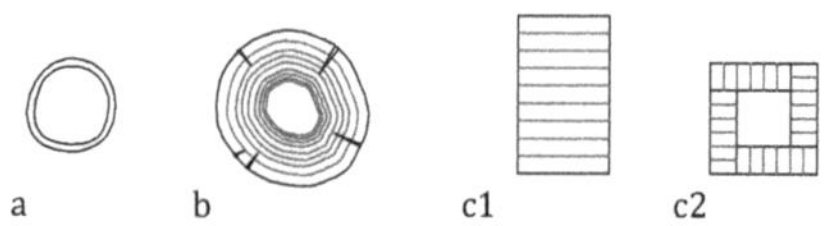

Figure 3. Bar element types. (**a**) bamboo bars; (**b**) solid wood longs; (**c1**) laminated timber section; (**c2**) laminated timber hollow section.

The first two groups belong to materials obtained directly from nature, and therefore, as in sawn timber construction, the shape and dimensions will not necessarily be optimized to the requirements of the project.

Regarding the third group, it is worth mentioning the contribution of the doctoral thesis of José Antonio Vázquez Rodríguez [12]. This work argues for the advantages of the shaping of bars by means of hollow sections of laminated timber (c2). The argument can be summarized in four points:

- The hollow section equals the material's value and offers less sensitivity to dimensional variations due to changes in environmental factors. Also, there is a reduction in risk against possible attacks of abiotic and biotic origin due to the fact that it is easier to carry out the integral treatment given its hollow character.
- There is exploitation of the advantages offered by wood products in general and laminated ones in particular. It is a material that, due to its production process, has fewer imperfections and, therefore, a higher resistance performance. The same could be said of microlam (VLV).
- The process of simple industrial execution consists of laminating rectangular pieces three or four times wider than the wall thickness of the hollow section so that, once the faces have been obtained by simple cutting, they can be glued against a master or guide piece and thus form the final piece. Consequently, the handling required is reduced, making the manufacturing cost competitive.
- The square or rectangular hollow section is considered suitable for any type of frame, although it seems to be particularly suitable for the semi-octahedron frame.

1.2.2. Joints

Depending on the type of bar used, the joint solution also varies. In timber construction, the critical part seems to be the connection between the timber bar and the (usually metallic) node. It is where the joining elements weaken the timber section and, at the same time, where the highest stress concentration occurs. In this regard, some of the research that has been developed (or is still in progress) is shown below:

- Bamboo connectors (Figure 4a): Ghavami and Moreira [13] developed a joint for the construction of double-layered bamboo space frames from octagonal plates welded to an octagonal base plate. During the research, the joint was tested on a double-layer bamboo space frame prototype with single supports every 4 m. The results were satisfactory. Currently, research is being continued with prototypes of larger spans.
- Solid wood log connectors (Figure 4b): One of the most well-known researchers in this field is Dr. Huybers [14]. Interested in the use of small-diameter logs, which, until then, had no use and were abundant in forestry waste, he developed a simple wire-tying method for attaching galvanized steel connector plates to the previously hollowed ends of roundwood poles.

Without going into much depth, some other joints that have been proposed are the KT-W joint by Katsuhiko Imai (University of Osaka) and the 14FTC-U joint by Alphose Zingoni (University of Cape Town).

- Connectors in glulam sections (Figure 4c): In the same thesis [12] in which hollow sections of glulam are developed as bars, the appropriate joint solution for them is also uncovered. A hollow sphere is proposed with a screwed cover that allows access to the interior and incorporates unthreaded holes for inserting the threaded rods and fixing them inside the sphere by means of nuts. At the opposite end, the bar is threaded into an octagonal nut. To attach the node to the wood, casting plates are fixed to the inside of the wooden element.
- Connectors by CNC fabrication (Figure 4d): So far, metal-to-wood joints have been presented. The node is usually made of a high strength steel, since the stress concentration is very high at that point. Steel is a material easily shaped to the desired geometry (spherical, cylindrical, etc.) and strong enough to resist this punctual concentration of forces.

A group of researchers from Victoria University of Wellington [15] proposed to work with plywood cut with the CNC router to create a wood–wood joint.

The results of the tests performed on the prototypes showed that this solution works with deformations of about 1/400 when overloads of use type A1 (residential zone housing) and B (administrative zones) are applied to it. Although the structure was not at its ultimate limit (collapse), it was concluded that the joint should be further stiffened.

Before moving on, we give one last note on the joints. Among the structures formed by bars, two subtypes can be distinguished: those in which the connection between bars is made through rigid nodes (that is, if the node undergoes a certain rotation, all the concurrent bars in it undergo the same rotation) or through articulated nodes (each concurrent bar can rotate independently of the others). This distinction, however, can often be overridden by the shape appearing when treating the nodes as rigid or as articulated, since the bending moments that necessarily appear in the first case are very small in comparison with the axial forces, which are of very similar magnitude in both cases. This consideration allows that it is not necessary to construct perfect articulated nodes and to use types of nodes that, in reality, are neither rigid nor articulated but are located, in general, in an intermediate zone [1,16].

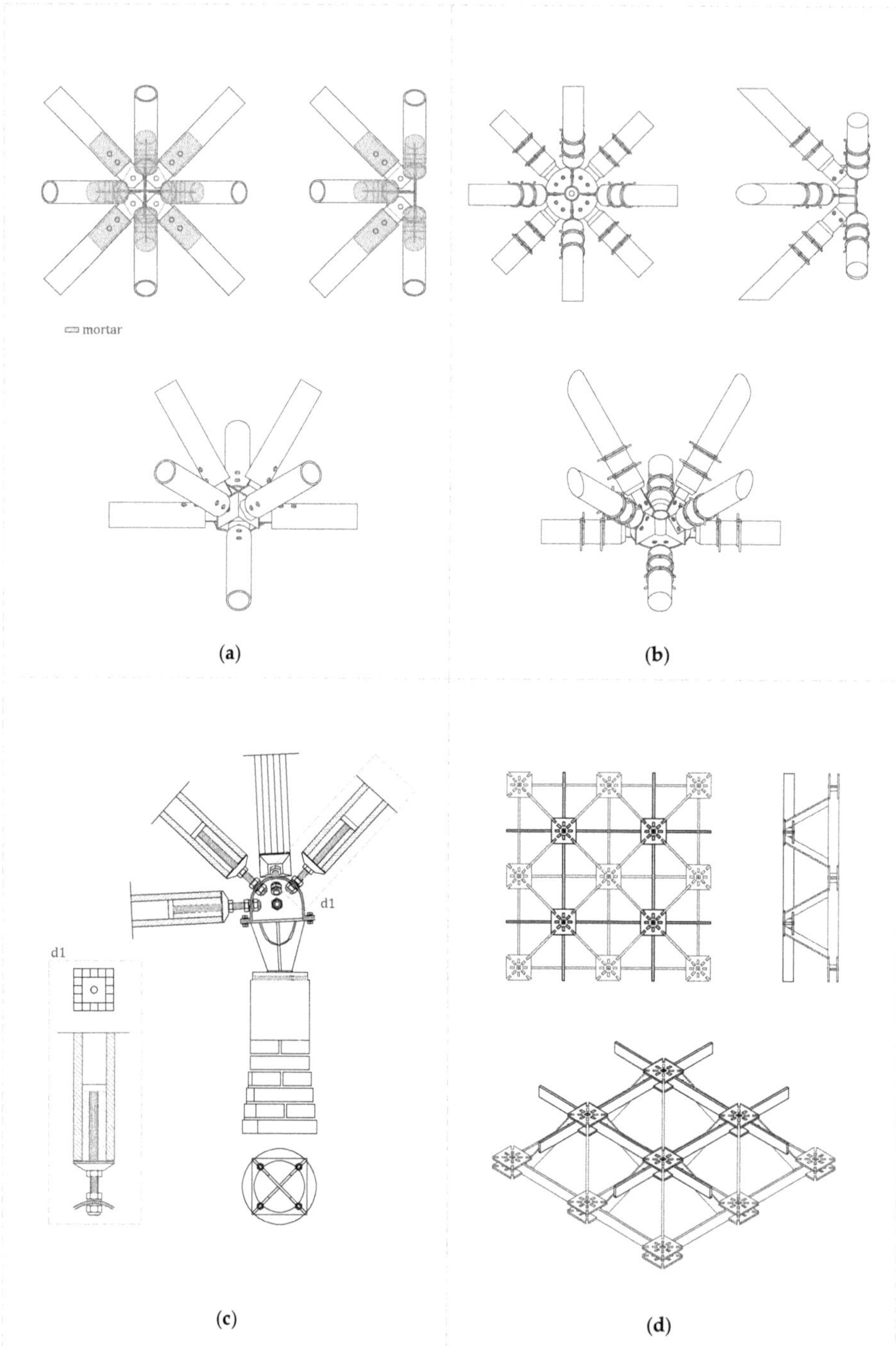

Figure 4. (**a**) Ghavami and Moreira's bamboo frame connector; (**b**) Hybers' solid wood log connector; (**c**) J.A. Vázquez's glulam hollow section joint; (**d**) Finch, Marriage, Gjerde and Pelosi's CNC fabrication frame with wood–wood joints.

1.2.3. Topology

The topology refers to the geometrical order of the bars and nodes. For greater stiffness of the flat spatial frames (indispensable in large spans), two layers are arranged connected to each other through the set of diagonal bars. Although the variety of typology is practically infinite, the present work focuses on those patented and commercialized for use in construction. The use of a custom-designed topology would also involve the design of the construction system of joints and bars, an area that is not the focus of this work. Based on the bibliography [12,17–21], the seven most mentioned and used typologies have been chosen. A classification has been proposed (Figure 5). To order them, each one has been named with a letter and a number to be able to identify them easily throughout the paper (Figure 6).

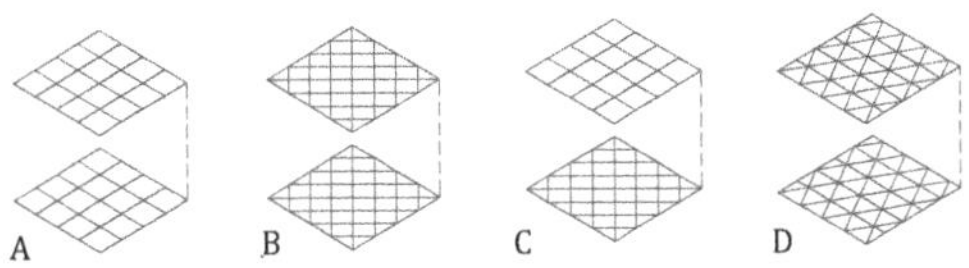

Figure 5. Frame topology classification. (**A**) group A; (**B**) group B; (**C**) group C; (**D**) group D.

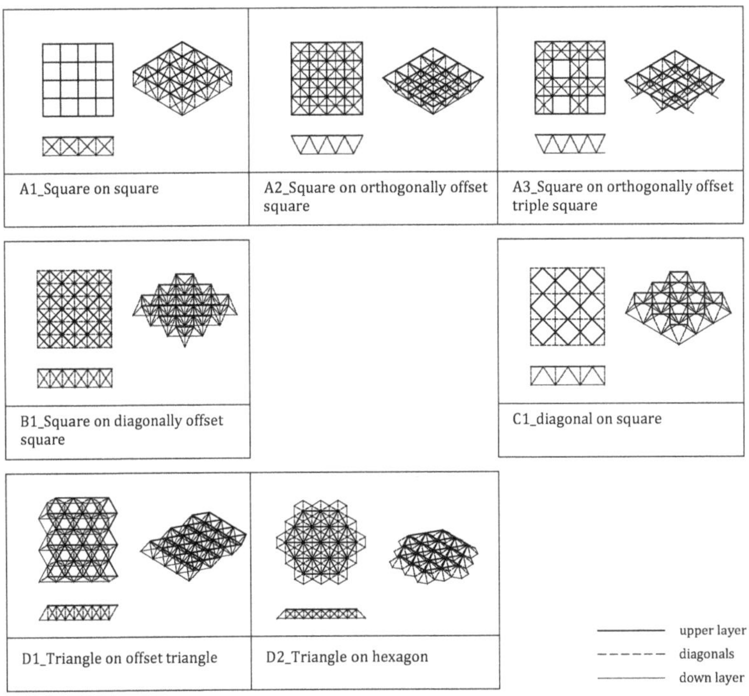

Figure 6. Analyzed topologies.

Group A: frames whose upper and lower layers are in parallel direction to the edges.

A1—Square on square: cubic module with vertical and diagonal bars. Absence of oblique diagonals.

A2—Square on orthogonally offset square: based on the pyramidal module with a square base. It is the most commonly used topology.

A3—Square on orthogonally offset triple square: based on the previous frame (A2), bars are suppressed in the lower layer in order to lighten it.

Group B: frames whose upper and lower layers are diagonal (45°) with respect to the edges.

B1—Square on diagonally offset square: pyramidal module frame with a square base arranged diagonally, at 45° to the edge.

Group C: frames in which one layer is parallel to the edges and the other diagonal (45°).

C1—Diagonal on square: created from the A1 frame, the top layer is given a 45° twist. It also has no oblique diagonals.

Group D: frames based on the triangular module.

D1—Triangle on offset triangle: created from the grouping of tetrahedral modules. It is the most rigid of the frames, as all the faces of the polyhedron are triangular, the undeformable geometry par excellence (Figure 7).

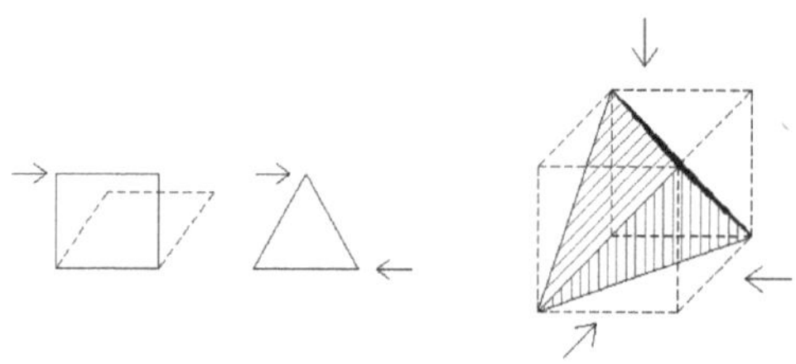

Figure 7. Rigidity of the tetrahedral module.

D2—Triangle on hexagon: based on the D1 frame, the lower layer is lightened by removing bars to create equilateral hexagons.

The different topologies behave differently depending on the applied loads, the position of the supports and the size of the module. Further variations can be introduced by changing the size of the top chord grid relative to the bottom chord grid. Wider geometries are usually possible in the bottom layer of a double-layer frame because the members are normally in tension. That is, the lower tension chords may be longer than the upper compression members (no buckling effect).

1.3. Computational Design

Visual programming software allow the creation of parametric scripts without requiring any computer programming knowledge. This helps architects in the design process by lightening data management, automating repetitive tasks and producing simulations [22]. It is all about data management; the principles of the model are parameters that can be changed anytime. The most illustrative example is probably the upside-down model of churches by Antonio Gaudi. The designer created intricate catenary arches by suspended weighted strings. By adjusting the position of the weights (the parameter in this case), the shape of the catenary arches changed. This parametric model enabled Gaudi to analyze in real time the structural stability of his project. Nowadays, the process has been digitalized, but the concept remains the same [23].

In this regard, computational design is highly suitable for the design and analysis of both simple and complex modular timber structures, such as the ones discussed in this paper. Some authors [24] argue that, despite the evident advantages of these programs to produce freeform and aesthetically pleasing designs, the challenge in their application lies in the computational cost associated with the optimization of their layout and limitations in their fabrication. In that aspect, this paper wants to contribute by proving that geometrically simpler flat space frames are as worthy of parametric analysis as the more complex ones. There is value in using pre-established flat topologies because they have an industry backing them up and are not as dominant—speaking about architectural composition—as a form-finding method of geometry can be. Architects need assistance in early design stages to develop those simpler projects as well.

1.4. Objective

To sum up the ideas that have been presented in the introduction, the aim of this paper is to test the utility of parametric tools in the early design and analysis stages of flat wooden spatial frames.

To do so, a case study will be developed. All the topologies to be used are those mentioned in the literature on space frames [20,21,25]. They have patented construction systems and are available on the market. Even so, it is not easy to find the technical/commercial documentation necessary to design spatial frames in the preliminary design phase (much less in wood), and this work intends to fulfill this need.

This is why, in this case, the parametric tools are not used to obtain a geometry but to analyze which of the geometries offered by the market is the most suitable for the specific project. The idea is to create a file that gathers the mentioned topologies in a parametric way to generate and pre-dimension them at the request of the designer. The computer process offers the results in real time, and it is up to the designer to choose the best fit [26–36]. The case study will be a simulation of this process.

2. Materials and Methods

The chosen software for this work is Grasshopper (build 1.0.0007) [37], an extension of Rhinoceros 3D (version 7.24.22297.11002) [38] (NURBS-based 3D modeling software) for algorithmic modeling that allows the creation and parametric edition of geometries. Grasshopper also has another plugin called Karamba 3D (version 2.2.0.16-220828) [39], which enables finite element analysis and pre-dimensioning of the parametrized geometry.

By combining the two software, it has been possible to create a "graphical calculation table" (Figure 8), which, by adjusting parameters such as dimensions and number of modules and choosing the topology (out of the 7 mentioned in Section 1.2.3 "Topology"), is able to compare results in real time and choose the combination that best suits the project. The geometry can also be imported into Rhino or AutoCAD to produce the graphical documentation.

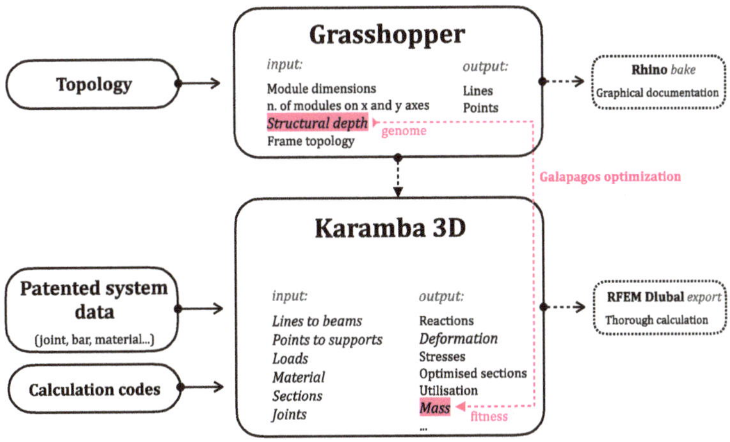

Figure 8. Workflow diagram.

Although it is not within the scope of this work, it is worth mentioning that add-ons can be installed to Grasshopper in order to enable the bidirectional data exchange between the pre-dimensioned parametric model and the FEM software (three-dimensional finite element analysis software) for a deeper calculation. It is an added value to be able to perform all design and calculation phases with the same parametric model.

2.1. Creation of the Parametric Model

All the frames are parameterized based on the parameters summarized in Figure 9. In the case of triangular and hexagonal modules, being equilateral, the sides (L) of the modules are defined. Further details on the Grasshopper parametrization process will not be developed, as it is not the subject of this work.

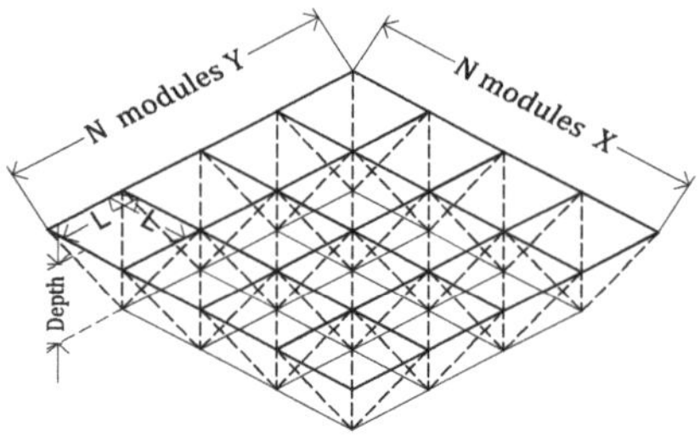

Figure 9. Parameters in the file.

Figure 10 shows, on the right, a screenshot of the completed Grasshopper file and on the left, the Rhino viewer. The model is organized in the following groups (numbered in Figure 10):

1. parametric definition of the geometry of the frame: the 7 chosen frame topologies and their respective parameterized supports.
2. grouping of the parameters: box grouping all the parameters to be chosen for the calculation, such as the frame topology, dimensions and number of modules and the depth.
3. transformation from geometrical model to structural model: by entering data (loads, supports, material, cross-sections and joints), the chosen geometry (in the parameters grouping, 2) is translated into a finite element model.
4. calculation and optimization: by means of Karamba 3D, the frame is calculated. Certain components (section optimization and genetic algorithms) are added to optimize the results.
5. visualization of results: a command is given to visualize the results in the Rhino viewer (6).

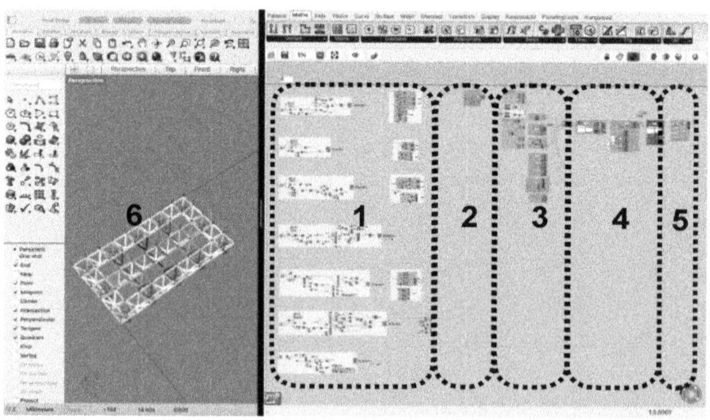

Figure 10. Screenshot of the Grasshopper file. (**1**) parametric definition; (**2**) parameters; (**3**) transformation from geometry to structure; (**4**) calculation; (**5**) visualization; (**6**) rhino viewer.

Before proceeding, a brief aside on the interoperability between FEM software and Grasshopper. A test has been performed using the interface of RFEM 6 Dlubal [40] (three-dimensional finite element analysis software) for Rhino and Grasshopper. Based on the existing Parametric FEM Toolbox plugin for RFEM 5, the RFEM 6 implements it by default. So by having the Grasshopper file and RFEM Dlubal opened side by side, and using the components that are shown in Figure 11, the connection is immediate. Changes in Grasshopper parameters are updated instantly in Dlubal as well. However, it must be emphasized that the transferred data are mainly geometrical (topology). The definition of sections and material must be performed using the exact terms of Dlubal for the program to process them correctly, and Karamba 3D components are not compatible. Therefore, this is a very useful feature in terms of handling the frame geometry (the main complexity of these structures) but somewhat limited for the rest (e.g., modeling hollow glulam bars). The remaining parameters for the complete calculation will be defined in the finite element program where the control of settings is considerably higher.

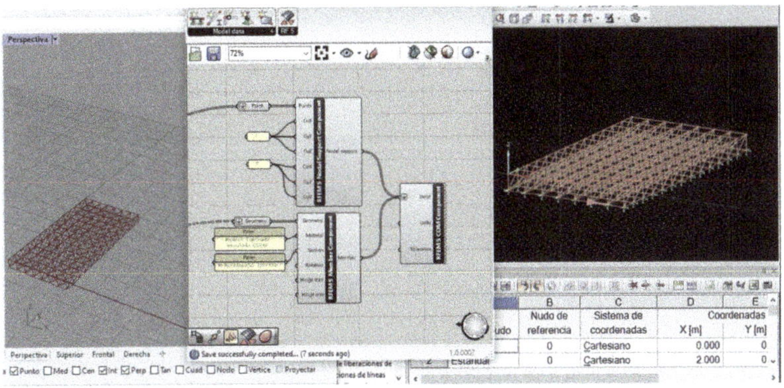

Figure 11. Screenshot of Rhino 7 (**left**), Grashopper (**center**) and RFEM 6 (**right**) interconnected.

2.2. *Optimization Components*

The potential of Karamba has been explored through the use of its optimization components. Each of them will be explained as follows:

2.2.1. Cross-Section Optimization

Using the optimize cross-section component, the previously calculated model has been connected, and the lists of possible cross-sections have been added so that the program chooses the optimum cross-section for each member. The 90% depletion rate and the deflection limit L/300 (limitation defined in Section 4.3.3.1 of the CTE-DB-SE, standards based on Eurocode 5 [41]) have been defined. This module provides the data used in the results section and also when defining the "fitness" of galapagos (evolutionary solver inside Grasshopper).

2.2.2. Genetic Algorithm

The genetic algorithm (galapagos) works with 2 outputs: one or several genome(s) and a single fitness. The latter is the value to be minimized or maximized, and for this purpose, the program finds the appropriate value of the genome(s). A galapagos has been considered: genomes: structural depth—fitness: mass.

2.3. *Data Entry*

In order to enrich the comparison and identify the limits of each topology, two variables have been considered: roof dimensions and distribution of supports.

Four roof dimensions have been established, all related to the usual dimensions for the uses for which such roofs are usually intended:

- 7 m × 12.5 m: In order to be able to analyze the behavior in relatively small spans, the 12.5 m dimension has been taken. This measure is the limit above which special transport is required. One of the advantages of space frames is that they are reduced to easily transportable bars and nodes regardless of the span they have to overcome.

- 12.5 m × 25 m: medium spans, very common in light covers for semi-Olympic swimming pools, frontons, children's playgrounds and so on.
- 20 m × 40 m: As the work has developed, it has been considered necessary to add a roof of intermediate measures between 25 and 50 m, since many of the topologies do not reach 50 m but 40 m (with the following limitations: GL24h wood and maximum depth of 3 m).
- 25 m × 50 m: large spans suitable for Olympic swimming pools, markets and so on.

In order to study how the distribution of supports affects the structural behavior, 3 options have been considered (Figure 12):

1. perimeter continuous support: The most stable of all. The situation may occur that the facade of the covered space is the supporting element, and in such cases, this is the option to go for.
2. three-sided support: All the roofs analyzed are rectangular in plan, and one of the longer sides has been freed to force the structure. This case is common in roofs for frontons, recreational areas, multi-purpose halls, etc., where the entire perimeter is closed except for one of the long sides.
3. four point support: The two previous options correspond to continuous supports (walls/columns in a row); this, however, is the option to analyze the performance with 4 punctual supports uniformly distributed. In order to generate numerical results [42,43], a system of hollow glulam bars and a generic articulated joint has been simulated.

Figure 12. Support distribution options. (**1**) perimeter continuous support; (**2**) three-sided support; (**3**) four point support.

As a summary, Table 1 with all data entries is attached:

Table 1. Data entry.

Sections (Figure 13)	Hollow section bars with a solid wood plug at both ends where a threaded tube is screwed. Values for b (cm) = between 10 y 22; values for e(mm) = 25, 30, 35, 40
Joints-Supports	Joints and supports are modeled as articulations.
Material	The GL24h class is chosen to see the performance of the frames with the lowest of the laminated classes (values from table E3 of the CTE-DB-SE-M).
from the current regulations (CTE-DB-SE, EA, M) are extracted:	
Loads	Dead weight, G1 overload, snow (<1000 m) and wind.
	5 hypothetical load combinations corresponding to light roofs (G1): H1_1.35 G + 1.5 Quse + (1.5 × 0.5 × Qsnow) H2_1.35 G + 1.5 Qsnow + (1.5 × 0.6 × Qwind pressure) H3_1.35 G + 1.5 Qwind pressure + (1.5 × 0.5 × Qsnow) H4_1.35 G + 1.5 Quse + (1.5 × 0.5 × Qsnow) + (1.5 × 0.6 × Qwind pressure) H5_1.35 G + 1.5 Qwind suction + (1.5 × 0.5 × Qsnow)
Service Class	2 (open roof)
Kmod	0.8 (service class 2 y medium duration)
Security coef.	1.35 (Table 4.1 -glued laminated timber)
Simultaneity rate	$\psi_{use\ class} = 0$; $\psi_{snow} = 0.5$; $\psi_{wind} = 0.6$ (Table 4.2)

Table 1. *Cont.*

Deflection limit	L/300 (Section "4.3.3.1 Flecha", case c)
Use case	2 (open roof) NP1 no specific requirements. All sides treated with insecticide and fungicide product.
of the current regulations (CTE-DB-SI) are extracted:	
Fire reaction	C-s2-d0 (SI 1-Table 4.1) by intumescent coating
Fire stability	R30 (SI 6-light roofs) by intumescent coating

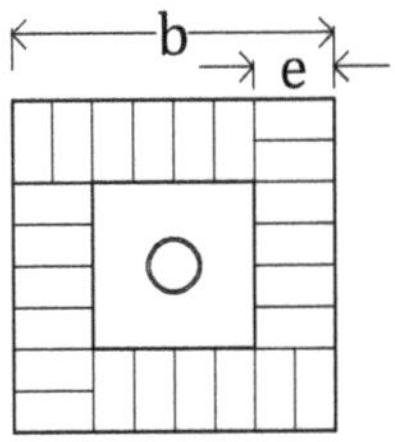

Figure 13. Bar section model. (b) section total length; (e) section side wall thickness.

3. Results and Discussion

The results presented below intend to demonstrate the utility of the program in providing architects with information on structural behavior, pre-dimensioning and metrics. The data that can be extracted from the model are extremely abundant. Keeping in mind that the objective of the paper is to create a working tool rather than a particular solution of these structures, the results shown below are only an abstract, an example, of all the information that can be handled with this parametric file (which can certainly be extended).

The results will be organized in two phases:

1. A first individual analysis, in which each frame is studied separately to obtain the optimal depth-to-span ratio by means of the genetic algorithm.
2. A second comparative phase, where the results and data on the performance of each topology are collected for a comparative analysis. This will enable the identification of the range of use and limitation of each frame.

Before analyzing the results, a clarification about the data entry is required: the file created is intended to pre-dimension roofs of variable dimensions and geometry. With so much uncertainty, simplifications have been made on the safety side that affect the results:

- Wind load: To avoid the extra parameterization of the pressure/suction areas defined by the CTE, the maximum load has been applied.
- Application of the loads: To calculate the point load, instead of differentiating between edge, corner and center nodes, the same area (that of the center being the largest) has been applied equally to all nodes.
- The joints are modeled as simple articulations, although the reality is that moments are always produced due to the lack of complete articulation. A developed calculation would involve nuancing the stiffness of the joint.

This parametric tool serves as a first draft of the structure, a quick way to discard options and arrive closer to the result. In the development of the specific project, other design programs should be used, for example, RFEM Dlubal (interfaced with the Karamba model), to further elaborate on the aspects that are not covered in this first step.

3.1. Individual Analysis

Optimal Depth for the Lowest Weight Possible

The galapagos algorithm has been used to determine the edge of each frame so that the mass is the minimum possible. Some authors [44] recommend an edge/modulus ratio between 0.5 and 1. They argue that lower ratios cause excessive stresses in the diagonal members, while higher ratios tend to increase the slenderness of these members forcing excessive buckling sizing. They conclude that, in general, in large-span structures with large stresses, it is desirable that the depth be close to the value of the modulus.

The results shown in Table 2 confirm the recommendation of the literature. All ratios are between 0.5 and 1, the average being 0.65. It can also be observed that as the module dimension increases, and therefore the span, the edge tends to increase and approaches the cubic ratio (edge = modulus value).

Table 2. Galapagos mass–depth results.

Frame	Number of Modules	Module Dimensions * (cm)	Galapagos Depth (cm)	Depth/Module Ratio
A1	x = 7, y = 13	200 300	100 210	0.50 0.70
A2	x = 7, y = 13	200 300	100 180	0.50 0.60
A3	x = 7, y = 13	200 300	110 210	0.55 0.70
B1	x = 5, y = 9	200 300	100 210	0.50 0.70
C1	x = 7, y = 13	200 -	140 -	0.70 -
D1	x = 14, y = 16	200 300	100 190	0.50 0.63
D2	x = 5, y = 9	150 250	100 170	0.67 0.68

* The module dimensions are decided based on two aspects: to have a reasonable number of modules to create the four types of covers and that all topologies have the same or similar dimensions to be able to compare them.

For the comparative results that will follow, a middle term has been established: the ratio (0.866) which means that all the bars (upper, lower and diagonal layers) are of the same length. Besides avoiding buckling problems (that are not considered at this stage of the numerical calculation), comparisons can be made under the same circumstances for all the frames without being particularly restrictive for any of them, and it also facilitates assembly logistics.

3.2. *Comparative Analysis*

The objective of this section is to analyze the behavior and limitations of each frame and to define its range of use. Three topics will be studied:

- Mass–deformation relationship: The relationship that summarizes the structural behavior of the frame. This shows how stiff the topology is and how much the section has to be increased to cover the required span (as it affects the mass).
- Exhaustion rates: This indicates how close the bars are to their maximum stress capacity. In addition, it also detects when the structure is being optimally utilized and when the geometry does not allow the bars to work efficiently.
- Metrics: Along with the structural behavior, the numbers of bars and nodes are collected for cost estimations.

3.2.1. Mass–Deformation Ratio

The mass–deformation relationship has been studied under the following parameters: support distribution, topology of the frame and module dimensions.

In general, the three-sided support is the worst performing. From 20 to 25 m, the deformations are too large (only the D1 frame resists). The perimeter support, of course, has the best values, followed, not far behind, by the 4-point supports.

The change can be seen in the lines drawn on the graphs of Figure 14: in small spans (7 × 12.5 m), the mass remains constant (because it does not require section optimization), and it is the deformation that changes. Therefore, the line is vertical. As the spans increase, the line starts to recline (the 20 × 40 graph is the most illustrative of this process). The inclination becomes dramatically more severe at the jump from 12.5 × 25 m to 20 × 40 m. This slope demonstrates that the program has enlarged the frame sections to satisfy the L/300 deflection limit (hence the increase in mass), and thereby, the usefulness of the section optimization component is demonstrated.

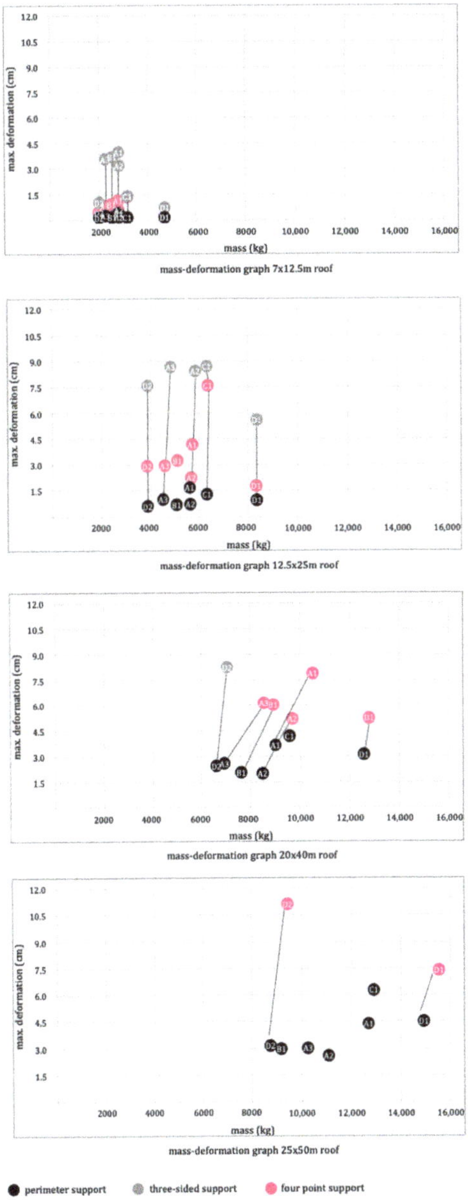

Figure 14. Mass–deformation ratio result graphs.

Looking at the topologies individually, the D1 frame is the one with the highest mass in all cases and, at the same time, one of the least deformed. A1 and C1 are the worst performers, as they need a lot of mass without deformations decreasing too much. The others have similar behaviors, although the A2 and D2 frames behave slightly better.

3.2.2. Exhaustion Rates

The exhaustion rate is the ratio between the stress supported by the bar and the maximum strength of the material. A maximum exhaustion of 90% has been defined in Karamba. All optimizations have been made on this basis. The columns that do not appear are those that do not meet the calculation requirements. Something noteworthy in the graphs (Figure 15) is the development of the meshes D1 and D2. In 7 × 12.5 m roofs, they are well below the others. As the span is increased, they start to work up to the 90% limit.

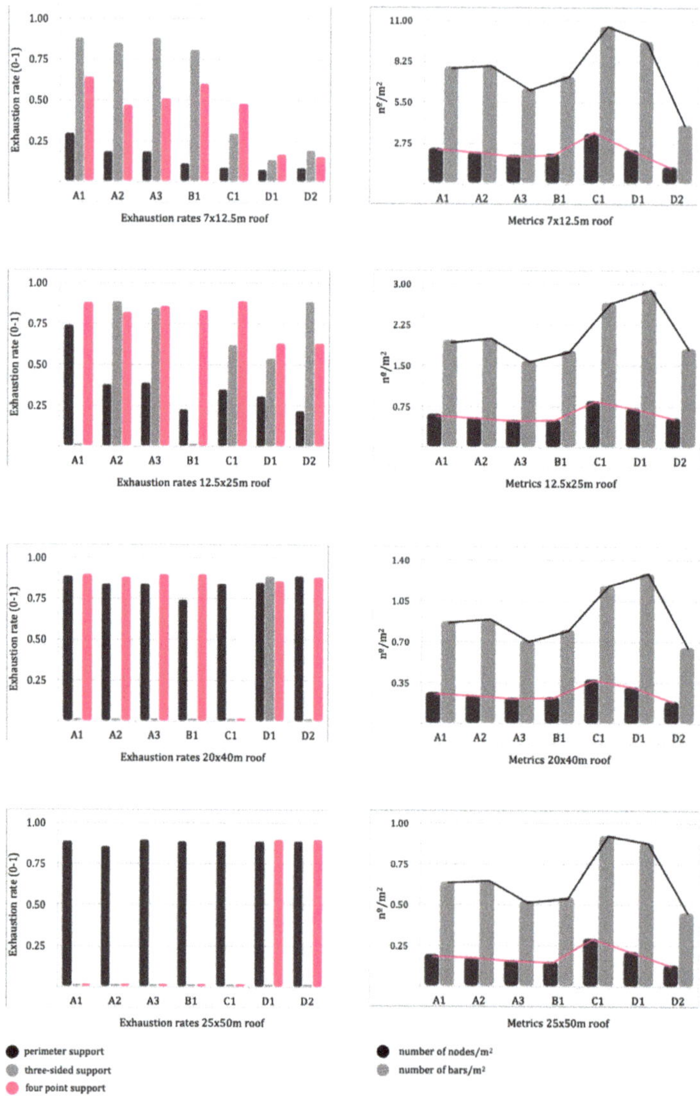

(a) (b)

Figure 15. (**a**) Utilization graphs; (**b**) metrics' graphs.

3.2.3. Metrics: Nodes and Bars

As a guideline, a breakdown by components of the total cost of the system is estimated. These are the approximate average values [44]: bars 30%, nodes 40%, protection and paint 10% and assembly 20%. From these data, it can be deduced that the cost of a space frame is closely related to the type and number of joints (40% nodes+30% bars). Therefore, in the absence of a budget, the number of nodes can be taken as a guide to cost. In such a case, economizing means that the number of nodes per square meter should be kept to a minimum.

The graphs from Figure 15 show that frames C1 and D1 are the most "expensive" ones, both in terms of the number of nodes and number of bars, while A3 and D2, the lightened topologies, are the most cost-efficient ones. Frame A2 is in the middle of both extremes. As the span of the structure increases, the two lines marked on the graphs maintain their outline because the number of modules remains constant. However, numerically speaking, the structure becomes "cheaper". The clearest example is to compare the values of the 7×12.5 and 25×50 roofs. Without going any further, the A1 mesh needs 2.46 nodes/m^2 to cover 91 m^2, while it only needs 0.2 nodes/m^2 for 1200 m^2.

By reducing the density of elements with longer bars, not only a greater load-bearing capacity is achieved, but also material and, therefore, cost is optimized. This is the advantage that space frames offer over other structural systems, and that is why they are so competitive in large spans.

4. Conclusions

4.1. Topologies

One of the objectives of the paper is to define the scope of each of the seven topologies, their strengths and limitations. After analyzing the results, the following conclusions are drawn:

- Group A (A1–A2–A3): These are the most balanced topologies, applicable in various situations. In this regard, it is only logical that they are the most common ones used. Of the three, A1 is the worst performer; its cubic module means that with the greatest number of nodes per square meter, it still has the greatest deformation.
- On the other hand, the semi-octahedral modulus provides A2 and A3 meshes with sufficient stiffness without affecting the mass too much. Between the last two, there is not that much difference; A3 turns out to be a good alternative to A2 when mass/economics is the determining factor, without serious detriment in deformations. In general, all three are relatively simple geometry topologies and easy to fit.
- Group D (D1–D2): their tetrahedral modulus makes them extremely rigid. Especially D1 is the preferred choice for large spans (over 40–50 m). From these spans on, the mesh starts to be competitive: the node-bar/m^2 ratio decreases drastically (which makes them more cost-effective) and the depletion rates go up (which indicates that the structure is working). A future research task would be to further analyze the limits of tetrahedral meshes in larger spans (above 50 m).
- The D2 mesh is presented as a good alternative to D1. Although the lower layer detaches a significant number of bars to form hexagons, it is still a rigid mesh, competitive for its lightness (with deformations lower than those of mesh A3, also lightened). One drawback that can arise is the difficulty in handling the hexagonal geometry. If the project does not contemplate the use of these characteristic shapes from the beginning, edges and joints can become an added complication.
- B1: It can be a good alternative to A2 mesh. The metrics (number of bars and nodes) are better, and for what is saved in elements, the deformations do not increase so much.
- The only doubt that may come up is (as in the D2) the geometry itself. As the bars are not arranged parallel to the edges, bars of special lengths arise to finish off the mesh, problems that group A can avoid.
- C1: Of the seven topologies, C1 is the one that behaves the worst in general. It tends to destabilize easily, and it is comfortable up to a 20–25 m span. Beyond this limit,

it becomes very limited. It is also the topology that needs more nodes and bars per square meter (slightly more than D1). In general, it seems that the geometry is not appropriate for generating a structural space frame.

It should be mentioned that in order to analyze the behavior of the topologies, the structure has been pushed to the limit. Some of the results collected in this paper, although complying with the standards, are quite tight (the deflections of the supports on three sides and the depletion indices). In the development of the calculation, the results should be handled more loosely, although for the purpose of this academic work, they are the appropriate ones.

Although the scope of the work has been flat meshes, the application of the system could be extrapolated from the plane to the curved surface and analyze, for instance, cylindrical or spherical meshes. This is a field to be explored in future projects.

4.2. Support Distributions

Another objective of the work is to analyze the repercussion of the different supports on the behavior of the structure. For this purpose, three cases have been considered: perimeter support, 3-sided support and four punctual supports. The following conclusions have been reached:

In general, the distribution of supports has a great influence on the structural behavior of the mesh, the change being less noticeable in the stiffer meshes (those of group D). Supports distributed in an orderly and symmetrical manner benefit the analyzed topologies. They do not tolerate mismatches and asymmetries well. Clearly, the perimeter continuous support is the most stable of all, although it is also limiting in the project. The four punctual supports offer the designer greater freedom.

The support on three sides is the most limiting. It is clear that the rectangular proportion (1/2) is not adequate in that case. Another future direction of the study would be to optimize the ratio of the roof to improve the values of the three-sided support. This would mean having to parameterize the lattices differently, so that changing the number of modules does not affect the overall dimensions of the roof. In this way, different ratios to 1/2 could be tested, using galapagos (fitness: arrow, genomes: number of modules on the x and y axis) to find the optimal ratio.

4.3. Using Grasshopper and Karamba

Understanding how information is handled between components is the most difficult part of Grasshopper, as it is a major shift from working with actual geometry to handling lists of data. Fundamental concepts such as "graft" and "flatten" become the cornerstone of modeling. The program is very powerful and useful. The genetic algorithm becomes an ally to obtain results without having to manually test the parameters, although the conditions must be well defined to avoid undesired results. It is also important to limit the number of variables so that the algorithm does not take too long to approach the answer (especially in these types of long files).

4.4. Final Conclusion

Although a workflow is proposed rather than specific numerical results, the case studies illustrate how much information this software can provide and thereby demonstrate their utility. In summary, a parametric analysis and pre-dimensioning tool has been created to project flat double-layer spatial meshes. A customizable and expandable "graphical spreadsheet" that accompanies the designer during the preliminary design phase.

Author Contributions: Conceptualization, M.M.-U., F.G.-Q. and J.A.B.-M.; Data curation, F.G.-Q. and J.M.R.-M.; Formal analysis, M.M.-U. and F.G.-Q.; Funding acquisition, F.G.-Q. and J.M.R.-M.; Investigation, M.M.-U.; Methodology, M.M.-U. and F.G.-Q.; Project administration, J.M.R.-M.; Resources, F.G.-Q., J.M.R.-M. and J.A.B.-M.; Software, M.M.-U. and F.G.-Q.; Supervision, F.G.-Q., J.M.R.-M. and J.A.B.-M.; Validation, F.G.-Q., J.M.R.-M. and J.B.A.; Visualization, M.M.-U., F.G.-Q.,

J.M.R.-M. and J.B.A.; Writing—original draft, M.M.-U.; Writing—review and editing, M.M.-U., F.G.-Q. and J.M.R.-M. All authors have read and agreed to the published version of the manuscript.

Funding: This research was funded by: 1. HAZI and the Basque Country Government, Vice-Ministry of Agriculture, Fisheries and Food Policy in the frame of the Timber Structures, Construction and Design Master. 2. University of the Basque Country. Architecture Department. 3. Fundación Tecnalia Research and Innovation.

Institutional Review Board Statement: Not applicable.

Informed Consent Statement: Not applicable.

Data Availability Statement: The data presented in this study are available in the article.

Conflicts of Interest: The authors declare no conflicts of interest.

References

1. Margarit, J.; Buxadé, C.; Buxadé i Ribot, C.; Margarit, J. *Las Mallas Espaciales en Arquitectura*; Gustavo Gili: Barcelona, Spain, 1972.
2. Xu, X.; You, J.; Wang, Y.; Luo, Y. Analysis and assessment of life-cycle carbon emissions of space frame structures. *J. Clean. Prod.* **2023**, *385*, 135521. [CrossRef]
3. Robotic Manufacturing of Timber Space-Frames—DigitalFUTURES. Available online: https://digitalfutures.international/workshop/robotic-manufacturing-of-timber-space-frames/ (accessed on 12 March 2023).
4. Huang, Y.; Garrett, C.R.; Ting, I.; Parascho, S.; Mueller, C.T. Robotic additive construction of bar structures: Unified sequence and motion planning. *Constr. Robot.* **2021**, *5*, 115–130. [CrossRef]
5. Finch, G.; Marriage, G. Reducing Building Waste through Light Timber Frame Design. In Proceedings of the PLEA 2018 HONG KONG, Hong Kong, China, 10–12 December 2018.
6. Koronaki, A.; Shepherd, P.; Evernden, M. Rationalization of freeform space-frame structures: Reducing variability in the joints. *Int. J. Archit. Comput.* **2020**, *18*, 84–99. [CrossRef]
7. Finch, G.; Marriage, G.; Pelosi, A.; Gjerde, M. Applications and Opportunities for Timber Space Frames in the Circular Economy. In Proceedings of the CIB World Building Congress, Hong Kong, China, 17–21 June 2019.
8. Bruun, E.P.G.; Adriaenssens, S.; Parascho, S. Structural rigidity theory applied to the scaffold-free (dis)assembly of space frames using cooperative robotics. *Autom. Constr.* **2022**, *141*, 104405. [CrossRef]
9. Baverel, O. *Nexorade: A Family of Interwoven Structures*; University of Surrey (United Kingdom): Guildford, UK, 2000.
10. Gutiérrez-Astudillo, N.C.; Flores, A.; Preciado, A. Reciprocal frame structures, a first academic approach to sustainable structures. In Proceedings of the IASS Annual Symposia, Tokyo, Japan, 26–30 September 2016. Available online: https://www.semanticscholar.org/paper/Reciprocal-frame-structures,-a-first-academic-to-Guti%C3%A9rrez-Astudillo-Flores/bad4f5510e8ebd89e009c3bd8d9762e3132616e5 (accessed on 25 April 2023).
11. Adriaenssens, S.; Block, P.; Veenendaal, D.; Williams, C. (Eds.) *Shell Structures for Architecture: Form Finding and Optimization*; Routledge: London, UK, 2014; ISBN 978-1-315-84927-0.
12. Vázquez, J.A. *Las Barras Huecas de Madera en la Construcción de Estructuras Espaciales*; Universidade da Coruña: La Coruna, Spain, 2001.
13. Ghavami, K.; Moreira, L.E. 61. Double-layer bamboo space structures. In *Space Structures 4*; Thomas Telford Publishing: London, UK, 1993; ISBN 978-0-7277-4941-3.
14. Huybers, P. Palos de madera como elementos de estructuras espaciales. *Rev. Digit. Cedex* **1990**, *79*, 99.
15. Finch, G.; Marriage, G.; Gjerde, M.; Pelosi, A. Experiments in Timber Space Frame Design: Fabrication, Construction and Structural Performance. In Proceedings of the 24th International Conference of the Association for Computer-Aided Architectural Design Research in Asia (CAADRIA), Hong Kong, China, 15–18 April 2019.
16. Lan, T.T. *Space Frame Structures*; CRC Press: Boca Raton, FL, USA, 1999.
17. Mupona, G.T. Development of Space Truss Systems in Timber. Master's Thesis, Faculty of Engineering and the Built Environment, University of Cape Town, Cape Town, South Africa, 2004. Available online: https://www.semanticscholar.org/paper/Development-of-space-truss-systems-in-timber-Mupona/211102da0a241e41447b23fcb658ca3a7b9fd99b (accessed on 12 March 2023).
18. Moggio, N. Glued Laminated Timber Space Truss Systems. Master's Thesis, Università di Trento, Trento, Italky, 2012. Available online: http://lup.lub.lu.se/student-papers/record/3131142 (accessed on 12 March 2023).
19. Álvarez Pablos, J. *Mallas Planas y Cilíndricas de Módulos Apilables de Madera Laminada*; Universidade da Coruña. Escola Técnica Superior de Arquitectura: La Coruna, Spain, 1998; ISBN 978-84-692-8073-7. Available online: https://ruc.udc.es/dspace/handle/2183/5561 (accessed on 12 March 2023).
20. Chilton, J. *Space Grid Structures*; Architectural Press: Oxford, UK, 2000; ISBN 978-0-7506-3275-1.
21. Eekhout, M. *Architecture in Space Structures*; Uitgeverij 010 Publishers: Rotterdam, The Netherlands, 1989; ISBN 978-90-6450-080-0.

22. Aksöz, Z.; Preisinger, C. An Interactive Structural Optimization of Space Frame Structures Using Machine Learning. In *Impact: Design with All Senses*; Gengnagel, C., Baverel, O., Burry, J., Ramsgaard Thomsen, M., Weinzierl, S., Eds.; Springer International Publishing: Cham, Switzerland, 2020; pp. 18–31.
23. Jormakka, K. *Basics Design Methods*; Birkhäuser: Basel, Switzerland, 2017; ISBN 978-3-0356-1147-2.
24. Koronaki, A.; Shepherd, P.; Evernden, M. Layout Optimization of Space Frame Structures. In Proceedings of the IASS Annual Symposium 2017, Hamburg, Germany, 25–28 September 2017.
25. Gordon, J.E. *Estructuras o Por qué las Cosas no se Caen*, 2nd ed.; Calamar: Madrid, Spain, 2004; ISBN 978-84-96235-06-9.
26. De Moraes, M.H.; Fraga, I.; Junior, W.; Christoforo, A. Comparative analysis of the mechanical performance of timber trusses structural typologies applying computational intelligence. *Rev. Árvore* **2022**, *46*, e4604. [CrossRef]
27. Apolinarska, A.A. Complex Timber Structures from Simple Elements: Computational Design of Novel Bar Structures for Robotic Fabrication and Assembly. Ph.D. Thesis, ETH Zurich, Zürich, Switzerland, 2018.
28. Andrén Jakobsson, N.; Bohman, S. *A Generative Design of Timber Structures according to Eurocode. Development of a Parametric Model in Grasshopper*; KTH Royal Institute of Technology: Stockholm, Sweden, 2019. Available online: http://urn.kb.se/resolve?urn=urn:nbn:se:kth:diva-255661 (accessed on 12 March 2023).
29. Almaraz, A. Evolutionary Optimization of Parametric Structures: Understanding Structure and Architecture as a Whole from Early Design Stages. Master's Thesis, University of A Coruña, La Coruna, Spain, 2015. Available online: https://ruc.udc.es/dspace/handle/2183/15965 (accessed on 12 March 2023).
30. Martínez Villarroya, D. Arquitectura Fractal: Optimización Topológica. Available online: https://oa.upm.es/55874/ (accessed on 12 March 2023).
31. Muñoz Ávila, D.A. Implementación de un Algoritmo de Optimización Multi—Objeto de una red de Barras Utilizando un Lenguaje de Programa-ción Visual en "Rhinoceros' y sus Extensiones. 2021. Available online: https://repositorio.uisek.edu.ec/handle/123456789/4519 (accessed on 12 March 2023).
32. Navarro Mateu, D. Procesos Naturales Aplicados a la Arquitectura Mediante Computación: Ciencia Evo-Devo y Modelado Algorítmico a Través de Grasshopper. Ph.D. Thesis, Universitat Internacional de Catalunya, Barcelona, Spain, 2017. Available online: https://www.tdx.cat/handle/10803/552427 (accessed on 12 March 2023).
33. Villar Monteagudo, I. Estudio Paramétrico Sobre Mallas de Doble Capa. 2015. Available online: https://ruc.udc.es/dspace/handle/2183/16171 (accessed on 12 March 2023).
34. Weldegiorgis, F.; Dhungana, A.R. Parametric Design and Optimization of Steel and Timber Truss Structures: Development of a Workflow for Design and Optimization Processes in Grasshopper 3D Environment. 2020. Available online: http://urn.kb.se/resolve?urn=urn:nbn:se:kth:diva-277901 (accessed on 12 March 2023).
35. Yan, X.; Bao, D.; Zhou, Y.; Xie, Y.; Cui, T. Detail control strategies for topology optimization in architectural design and development. *Front. Archit. Res.* **2022**, *11*, 340–356. [CrossRef]
36. Preisinger, C. Linking Structure and Parametric Geometry. *Archit. Des.* **2013**, *83*, 110–113. [CrossRef]
37. Network, Scott Davidson Created This Ning Network. Grasshopper. Available online: https://www.grasshopper3d.com/ (accessed on 29 April 2023).
38. Associates, Robert McNeel & Associates. Rhinoceros 3D. Available online: https://www.rhino3d.com/es/ (accessed on 29 April 2023).
39. Karamba3D. Available online: https://karamba3d.com/ (accessed on 29 April 2023).
40. Software Para Análisis Estructural y Dimensionado | Dlubal Software. Available online: https://www.dlubal.com/es (accessed on 29 April 2023).
41. UNE-EN 1995-1-1:2016 Eurocódigo 5. Proyecto de Estructuras de Madera. Parte 1-1: Reglas Generales y Reglas para Edificación. Available online: https://www.une.org/encuentra-tu-norma/busca-tu-norma/norma?c=N0056510 (accessed on 1 May 2023).
42. Muñoz-Vidal, M. *Generación y Cálculo de Mallas Espaciales*; Universidad, Departamento de Tecnolog'ia de la Construcción: La Coruña, Spain, 1993.
43. Navarro Carrasco, S. Cálculo Estructural de Una Gran Marquesina Espacial Ejecutada Íntegramente Mediante una Malla Espacial de Doble Capa de Acuerdo con los Eurocódigos que Disponen de Anejo Nacional; Cartagena, Spain. 2015. Available online: https://repositorio.upct.es/handle/10317/4912 (accessed on 12 March 2023).
44. Cavia-Sorret, P. Las mallas espaciales y su aplicación en cubiertas de grandes luces. *Re. Rev. De Edif.* **1993**, *15*, 7–15. [CrossRef]

applied sciences

MDPI

Article

Experimental and Numerical Research on the Splitting Capacity of European Beech Beams Loaded Perpendicular to the Grain by Connections: Influence of Different Geometrical Parameters

José Luis Gómez-Royuela [1], Almudena Majano-Majano [1], Antonio José Lara-Bocanegra [1], José Xavier [2,3,*] and Marcelo F. S. F. de Moura [4]

1 Department of Building Structures and Physics, ETS of Architecture, Universidad Politécnica de Madrid (UPM), Avda. Juan de Herrera 4, 28040 Madrid, Spain; joseluis.gomez.royuela@upm.es (J.L.G.-R.); almudena.majano@upm.es (A.M.-M.); antoniojose.lara@upm.es (A.J.L.-B.)
2 UNIDEMI, Department of Mechanical and Industrial Engineering, NOVA School of Science and Technology, Universidade NOVA de Lisboa, 2829-516 Caparica, Portugal
3 Laboratório Associado de Sistemas Inteligentes, LASI, 4800-058 Guimarães, Portugal
4 Department of Mechanical Engineering and Industrial Management, Faculty of Engineering, University of Porto, Rua Dr. Roberto Frias, 4200-465 Porto, Portugal; mfmoura@fe.up.pt
* Correspondence: jmc.xavier@fct.unl.pt

Abstract: In the present work, single- and double-dowel joints following different geometric configurations are experimentally and numerically investigated to derive the splitting behaviour of beech wood (*Fagus sylvatica* L.), one of the most widespread hardwood species in Europe for structural purposes. The influence of the spacing between dowels, their distance to the supports, and the slenderness of the beams is analysed. The correlation of the experimental failure loads with those predicted numerically by cohesive zone finite element-based models using the fracture properties of the species is discussed. The experimental results are also compared with those obtained from the normative expression included in Eurocode 5 and two other design models reported in the literature. The splitting failure loads predicted by both the analytical and numerical models were found to be conservative, the latter being closer to the experimental values.

Keywords: dowel connection; hardwood; numerical model; fracture mechanics; timber structures; Fagus sylvatica; cohesive zone modelling

1. Introduction

In recent decades, there has been significant growth in the development of renewable and environmentally sustainable building materials, spurred by the imperative to address the impacts of climate change. The building sector is a significant consumer of primary energy, contributing significantly to the release of substantial amounts of carbon dioxide (CO_2) into the atmosphere [1,2]. In this context, wood and wood-based products have attracted great interest in architecture and civil engineering because of their renewable resource potential when sustainably managed [3], their savings in primary energy and CO_2 emissions during the manufacturing process [1,4], and the added value they bring to buildings. Furthermore, among the main structural building materials, wood is the only one composed of substantial amounts of carbon captured from the atmosphere during the tree growth process (around 50% of its dry weight) [5]. Currently, most structural timber products are made from softwoods, but in recent years, there has been growing interest in the use of hardwoods for structural purposes. The reasons for this are multiple and may be due, among others, to the under-use of aged hardwood forests and the high mechanical properties compared to softwoods [6,7].

In this respect, *Fagus sylvatica* L., or European beech, is one of the most important and widespread hardwood species in Europe (from southern Scandinavia to northern Sicily and from northwestern Turkey to northwestern Spain), with an average height of about 30–40 m. [8,9]. The large beech reserves available in European forests have aroused the interest of the scientific community. Several investigations have been carried out on this material for structural purposes, both on solid wood in terms of determining its elastic constants [10–13], mechanical properties [14–16], or fracture behaviour [17,18], as well as on structural products made of beech such as glulam [19] or laminated veneer lumber (LVL) [20]. European beech also offers good strength and stiffness compared to softwoods, which makes it an excellent choice for the design of timber structures, especially in the assessment of high-performance scenarios, such as connections with fasteners or large span elements, allowing for a reduction in the cross-section of the main members.

Mechanical connections with steel dowel-type fasteners are among the most widely used connections in timber structures [21–23] because of their simplicity of manufacture and assembly and the possibility of being designed to have a ductile connection failure. However, there are certain situations where brittle failure of the timber member may occur at load levels below the bearing capacity for desirable ductile behaviour, especially in designs with connections loaded at an angle to grain where tension stresses perpendicular to the grain are induced, for which timber inherently exhibits relatively low strength and stiffness. This brittle splitting failure is, therefore, one of the most critical in timber structures and deserves particular concern in design.

Eurocode 5 [24] includes the so-called European Yielding Model (EYM) originally formulated by Johansen [25], which is used to predict the load carrying capacity of dowel-type connections subjected to shear force. This model always considers the plastic behaviour of timber and also the possible plastic behaviour of the dowel depending on the failure mode of the connection. However, this model can dangerously overestimate the ultimate load in connections if splitting failure occurs. Concerning the prediction of the splitting capacity of connections loaded at an angle to the grain, Eurocode 5 provides a relatively simple relationship developed theoretically by Van der Put and Leijten [26,27] following an energetic approach in the framework of linear elastic fracture mechanics (LEFM). The expression only applied to softwood and considered the timber member width, the distance from the fastener to the loaded edge of the member, and the ratio of the latter to the member depth (commonly known as the α parameter). It is widely demonstrated and specified in some design codes that when $\alpha \geq 0.7$, splitting may be neglected. The expression also includes a constant value arising from the fracture energy calibration parameter. However, the scientific literature suggests that this value is a considerable overestimation for some wood species commonly used in Europe [28]. The expression disregards the influence of other parameters, such as the number and geometry of groups of fasteners.

Further experimental and numerical research has been carried out to propose alternative analytical formulations to the normative expression, taking into account the effect of different geometrical parameters of the connection and material properties. Most of these experimental studies have been conducted on softwoods, e.g., the first studies reported in [29–32] used to calibrate the analytical model proposed by Van der Put and Leijten or later relevant studies on sawn spruce and glulam [33–36], radiata pine LVL [37], southern pine LVL [38,39], and southern pine MSR lumber [38]. However, very little research has focused on the splitting behaviour of dowel connections in hardwood species. Recent studies can be found on *Eucalyptus globulus* sawn wood [40] and yellow poplar PSL [38]. There is a work on beech beams loaded perpendicularly to the grain by connections, although the α values considered were higher than 0.7 [23].

In terms of numerical investigations, several authors have developed finite element models (FEMs) to realistically replicate the behaviour of structural timber elements with fastener connections [22,34,41–43]. One method involves utilising cohesive zone models (CZMs) incorporating traction–separation laws, which are treated as inherent material properties. These models require predefining possible damage paths. Strength criteria based on limit stresses are applied to indicate the onset of the damage. Damage evolves as described by the traction–separation law following the principles of fracture mechanics.

This method was first applied to wood in [44], and its capability has been sufficiently demonstrated in subsequent studies [41,42,45], as well as in previous work by the authors focusing on the simulation of crack propagation in double cantilever beams (DCBs) [17] and end-notched flexure (ENF) tests [18] in beech wood, offering good agreement with experimental results.

The main objective of the present research is to determine the splitting capacity of *Fagus Sylvatica* L. (European beech) solid wood beams loaded perpendicular to the grain by connections, considering α values lower than 0.7. An experimental campaign is carried out on connections consisting of one and two dowels placed at various positions along beams of different lengths and depths aiming to study the influence of these geometrical parameters. Cohesive zone finite element-based models using the fracture properties of the species are also developed, which are capable of reproducing crack development. Finally, the results are compared with the theoretical splitting failure loads obtained by applying the analytical expression included in Eurocode 5 [24] and two other alternative design models reported in the literature [46,47], which offer a more complete formulation regarding the consideration of different geometrical parameters of the connection.

2. Materials and Methods

2.1. Experimental Test

European beech (*Fagus sylvatica* L.) solid wood from Austria was used in the experimental tests. The boards were obtained from heartwood and cut in the tangential direction. Before testing, the boards were conditioned in a climatic chamber at 20 °C and 65% relative humidity until equilibrium moisture content (MC) was reached. After conditioning, all boards were visually graded according to the German standard DIN 4074-5:2008 [48]. The specimens were free from knots and cracks. All specimens met the requirements for the LS13 strength class [48], which would allow a strength class assignment of D40 according to EN 1912:2012 [49]. MC was determined using the oven-drying method using a section taken from each board following EN 13183-1:2002 [50]. This resulted in an average value of 10.3%. The density (ρ) was also measured using a section taken from each board according to EN 408:2012 [51] and adjusted to the reference moisture content of 12% [52], giving a mean value of 732 kg/m^3.

The beams were subjected to 3- or 4-point bending tests depending on the connection layout studied (Figure 1), at a constant crosshead displacement rate set between 0.75 and 3.50 mm/min, so that failure took place in approximately 5 min.

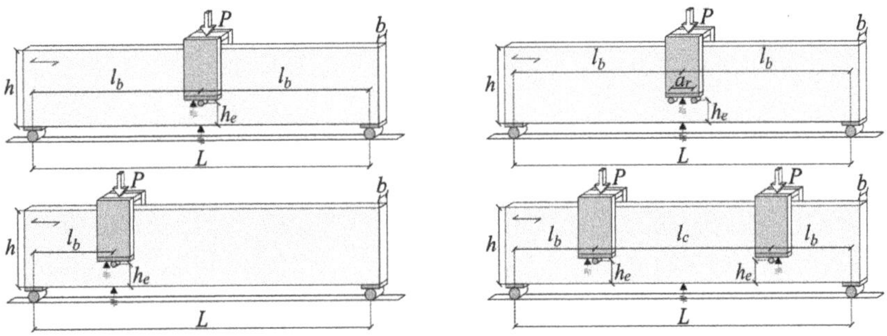

Figure 1. Bending test set-ups.

Displacements (δ) were measured using linear variable differential transformers (LVDTs) and recorded at a frequency of 1 Hz during the tests. To analyse joint behaviour and dowel sliding, LVDTs were placed both at the bottom of the loading plates, as close as possible to the dowel, and at the bottom of the beam (black triangles in Figure 1). Holes of the same diameter as the dowels were drilled in the timber beams so that there was no clearance between them.

The experimental programme was divided into two groups of specimens according to the beam depth (h): group 1 with h = 100 mm and group 2 with h = 200 mm. The beam width (b) was 48 mm for all specimens. In all tests, dowels with a diameter (d) of 16 mm and 120 mm length made of S355-grade steel were used. With this diameter, the dowel slenderness (b/d) was small, and steel yielding was avoided.

The distance to the loaded edge (h_e) was set at $4d$ (minimum distance established in Eurocode 5 [24]), remaining constant in all tests, resulting in joints with two different values of the α parameter ($\alpha = h_e/h$), namely 0.64 for group 1 and 0.32 for group 2. In each of the groups, beams with one dowel (1D) and two dowels (2D) were tested in order to analyse the influence of the number of dowels and connections on the splitting capacity.

As a reference configuration, a span L of 950 mm (1000 mm specimen length) and one connection with 1D and 2D (spaced $a_r = 4d$ apart) located at mid-span were established. In order to analyse the influence of the distance l_b from the connection to the support and the distance l_c between the connections on the splitting capacity, different l_b values were considered (360 mm, 240 mm, and 120 mm), which corresponded to support spacings of approximately 0.38, 0.25, and 0.13 times the span. In order to also analyse the influence of the slenderness of the beam on the splitting capacity, beams with the connection at mid-span but with L of 450 mm (500 mm specimen length) were also tested. Figure 2 illustrates the set of reference configurations and their variations.

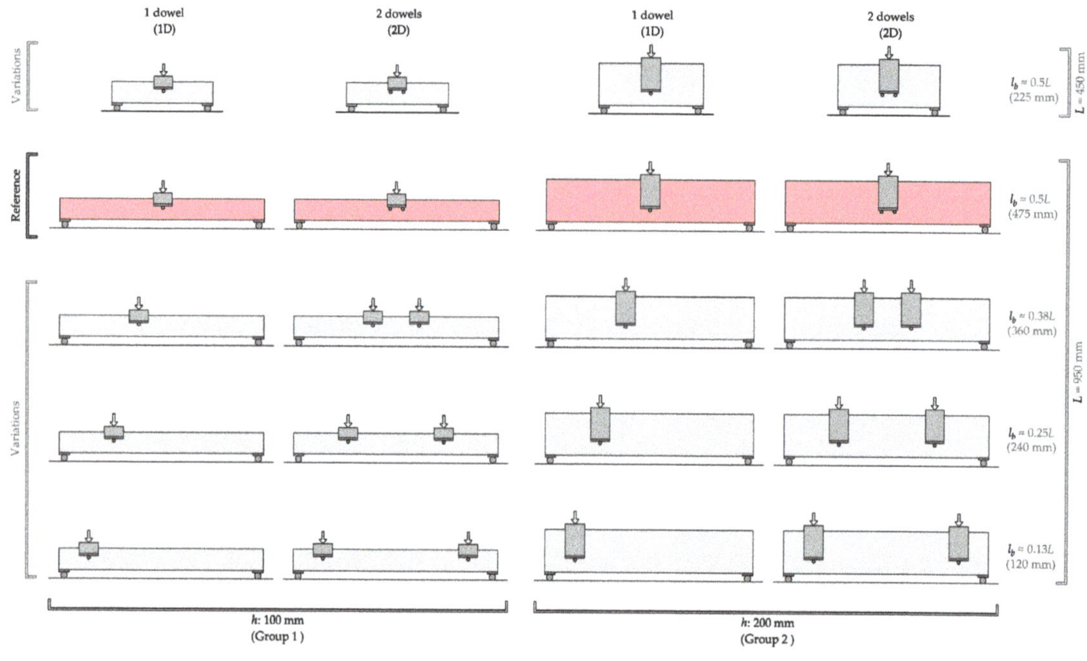

Figure 2. Experimental configurations tested: reference configurations (in red) with one- and two-dowel connections placed at mid-span for each beam depth group and variations (in grey) regarding connection position and beam slenderness.

Table 1 summarises, for each series of tests, the geometrical parameters of the beam (span and depth), as well as the density and moisture content, the position of the connection in terms of distance to the support l_b and the spacing between connections l_c (in case of two connections), the position of the connection in relation to the beam depth by means of the α parameter, and the number of replicates tested.

2.2. Numerical Analysis

Numerical finite element analyses of all configurations studied in the experimental campaign (Figure 2) were carried out using ABAQUS v2021 software [53]. Brittle splitting failure is addressed as simply as possible and in a way that provides sufficiently safe results from a structural design point of view.

Since the dowel had low slenderness, it showed very small deformations, and the authors did not observe any yield hinges, so a simplified 2D model was developed in which a cohesive zone model (CZM) was implemented to reproduce the crack growth. The timber beam and the steel elements (supports and dowels) were defined by 8-node quadratic plane stress elements (CPS8R). The model mesh was refined in the crack propagation region, as well as in the dowel and support locations (Figure 3a,b), using a minimum and maximum mesh size of 1 mm × 1 mm and 4 mm × 4 mm, respectively. To simulate crack growth and propagation, 6-node quadratic cohesive surfaces were used together with the Newton–Cotes integration scheme [54]. To reduce the computational time and in accordance with the findings in [55], only one fracture plane at mid-height of the dowel was modelled. The contact between the steel elements and wood was introduced using 6-node quadratic contact surfaces with a friction coefficient of 0.15 [56]. The analyses were carried out considering a non-linear geometrical behaviour of the beams. The Newton–Raphson convergence method was applied to solve the systems of non-linear equilibrium equations. An isotropic material was considered for steel elements and an orthotropic material for wood. The yielding behaviour of wood was neglected for the sake of the simplification intended by the model. Consequently, a linear elastic constitutive response of the materials was assumed. The application of the load was performed by a uniform vertical displacement imposed at the central node of the dowels in the range of 6 and 20 mm introduced in increments of no more than 0.01 mm to reproduce stable crack growth and to obtain accurate results.

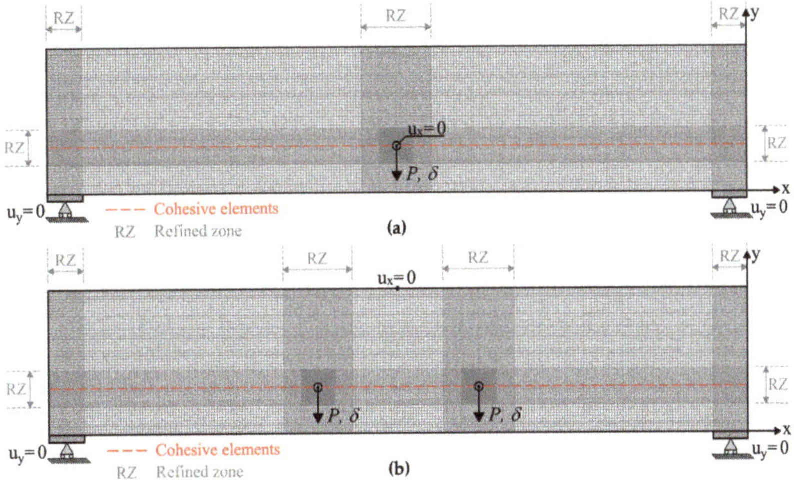

Figure 3. Numerical model: (**a**) single connection; (**b**) double connection.

Table 1. Experimental test program.

	Number of Dowels (no.dowels)	Span L (mm)	Connection–Support Distance l_b (mm)	Series h/no.dowels/L/l_b	Number of Specimens	Density ρ (kg/m^3)	Moisture Content (%)	Connection Spacing l_c (mm)	α h_e/h
Group 1 (h: 100 mm)	1D	450	225	100/1D/450/0.5L	4	747	9.7	-	0.64
		950	475	100/1D/950/0.5L	4	734	11.1		
			360	100/1D/950/0.38L	3	708	10.4		
			240	100/1D/950/0.25L	3	727	10.4		
			120	100/1D/950/0.13L	3	730	10.1		
	2D	450	225	100/2D/450/0.5L	4	727	10.2	-	0.64
		950	475	100/2D/950/0.5L	4	710	10.7		
			360	100/2D/950/0.38L	3	748	10.7	230	
			240	100/2D/950/0.25L	3	738	11.5	470	
			120	100/2D/950/0.13L	3	717	10.9	710	
Group 2 (h: 200 mm)	1D	450	225	200/1D/450/0.5L	4	756	9.7	-	0.32
		950	475	200/1D/950/0.5L	3	744	10.0		
			360	200/1D/950/0.38L	3	716	9.7		
			240	200/1D/950/0.25L	3	730	10.1		
			120	200/1D/950/0.13L	3	758	9.9		
	2D	450	225	200/2D/450/0.5L	4	730	9.6	-	0.32
		950	475	200/2D/950/0.5L	4	713	10.0		
			360	200/2D/950/0.38L	3	741	10.0	230	
			240	200/2D/950/0.25L	3	742	10.9	470	
			120	200/2D/950/0.13L	3	716	10.3	710	

A bilinear cohesive law model with a linear softening relationship was used to reproduce the fracture process in mixed-mode I + II (Figure 4) [57,58].

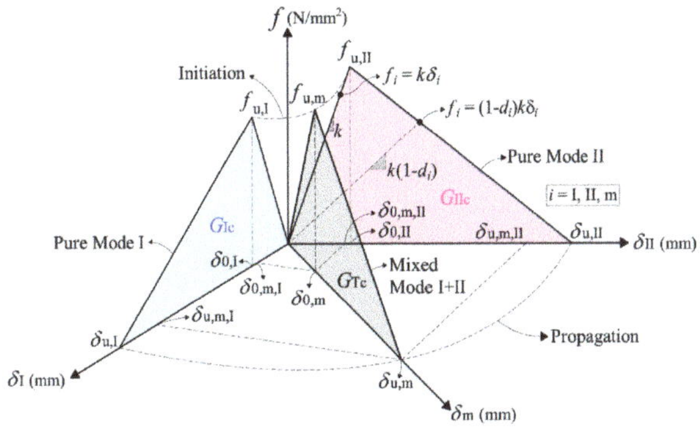

Figure 4. Bilinear cohesive law in mixed−mode I + II implemented in ABAQUS. Separation (δ) and traction (f).

The proposed damage model is characterised by two branches. The first one describes the elastic behaviour of the undamaged material until the stress limit ($f_{u,i}$, i = I, II, m) is reached, and the second one defines the softening behaviour under damage (Figure 4). Under mixed−mode I + II, an equivalent relative displacement combining the two loading modes is defined (Figure 4) according to Equation (1):

$$\delta_{\mathrm{m}} = \sqrt{\delta_{\mathrm{I}}^2 + \delta_{\mathrm{II}}^2} \tag{1}$$

which can be written as

$$\delta_{\mathrm{m}} = \delta_{\mathrm{I}}\sqrt{1 + \beta^2} \tag{2}$$

being

$$\beta = \frac{|\delta_{\mathrm{II}}|}{\delta_{\mathrm{I}}} \tag{3}$$

In the first branch, the constitutive relationship is given by

$$f_{\mathrm{m}} = k\delta_{\mathrm{m}} \tag{4}$$

where f_{m} is the resulting mixed−mode I + II traction and k is the initial interface stiffness usually known as the penalty parameter which is set by the user. In this study, the value of 10^6 N/mm^3 [54,59] was adopted in all modes (I, II, mixed) to avoid undesirable interpenetration of the crack faces and numerical problems. A quadratic stress criterion is assumed to simulate damage onset under mixed-mode loading:

$$\left(\frac{f_{\mathrm{I}}}{f_{\mathrm{u,I}}}\right)^2 + \left(\frac{f_{\mathrm{II}}}{f_{\mathrm{u,II}}}\right)^2 = 1 \tag{5}$$

where (f_{I}, f_{II}) are the current mode I and II loading components and ($f_{\mathrm{u,I}}$, $f_{\mathrm{u,II}}$) are the corresponding local strengths (Figure 4). After some algebraic manipulation involving Equations (1)–(5), the relative displacement at damage onset becomes

$$\delta_{0,\mathrm{m}} = \frac{\delta_{0,\mathrm{I}}\delta_{0,\mathrm{II}}\sqrt{1+\beta^2}}{\sqrt{\beta^2\delta_{0,\mathrm{I}}^2 + \delta_{0,\mathrm{II}}^2}} \tag{6}$$

where $\delta_{0,i}$ (i = I, II) are the relative displacements at damage onset under pure mode loading (Figure 4). During propagation, the traction–separation relationship can be directly defined as

$$f_{\mathrm{m}} = (1 - d_{\mathrm{m}})k\delta_{\mathrm{m}} \tag{7}$$

where d_{m} is the damage parameter defined from the linear softening relationship between $\delta_{0,\mathrm{m}}$ and the ultimate relative displacement $\delta_{\mathrm{u,m}}$ according to

$$d_{\mathrm{m}} = \frac{\delta_{\mathrm{u,m}}(\delta_{\mathrm{m}} - \delta_{0,\mathrm{m}})}{\delta_{\mathrm{m}}(\delta_{\mathrm{u,m}} - \delta_{0,\mathrm{m}})} \tag{8}$$

The ultimate relative displacement $\delta_{\mathrm{u,m}}$ is calculated by means of the linear energetic criterion used to simulate damage propagation under mixed-mode I + II loading:

$$\frac{G_{\mathrm{I}}}{G_{\mathrm{Ic}}} + \frac{G_{\mathrm{II}}}{G_{\mathrm{IIc}}} = 1 \tag{9}$$

where G_{ic} (i = I, II) are the critical fracture energy and G_i (i = I, II) are the energy components of the current mode mixity given by

$$G_i = \frac{1}{2}k\delta_{0,\mathrm{m},i}\delta_{\mathrm{u,m},i} \tag{10}$$

with $\delta_{0,\mathrm{m},i}$ and $\delta_{\mathrm{u,m},i}$ representing the critical mode i components of the relative displacements at damage onset and at failure, respectively. From Equations (3) and (10), the strain energy release rate under mixed-mode loading can be written as

$$G_{\mathrm{Tc}} = G_{\mathrm{I}} + G_{\mathrm{II}} = G_{\mathrm{I}}\left(1 + \beta^2\right) \tag{11}$$

Combining Equations (11) and (9) yields

$$G_{\mathrm{Tc}} = \frac{G_{\mathrm{Ic}}G_{\mathrm{IIc}}(1 + \beta^2)}{G_{\mathrm{IIc}} + \beta^2 G_{\mathrm{Ic}}} \tag{12}$$

Knowing the critical strain energy release rate (G_{Tc}), $\delta_{\mathrm{u,m}}$ can be determined as follows:

$$\delta_{\mathrm{u,m}} = \frac{2G_{\mathrm{Tc}}}{k\delta_{0,\mathrm{m}}} \tag{13}$$

which makes it possible to define the damage parameter (Equation (8)) that represents the progressive propagation of damage under mixed-mode I + II loading. It should be noted that this general formulation also addresses the pure mode loading cases (I or II) since they are particular cases of the mixed-mode loading scenario.

The mean values of the fracture parameters of European beech defining the mode I and II cohesive laws in the TL crack propagation system, as well as the elastic constants of the material, were obtained from previous work by the authors [10,17,18] and are compiled in Tables 2 and 3. For the dowel steel, a modulus of elasticity of 210,000 N/mm^2 and a Poisson's ratio of 0.3 were considered. These material properties were used as input data in the numerical models.

Table 2. Average values and standard deviation of the elastic constants of European beech [10].

	E_L (N/mm^2)	E_R (N/mm^2)	E_T (N/mm^2)	ν_{RL} (-)	ν_{TL} (-)	ν_{RT} (-)	G_{LR} (N/mm^2)	G_{LT} (N/mm^2)	G_{RT} (N/mm^2)
Average	13,811	1590	832	0.51	0.44	0.32	1108	706	349
St. Dev.	1323	541	115	0.030	0.015	0.041	202	139	53

Table 3. Numerical parameters of the bilinear cohesive law: maximum traction (f_u) and critical fracture energy (G_c) in fracture modes I and II and TL crack propagation system for beech [17,18].

	$f_{I,u}$ (N/mm^2)	$f_{II,u}$ (N/mm^2)	G_{Ic} (N/mm)	G_{IIc} (N/mm)
Average	8.74	18.88	0.46	1.17
St. Dev.	1.97	3.34	0.10	0.26

2.3. Design Models

Several design proposals can be found in the literature to predict the splitting capacity of timber connections loaded perpendicular to the grain. However, there is much discussion on the parameters influencing the splitting behaviour, and no solid proposal has yet been developed that is widely accepted by the research community. A comprehensive review of most of the existing models, their theoretical basis, and their scope in the design of timber connections can be found in [34,60]. A summary of simple analytical models and the relationships between them is compiled in [61]. The three design models used in this work to predict the splitting failure load of the European beech are described below.

The first is the model formulated by Van der Put [26]. It is one of the first analytical models to predict the maximum shear force for the splitting failure of a simply supported beam. The model was developed using an energy approach in the framework of linear elastic fracture mechanics (LEFM). A single force acting at the mid-span of the beam was considered a starting point regardless of the type of fastener used to transfer the load to the beam. Subsequently, Van der Put and Leijten [27] calibrated the original model [26] for softwood using experimental results available in the literature. Some other considerations for this model can be found in [62]. The proposal has been widely discussed, mainly because of its simplicity. Jensen [63,64] published an extended model without any simplifying assumptions and including the contribution of normal forces in the section equilibrium, in which the original model of Van der Put appears as a special case. In addition, following the same approach, models for beams with two connections were developed. Also based on the original proposal, Ballerini [65] presented a semi-empirical expression taking into account the geometry of the connection. But despite successive proposals, the original model of Van der Put forms the basis of the design expression given in Eurocode 5 [24] (Equations (14)–(17)), which will be used to calculate the theoretical failure loads for beech in the present work:

$$F_{v,\mathrm{Ed}} \leq F_{90,\mathrm{Rd}} \tag{14}$$

$$F_{v,\mathrm{Ed}} = \max\begin{cases} F_{v,\mathrm{Ed},1} \\ F_{v,\mathrm{Ed},2} \end{cases} \tag{15}$$

$$F_{90,\mathrm{Rd}} = \frac{F_{90,\mathrm{Rk}}}{\gamma_\mathrm{M}} k_\mathrm{mod} \tag{16}$$

$$F_{90,\mathrm{Rk}} = 14b\, w \sqrt{\frac{h_e}{1-\frac{h_e}{h}}} \tag{17}$$

where $F_{v,Ed,1}$ and $F_{v,Ed,2}$ are the shear forces in the main beam on both sides of the connection, $F_{90,Rk}$ is the characteristic splitting capacity, k_{mod} is the modification factor considering load duration and moisture content, γ_M is the partial factor for material properties, b is the width of the beam, h is the depth of the beam, h_e is the distance from the fastener to the loaded edge of the beam, and w is a factor depending on the type of fastener (1 for dowels). It is important to underline that Equation (17) includes also a constant 14 value. According to the original work of Van der Put and Leijten [27], this 14 value is derived from the expression $C_1 = (GG_c/0.6)^{0.5}$, where the so-called *apparent fracture parameter* $(GG_c)^{0.5}$ was used as a fitting parameter by taking test results from a limited number of sources in the literature with different connection types. A value of C_1 = 10 N/mm$^{1.5}$ was suggested for use in the code design expression, somewhat lower than the C_1 = 14 N/mm$^{1.5}$ finally adopted. Since the expression of the code is only valid for softwoods, in the present work it was decided to apply the original relationship and determine the C_1 value for European beech instead of C_1 = 14 N/mm$^{1.5}$, by means of the shear modulus G and critical fracture energy G_c of the material obtained experimentally, resulting in C_1 = 23.27 N/mm$^{1.5}$.

The second design model that was applied to estimate European beech's carrying capacity in the present work is the model formulated by Jensen [46]. It relies on Timoshenko's beam theory for elastic foundations (BEFs) and the quasi-non-linear fracture mechanics. For its development, a beam is considered with a cracking layer modelled by springs to which the material fracture properties are assigned. Upon cracking, the beam below this layer is considered to be resting on elastic Wrinkler springs connected to the upper part, which is assumed to be infinitely rigid. For a single load acting far from the beam end and small crack lengths, Equations (18)–(22) result:

$$P_u = \gamma\, P_{u,LEFM} \tag{18}$$

$$P_{u,LEFM} = 2bC_1\sqrt{\frac{h_e}{1-\frac{h_e}{h}}} \tag{19}$$

$$\gamma = \frac{\sqrt{2\zeta+1}}{\zeta+1} \tag{20}$$

$$\zeta = \frac{C_1}{f_t}\sqrt{10\frac{G}{E}\frac{1}{h_e}} \tag{21}$$

$$C_1 = \sqrt{\frac{5}{3}GG_f} \tag{22}$$

with P_u being the splitting load of the connection, E the longitudinal modulus of elasticity, G the shear modulus of elasticity, and G_f the fracture energy. One of the differentiating features of this model compared to previous proposals is that it takes into account the perpendicular tensile strength of the timber, f_t.

The third of the design approaches selected to estimate the theoretical splitting failure load of European beech is the one postulated by Franke and Quenneville [47] given by Equations (23)–(26). The design approach was developed through experimental investigations and finite element numerical simulations considering numerous connection arrangements, in the framework of non-linear fracture mechanics. It is one of the most comprehensive models in the literature, which for the first time considers mixed-mode I + II (tension and shear) failure criteria that represent a more realistic situation and not only mode I, as most of the previous models conservatively do:

$$F_{90} = \frac{b\cdot 10^3}{\left(\frac{G_{\mathrm{I,\,norm}}}{G_{\mathrm{Ic}}}+\frac{G_{\mathrm{II,\,norm}}}{G_{\mathrm{IIc}}}\right)}\cdot k_r \tag{23}$$

$$G_{\mathrm{I,\,norm}} = e^{\left(h^{-1}\left(200-10h_e\cdot h^{-0.25}-a_r\right)\right)} \tag{24}$$

$$G_{\mathrm{II,\,norm}} = \left(0.05+0.12\frac{h_e}{h}+1\cdot 10^{-3}a_r\right) \tag{25}$$

$$k_r = \begin{cases} 1 & \text{for } n=1 \text{ dowel} \\ 0.1+(\arctan(n))^{0.6} & \text{for } n>1 \text{ dowels} \end{cases} \tag{26}$$

with F_{90} being the splitting load of the connection, G_{Ic} and G_{IIc} the critical fracture energy of the material in mode I and II, respectively, and $G_{I,\,norm}$ and $G_{II,\,norm}$ the normalised fracture energy in mode I and II, respectively. The model also includes other parameters concerning the connection layout, such as the width of the connection (a_r) and the number of rows of fasteners parallel to the grain through the k_r factor.

The material properties of the European beech that are used as input data in the three above-mentioned models to calculate the corresponding theoretical splitting ultimate loads were obtained from previous work of the authors and are presented in Tables 2 and 3. A value of 8.9 N/mm^2 is used for the tensile strength perpendicular to the grain (f_t) [13].

3. Results and Discussion

3.1. Experimental Results

3.1.1. Splitting Loads and Failure Modes

The results on ultimate loads and failure modes from the experimental work on European beech are presented here.

Splitting was the type of failure that governed the collapse of most of the beams tested. However, some specimens failed in bending. Table 4 shows the percentage of failures due to splitting or bending per test series, as well as the mean values of the connection failure load (P_{exp}) and the maximum shear force corresponding to P_{exp} ($V_{shear,max}$) for each of these two failure modes. The location of P_{exp} is represented in Figure 1.

Table 4. Experimental results: percentage of beams with splitting/bending failure per series (%); average values of the connection failure load (P_{exp}) and the maximum shear force corresponding to the P_{exp} ($V_{shear,max}$) for each failure mode; coefficient of variation (COV); ratio between the distance of the connection to the support (l_b) and the beam depth (h).

	Series	Failure Mode				
		Splitting/Bending				
	h*/no.dowels/*L*/*l_b	**%**	**P_{exp} (kN)**	**$V_{shear.max}$ (kN)**	**COV (%)**	**l_b/h**
Group 1 (*h*: 100 mm)	100/1D/450/0.5L	100/-	32.5/-	16.3/-	5.1/-	2.25
	100/1D/950/0.5L	25/75	31.0/30.3	15.5/15.2	-/4.1	4.75
	100/1D/950/0.38L	67/33	26.7/33.3	16.6/20.7	7.4/0	3.60
	100/1D/950/0.25L	67/33	26.3/28.3	19.7/21.1	5.7/0	2.40
	100/1D/950/0.13L	100/-	23.9/-	20.9/-	6.0/-	1.20
	100/2D/450/0.5L	100/-	49.0/-	24.5/-	10.5/-	1.93
	100/2D/950/0.5L	-/100	-/37.6	-/18.8	-/8	4.43
	100/2D/950/0.38L	33/67	23.7/23.9	23.7/23.9	0/0.4	3.60
	100/2D/950/0.25L	67/33	27.1/21.5	27.1/21.5	8.7/0	2.40
	100/2D/950/0.13L	100/-	25.0/-	25.0/-	10.1/-	1.20
Group 2 (*h*: 200 mm)	200/1D/450/0.5L	100/-	27.2/-	13.6/-	7.8/-	1.13
	200/1D/950/0.5L	100/-	30.8/-	15.4/-	11/-	2.38
	200/1D/950/0.38L	100/-	24.3/-	15.1/-	10.0/-	1.80
	200/1D/950/0.25L	100/-	20.5/-	15.3/-	14.2/-	1.20
	200/1D/950/0.13L	100/-	27.4/-	23.9/-	6.5/-	0.60
	200/2D/450/0.5L	100/-	29.2/-	14.6/-	20.5/-	0.97
	200/2D/950/0.5L	100/-	33.6/-	16.8/-	5.1/-	2.22
	200/2D/950/0.38L	100/-	22.8/-	22.8/-	20.1/-	1.80
	200/2D/950/0.25L	100/-	26.9/-	26.9/-	18.9/-	1.20
	200/2D/950/0.13L	100/-	28.5/-	28.5/-	24.1/-	0.60

It should be noted that all the tested connections have the same distance from the loaded edge (h_e) but different values of α. This leads to the fact that all tested series in group 2 failed by splitting but not the series in group 1. In particular, all beams in the reference series of group 1 failed in bending, except for one beam in the reference series with one dowel. In addition, in the series of group 1 with the off-centre connection, some beams also failed by bending, except in the series with the connection closer to the support, where all beams failed by splitting. The closer the connection is to the supports, the more predominant the splitting failure. This trend is logical since the moment generated by the load decreases with the proximity of the connection to the support. It is worth noting that bending failure only occurred in the beams with the greatest slenderness and α ratio of 0.64, where the ultimate loads shown in Table 4 caused stresses in the outermost wood fibres of around 90 N/mm^2, which are in line with the limit values reported in the literature [15].

It was observed that the experimental tests reached the maximum load by developing two different fracture patterns: one completely brittle (pure splitting) and the other in a combination of failure modes (a non-negligible plastic behaviour of the wood develops before splitting occurs), as shown in Figure 5. The behaviour of the combined failure mode is directly related to the formation of small cracks in the wood at the bottom of the dowel and, in some cases, to the yielding of the wood around the connection, as these areas are singular points subjected to high stress concentrations perpendicular to the grain.

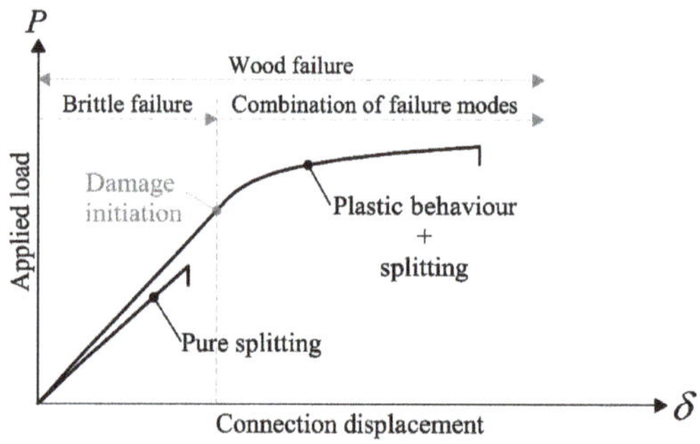

Figure 5. Governing failure modes in the experimental tests.

A summary of the load–displacement curves (P-δ) for each of the tested specimens is shown in Figures 6 and 7. In these graphs, the load and displacement of the connection are plotted, according to the placement of LVDTs depicted in Figure 1. However, in Figure 7b, the total load on the beam ($2P$) is plotted in order to compare the results between the graphs depicted in Figure 7.

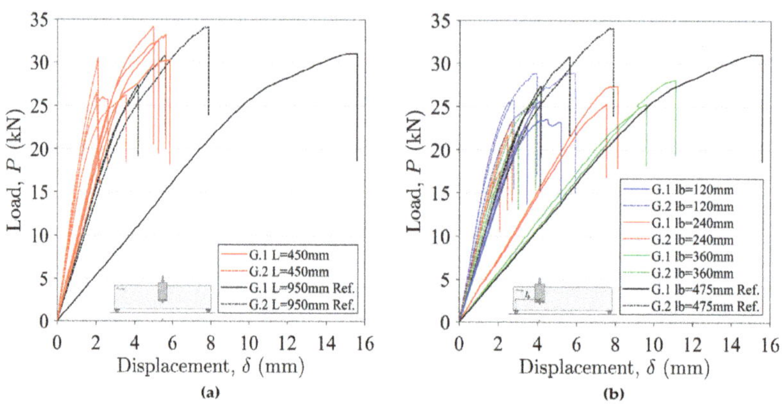

Figure 6. Experimental P-δ curves of beams with one dowel: (**a**) 1D with connection at mid-span; (**b**) 1D with the connection placed at off-centre of the beam.

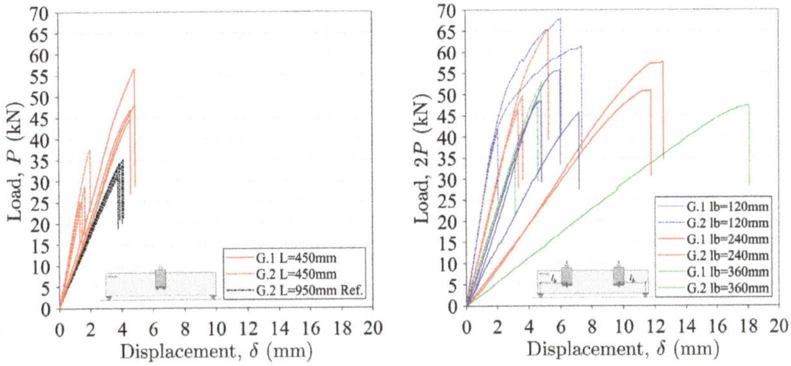

(a) (b)

Figure 7. Experimental P-δ curves of beams with two dowels: (**a**) 2D with connection at mid-span; (**b**) 2D with the connection placed at off-centre of the beam.

As can be seen, a single-dowel-type connection tends to develop a mixed-mode behaviour, but a double-dowel-type connection tends to collapse in a brittle manner. In particular, all beams with two dowels located in the centre of the span and those with two off-centre connections closer to the centre of the span break in a completely brittle manner. However, the rest of the beams showed a mixed-mode behaviour. This difference in behaviour depends on the configuration of the connection and is in line with the results of other research in the literature for other wood species [29,34,38,40,66].

In the series investigated, the splitting failure was characterised by the propagation of a main crack, which extended along the beam length and through the full entire thickness. This failure mode appeared suddenly and in a brittle manner. In some tests, the splitting failure was preceded by the formation of small cracks around the dowel, but eventually only one crack propagated and in a forceful manner. In some specimens, the crack shape that developed along the beam was very straight and continuous (see Figure 8a), but in other cases, the shape was oblique and/or discontinuous (see Figure 8b). The crack propagation patterns across the width of the beam also exhibited similar differences between the specimens (see Figure 9). However, the results showed that the less straight and continuous the crack propagation was, both longitudinally and transversely, the higher the splitting capacity.

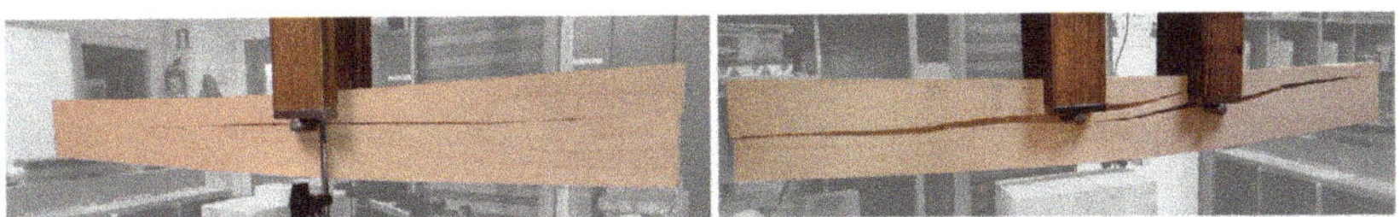

(a) (b)

Figure 8. Typical failures: (**a**) straight and continuous crack; (**b**) oblique and discontinuous crack.

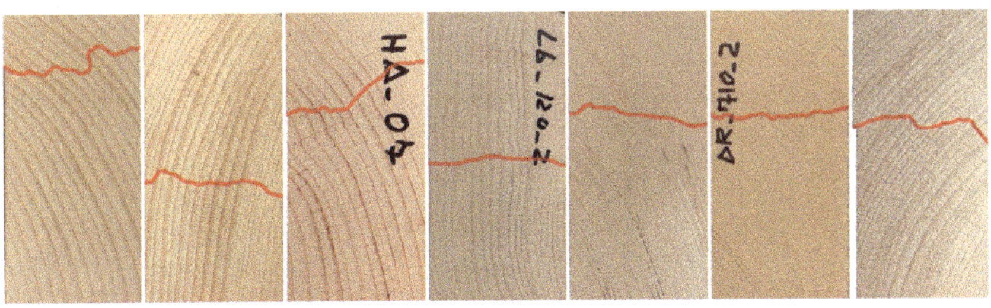

Figure 9. Typical crack patterns.

In all the series with two dowels and a span of 950 mm in group 2, the crack was propagated to the end of the beam, splitting it into two parts. In addition, this also happened in the beams of group 1 with a span of 450 mm and in all the series with two connections except for the one where the connections are closer to the supports (100/2D/950/0.13L). However, this did not happen in the rest of the series, where the crack did not reach the ends. Some of these crack patterns are shown in Figure 10.

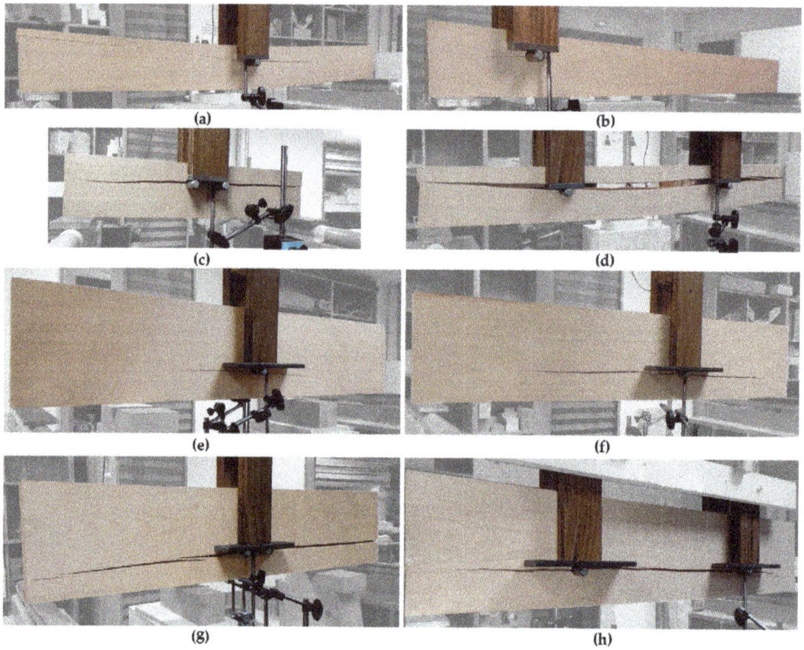

Figure 10. Typical splitting failure of series with h = 100 mm (group 1 (**a–d**)) and h = 200 mm (group 2 (**e–h**)): (**a**,**e**) 1D with connection at mid-span; (**b**,**f**) 1D with the connection placed at off-centre of the beam; (**c**,**g**) 2D with connection at mid-span; (**d**,**h**) 2D with the connections placed at off-centre of the beam.

3.1.2. Influence of the Number of Dowels at the Connection

To study the influence of the number of dowels on the connection, the series with the connection located at mid-span with one and two dowels and beam spans of 450 mm and 950 mm were considered.

Table 5 summarises the results of the series studied, including the mean value of the splitting failure load, the parameter α, the slenderness of the beam, and the percentage load increase in the connections with two dowels with respect to their counterparts with one dowel.

As can be seen, the load gain of 2-dowel connections relative to single-dowel connections never reaches twice the value.

The influence of the parameter α can be analysed by comparing the series with different values of α and similar slenderness (4.75 slenderness for α = 0.32 and 4.5 slenderness for α = 0.64). The results show that the load gain for α = 0.32 is very small, only 9%. However, for the series with α = 0.64, the load gain attained is much higher, 51%, although still far from the expected doubling. Note that in this case α is doubled, but the load gain is more than five times higher.

Table 5. Load increase in series with two-dowel connections compared to one-dowel connections.

Number of Dowels (no.dowels)	Series h/no.dowels/L/l_b	L (mm)	Slenderness (L/h)	h_e (mm)	α h_e/h	P_{exp} (kN)	Load Increase (%)
1D 2D	100/1D/450/0.5L 100/2D/450/0.5L	450	4.50	64	0.64	32.5 49.0	51%
1D 2D	200/1D/450/0.5L 200/2D/450/0.5L	450	2.25	64	0.32	27.2 29.2	7%
1D 2D	200/1D/950/0.5L 200/2D/950/0.5L	950	4.75	64	0.32	30.8 33.6	9%

The effect of the span can be analysed by comparing the results of the series with the same depth (h = 200 mm) but different spans (450 mm and 950 mm). It can be observed that, as the span increases, the load gain for using two dowels instead of one is almost negligible, from 7% for the series with a 450 mm span to 9% for the series with a 950 mm span. These results show that span has hardly any influence, at least for low α values such as the one studied (α = 0.32).

The increase in load obtained as a function of the number of dowels in the connection is consistent with other research [34,40,67], where increasing the number of dowels led to an increase in splitting failure load.

It is worth mentioning that the load carrying capacity of the connection increases slightly with increasing beam span. Figure 11 shows the ultimate splitting loads in relation to the beam span for the series of single- and double-dowel connections at mid-span for h = 200 mm.

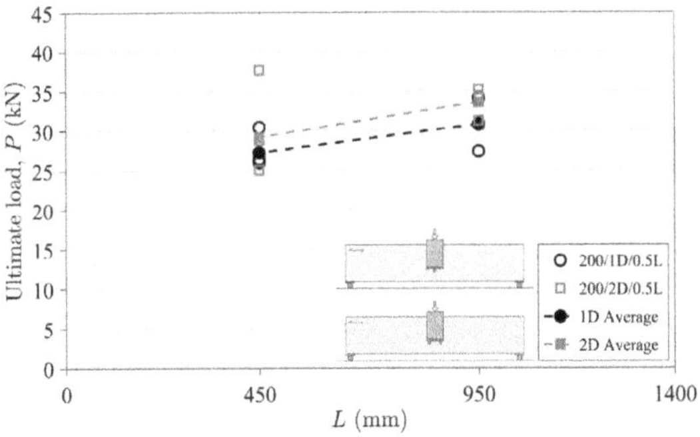

Figure 11. Splitting failure loads for one- and two-dowel connections in relation to beam span, for h = 200 mm.

According to the results, it follows that the larger the span, the higher the splitting loads (13% for 1D and 15% for 2D). In terms of beam slenderness (L/h), the higher the slenderness, the higher the load carrying capacity, which is in agreement with the results for spruce glulam available in [34,68]. In the case of the series with h = 100 mm, this comparison is not possible because the beams with larger spans failed in bending.

3.1.3. Influence of Distances l_b and l_c

The evolution of the maximum splitting load for the beams with one-dowel connections placed at different distances from the nearest support (l_b) is shown in Figure 12a. Figure 12b shows the load evolution for beams with two-dowel connections as a function of the spacing between the dowels (l_c). The results in both graphs are given for beams with h = 100 mm and h = 200 mm.

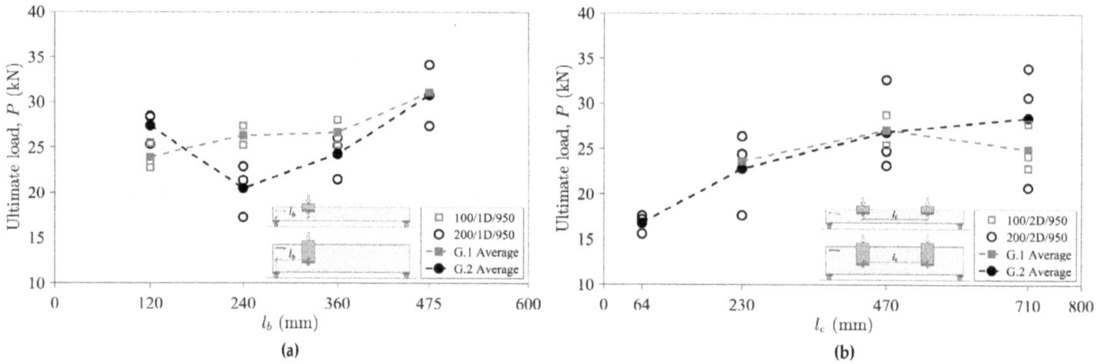

Figure 12. (**a**) Splitting failure loads of single-dowel connections versus the distance between the connection and the support l_b, for the two different beam depths; (**b**) splitting failure loads of two-dowel connections versus dowel spacing l_c, for the two different beam depths.

When looking at the evolution of the splitting capacity as a function of l_b for beech beams with one-dowel connections (Figure 12a), the series of h = 200 mm shows a decrease in splitting failure loads with the proximity of the dowel to the support, up to the distance l_b = 240 mm. However, the load increases again when the dowel is at the closest distance to the support of those studied (l_b = 120 mm). The specific l_b distance at which this trend change occurs is difficult to determine with accuracy as many tests would be necessary. However, a value of $l_b \approx h$ seems a good approximation. This apparent increase in the splitting capacity of the beam may be due to the fact that, for distances of $l_b \approx h$, part of the load is transferred directly to the supports [62]. However, the splitting failure is not negligible even if $l_b < h$.

A similar trend is observed for beams with h = 100 mm, with a decrease in ultimate load with decreasing l_b, although in this case with no increase in load for the lowest l_b. This observation may be related to the fact that the smallest value of l_b tested is greater than h and therefore the dowel is not close enough to the support to transfer the load directly to it.

There are not many experimental studies in the literature focusing on the influence of l_b. Some literature sources report that splitting capacity is independent of l_b [35,61], contrary to the results of the present study, where the influence was evident. Other studies with off-centre dowel connections in sawn spruce, spruce glulam, Douglas fir glulam, or Radiata pine LVL can be found in [28,34,69]. However, due to the limited results and geometries addressed, it is difficult to draw clear conclusions.

When comparing the test results between the two beam depths, it can be seen that the h = 100 mm series tends to attain higher splitting loads than the h = 200 mm series. This gain in load may be explained, to some extent, by the difference in the α ratios, so that the higher the α, the higher the ultimate load. This influence of the parameter α is in line with the findings described in Section 3.1.1.

Regarding the evolution of the ultimate splitting loads as a function of l_c for beech beams with two-dowel connections (Figure 12b), it can be seen that the loads increase with the dowel spacing, which is in agreement with the test results reported in [34,36,66,67,70].

The results from the h = 200 mm series show that the ultimate loads measured for beams with small dowel spacings (l_c = 4d) do not differ significantly from the loads reached for beams with a single dowel placed at mid-span, and the larger the dowel spacing, the higher the splitting load. Small increases in the l_c spacing lead to load increases, but this tends to stabilise at large l_c spacings. A similar trend is observed for the series with h = 100 mm, although many of the specimens in this series failed in bending (Table 4). It should be noted that the average values of the loads reached are practically identical between the two groups, except for the series with an l_c spacing of 710 mm, where the specimens with h = 100 mm showed lower splitting capacity. In this case, it seems that α

does not influence the ultimate load, even though the series with h = 100 mm have twice the α ratio of those with h = 200 mm.

It is interesting to note that series with $l_b/h < 2.4$ failed by splitting and those with $l_b/h \geq 2.4$ were susceptible to fail by bending. Furthermore, when $l_b/h \geq 4.40$, the governing failure mode appears to be primarily bending. However, these values are influenced by the number of dowels with which the connection is loaded, as beams with two dowels can transfer more load to the beam before splitting occurs. This favours bending as the dominant failure mode.

3.1.4. Influence of the Number of Connections

The effect of placing another connection symmetrically to the centre of the beam is discussed here. Figure 13 shows the evolution of the load per connection as a function of l_b for beams with single (1D) and double symmetrical (2D) connections, for the two groups of beam depths tested (100 and 200 mm). It is observed that the load of beams with a single connection is higher than that of beams with two connections in the case of l_b of 360 mm, i.e., a connection spacing l_c of 230 mm. Considering the linear relationship plotted in Figure 13, the value of l_c required for a multiconnection beam to exceed the load corresponding to its single-connection counterpart would be approximately 420 mm and 280 mm for h = 100 mm and h = 200 mm, respectively ($l_c \approx 4.2h$ and $1.4h$, respectively). This could be due to an influence of the α parameter, suggesting that a higher α requires higher l_c connection spacing.

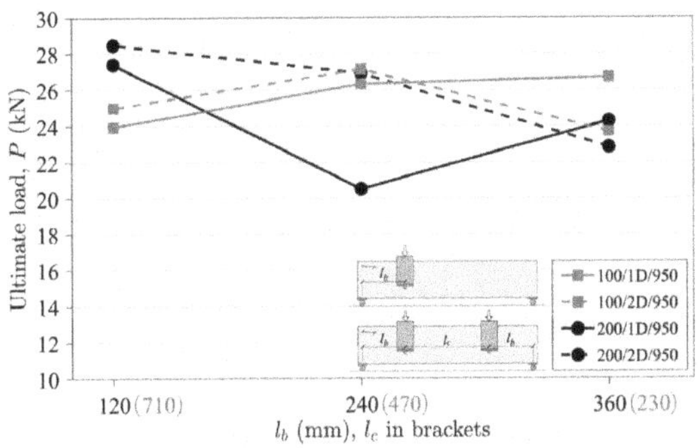

Figure 13. Splitting failure load per connection for single and double symmetrical connections as a function of the distance to the support l_b, for the two different beam depths.

In the case of beams with h = 100 mm, it is observed that the load per connection in beams with multiple connections corresponding to l_b distances of 120 mm and 240 mm ($l_c \geq 470$ mm) is approximately 4% higher than the equivalent case with a single connection. However, in the case of beams with h = 200 mm, this increase does not remain constant and amounts to 4% and 31% for l_b distances of 120 mm and 240 mm, respectively. In this case, there is a significant gain for l_b = 240 mm. It should be noted that this distance l_b corresponds to the geometry for which the most critical case of its equivalent with one connection occurs. In both groups of beam depths, for a certain distance between connections (l_c), there is an inflexion point at which beams with multiple connections reach more load than their counterparts with a single connection. Therefore, a beam with multiple connections arranged symmetrically to the centre of the beam reaches more than twice the splitting load than its counterpart with a single connection.

According to the findings in Section 3.1.3, the total load of a beam with two dowels located at off-centre of the beam does not double the load of the reference connection, i.e., a connection with two dowels located at mid-span. However, when comparing similar geometries, there is a certain connection spacing (l_c) where the stresses introduced into the beam by each connection do not interfere with each other and, in such cases, the total load of a beam with multiple connections can be more than twice the load of its counterpart with a single connection.

3.2. Comparison between Experimental, Numerical, and Theoretical Failure Loads

The adequacy of the numerical and design models described in Sections 2.2 and 2.3 for the prediction of the splitting failure loads of beech beams loaded perpendicular to the grain by the connections is discussed here in relation to the experimental results. The material properties of the European beech used as input values in the models were also presented in Section 2.2, which corresponded to the tangential direction of the wood as this was the main orientation of the cross-section of the beams.

Figure 14 shows representative results of the experimental (grey lines) and numerical (black lines) load–displacement curves, depicting both the dowel displacement (dashed lines) and the beam bottom face displacement (solid lines). As can be seen, the numerical results agree well with the experimental results for the main parameters, such as stiffness and ultimate load. Still, the numerical curves were able to better reproduce the experimental behaviour of the beams with a brittle failure than those with a mixed-mode failure pattern (see Figure 5). The good agreement between the experimental and numerical initial stiffness shows that the average values of the elastic constants used are representative of the material. This consistency between the numerical and experimental initial stiffness is more evident in the curves plotted considering the displacement of the bottom of the beam than in those plotted with the displacement under the dowel, where the numerical ones show a higher stiffness. The maximum splitting capacity of the geometries of group 1 (h = 100 mm) matches the experimental results, but in the case of group 2 (h = 200 mm), the predicted loads are always lower than the experimental ones.

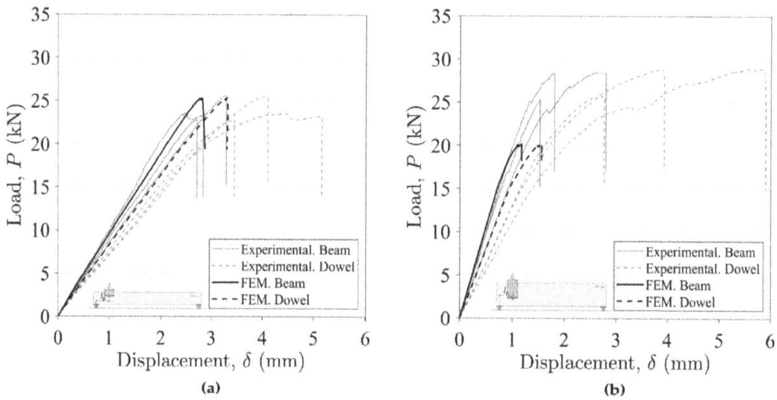

Figure 14. Representative experimental and numerical P-δ curves: (**a**) series with h = 100 mm (100/1D/950/0.13L); (**b**) series with h = 200 mm (200/1D/950/0.13L).

It should be highlighted that the expression in Eurocode 5 [24] focuses on the prediction of the maximum shear force, so the maximum splitting capacity of the connection has been obtained by applying beam theory. Thus, the maximum splitting capacity foreseen by Eurocode 5 [24] is twice the value of the shear force (Equations (14)–(17)) for all series, except for series 1D with the connection located at off-centre of the beam, where the maximum load carrying capacity was determined according to the following expression:

$$P_{\text{pre}} = V_{s,\max}\frac{L}{L - l_b} \tag{27}$$

where $V_{s,\max}$ is the maximum shear force obtained directly from Equation (14), L is the beam span, and l_b is the distance from the connection to the nearest support. It should be noted that only Eurocode 5 [24] and the numerical models take into account the position of the connection with respect to the beam.

The results of the predicted failure loads from the numerical finite element analysis and the three design models studied are presented in Table 6. The predicted values normalised to the experimental results for all cases are also included. These ratios are also shown graphically in Figure 15. A ratio ≤ 1 represents a safe prediction and a ratio > 1 an unsafe prediction.

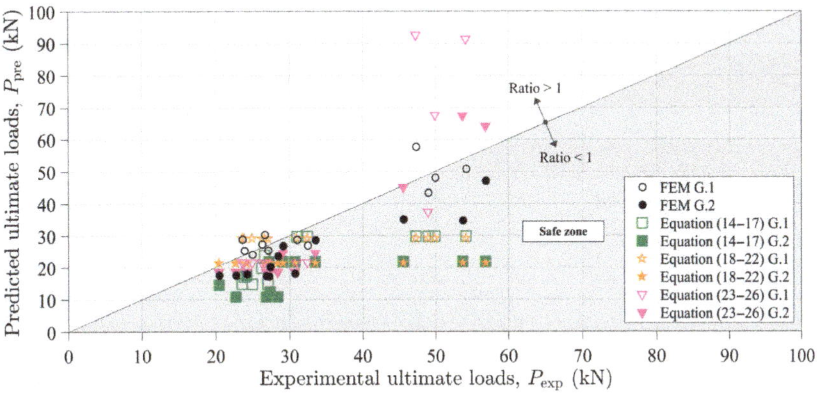

Figure 15. Experimental failure loads versus values predicted by FEM (black symbols) and by the design models (symbols in colours) for the European beech.

As can be seen from the results, in general terms, the best theoretical predictions were found for low experimental loads (between 20 and 35 kN), and there was more dispersion at higher experimental loads (between 45 and 60 kN).

When the connection is located at the centre of the beam (l_b = 0.5L), all models provide safe predictions for all the cases studied (one and two dowels and the two beam edges). Specifically, in the case of one dowel (1D), the Eurocode 5 base model and the Jensen model similarly give the best predictions. However, when there are two dowels (2D), it is FEM that stands out with theoretical loads closer to the experimental ones when applying the three design models, and it is precisely the Eurocode 5 base model and Jensen that offer the worst predictions this time, contrary to the previous case. This is consistent with the models that do not consider the number of fasteners in their formulation.

When the connection is off-centre (l_b = 0.38L, 0.25L, and 0.13L) and considering 1D, the Eurocode 5 model is the one that gives the predictions least adjusted to the experimental values in most cases, although always on the safe side. The closer the dowel is to the support, the more conservative it is. It should be noted that the C_1 value used in the expression corresponds to beech. If the fixed value C_1 = 14 given in the Eurocode 5 expression (Equations (14)–(17)) for softwoods had been used, the predicted loads would have been even more conservative.

In the case of off-centre connection, but focusing now on 2D, two assumptions have been considered for the analysis of results: each dowel as part of a separate connection (i.e., the beam would have two connections with 1D each) and the two dowels as part of the same connection. The results of this second assumption are marked with * in Table 6.

According to the first scenario of two separate connections, the Eurocode 5 model seems to be too conservative again, this time with predictions below 63% of the experimental load. The FEM appears to give theoretical values closer to the experimental ones.

Table 6. Experimental (P_{exp}) vs. predicted (numerical and theoretical, P_{pre}) splitting capacity.

Series		Exp. Test	Numerical Analysis		Design Models: Eurocode 5 [24] Equations (14)–(17)		Design Models: Jensen [46] Equations (18)–(22)		Design Models: Franke and Quenneville [47] Equations (23)–(26)	
	h/no.dowels/L/l_b	P_{exp} (kN)	P_{pre} (kN)	P_{pre}/P_{exp}	P_{pre} (kN)	P_{pre}/P_{exp}	P_{pre} (kN)	P_{pre}/P_{exp}	P_{pre} (kN)	P_{pre}/P_{exp}
G.1	100/1D/450/0.5L	32.5	27.0	0.83	29.8	0.92	29.2	0.90	21.6	0.66
	100/1D/950/0.5L	31.0	28.8	0.93	29.8	0.96	29.2	0.94	21.6	0.70
	100/1D/950/0.38L	26.7	30.2	1.13	24.0	0.90	29.2	1.10	21.6	0.81
	100/1D/950/0.25L	26.3	27.3	1.04	19.9	0.76	29.2	1.11	21.6	0.82
	100/1D/950/0.13L	23.9	25.3	1.06	17.0	0.71	29.2	1.22	21.6	0.90
	100/2D/450/0.5L	49.0	43.4	0.89	29.8	0.61	29.3	0.60	37.4	0.76
	100/2D/950/0.5L	-	-	-	-	-	-	-	-	-
	100/2D/950/0.38L	23.7	28.8	1.22	14.9	0.63	29.3	1.24	21.6	0.91
	100/2D/950/0.25L	27.1	25.4	0.94	14.9	0.55	29.2	1.08	21.6	0.80
	100/2D/950/0.13L	25.0	24.0	0.96	14.9	0.60	29.2	1.17	21.6	0.86
G.1 * ($P = 2P$)	100/2D/950/0.38L	47.3	57.6	1.22	29.8	0.63	29.3	0.62	92.6	1.96
	100/2D/950/0.25L	54.2	50.7	0.94	29.8	0.55	29.2	0.54	91.4	1.69
	100/2D/950/0.13L	50.0	48.0	0.96	29.8	0.60	29.2	0.58	67.6	1.35
G.2	200/1D/450/0.5L	27.2	17.2	0.63	21.8	0.80	21.4	0.79	18.6	0.68
	200/1D/950/0.5L	30.8	18.1	0.59	21.8	0.71	21.4	0.70	18.6	0.60
	200/1D/950/0.38L	24.3	17.9	0.74	17.6	0.72	21.4	0.88	18.6	0.77
	200/1D/950/0.25L	20.5	17.5	0.85	14.6	0.71	21.4	1.05	18.6	0.91
	200/1D/950/0.13L	27.4	20.2	0.74	12.5	0.46	21.4	0.78	18.6	0.68
	200/2D/450/0.5L	29.2	26.8	0.92	21.8	0.75	21.5	0.74	24.7	0.85
	200/2D/950/0.5L	33.6	28.6	0.85	21.8	0.65	21.4	0.64	24.6	0.73
	200/2D/950/0.38L	22.8	17.5	0.77	10.9	0.48	21.5	0.94	18.6	0.82
	200/2D/950/0.25L	26.9	17.3	0.64	10.9	0.41	21.5	0.80	18.6	0.69
	200/2D/950/0.13L	28.5	23.5	0.83	10.9	0.38	21.4	0.75	18.6	0.65
G.2 * ($P = 2P$)	200/2D/950/0.38L	45.6	35.0	0.77	21.8	0.48	21.5	0.47	45.1	0.99
	200/2D/950/0.25L	53.8	34.7	0.64	21.8	0.41	21.5	0.40	67.4	1.25
	200/2D/950/0.13L	56.9	47.1	0.83	21.8	0.38	21.4	0.38	64.1	1.13

* This comparison is made considering that the beam has only one connection, i.e., $P = 2P$.

Under the second assumption of considering the two dowels as part of the same connection, the theoretical failure loads predicted by the FEM and by the Eurocode 5 model double their value, but the ratio between these and the experimental ones is the same as that obtained under the first assumption, thus making the Eurocode 5 model quite conservative, as mentioned above.

However, the load ratios resulting from applying the models of Jensen (Equations (18)–(22)) and Franke and Quenneville (Equations (23)–(26)) differ between the two assumptions. Specifically, Jensen's model leads to the same theoretical loads for both scenarios, and therefore, the ratios are halved when only one connection is considered, which means overly conservative predictions with theoretical values below 62% of the experimental load, similar to those obtained by applying the Eurocode 5 model (it should be remembered that both models disregard the number of dowels in the connection in their formulation). On the other hand, if the connections were considered independent, Jensen's model gives results closer to the experimental ones but sometimes in a non-conservative way, as was the case for beams of h = 100 mm.

Focusing on the Franke and Quenneville model, the difference in behaviour between both assumptions is the reverse: the consideration of independent connections is the one that gives conservative results, while the second assumption of considering a single connection with two dowels results for most of the cases unsafe, with some theoretical values far away from the experimental ones, especially in the beams of h = 100 mm. Therefore, the Franke and Quenneville design model considering two independent connections, as well as the FEM model, would be the most appropriate when off-centre multidowel connections are presented.

4. Conclusions

The numerical and experimental splitting capacity results for European beech (*Fagus sylvatica* L.) beams loaded perpendicular to the grain by single- and double-dowel connections with different locations along the beam span are provided.

The use of an additional dowel slightly increases the splitting capacity at low α (h_e/h) values (0.32) but significantly increases it at high α values (0.64), by about 50%. For the same test configuration, the larger the beam span, the higher the splitting load. For single-dowel connections, the position of the dowel along the beam span influences the splitting capacity, which is lower the closer the connection is to the supports. For double-dowel connections, the greater the distance between the dowels, the greater the splitting load, but in no case will the increased load from an additional dowel double the ultimate load of a single-dowel connection. Beams with $l_b/h < 2.4$ fail by splitting, and those with an l_b/h ratio ≥ 2.4 are susceptible to fail by bending, but these values are influenced by the number of dowels loaded into the connection, as beams with two dowels can transfer more load to the beam before splitting occurs. This fact favours bending as the dominant failure mode.

A 2D finite element model was developed for each geometry in the experimental programme, using a cohesive element model and a bilinear cohesive law to reproduce the mode I and II fracture process.

The proposed finite element model successfully replicated the behaviour of beams exhibiting purely brittle failure patterns more effectively than those displaying mixed-mode failure.

The numerical P-δ curves are in good agreement with the experimental ones. The proposed numerical model demonstrates its efficacy as a powerful tool for predicting the splitting capacity of timber connections with various arrangements. Moreover, it proves to be structurally safe from a design standpoint.

Of the three design models based on fracture mechanics (including the Eurocode 5 expression), the models based on elastic and linear fracture energy generally give good load predictions for single-dowel arrangements. For two-dowel layouts, where the two dowels are considered as part of the same connection, design models that include fracture properties only in mode I provide too conservative values. The model that considers mode

I and II fracture and the influence of joint width (l_c) predicts overestimated values for high l_c configurations.

Further experimental research on the splitting behaviour considering other connection geometries and hardwood species would be desirable in order to obtain a general expression valid for all of them.

Author Contributions: Conceptualisation, J.L.G.-R., A.J.L.-B. and A.M.-M.; methodology, J.L.G.-R., A.J.L.-B., A.M.-M., J.X. and M.F.S.F.d.M.; validation, J.L.G.-R.; formal analysis, J.L.G.-R.; investigation, J.L.G.-R., A.J.L.-B. and A.M.-M.; writing—original draft preparation, J.L.G.-R.; writing—review and editing, J.L.G.-R., A.J.L.-B., A.M.-M., J.X. and M.F.S.F.d.M.; visualisation, J.L.G.-R.; supervision, A.M.-M., J.X. and M.F.S.F.d.M.; funding acquisition, A.M.-M., J.X. and M.F.S.F.d.M. All authors have read and agreed to the published version of the manuscript.

Funding: Part of the work was undertaken during a short-term scientific stay by the first author at the Faculty of Engineering (University of Porto) in 2021, with the financial support provided by Programa Propio de I+D+i 2021 de la Universidad Politécnica de Madrid. This work is part of the R&D&I Project PID2020-112954RA-I00 funded by MCIN/AEI/10.13039/501100011033. The authors gratefully acknowledge Fundação para a Ciência e a Tecnologia (FCT-MCTES) for the financial support of the Laboratório Associado de Energia, Transportes e Aeronáutica (LAETA) by the project UID/EMS/50022/2020 and the Research and Development Unit for Mechanical and Industrial Engineering (UNIDEMI) by the projects UIDB/00667/2020 and UIDP/00667/2020.

Institutional Review Board Statement: Not applicable.

Informed Consent Statement: Not applicable.

Data Availability Statement: The original contributions presented in the study are included in the article material, further inquiries cab be directed to the corresponding author/s.

Conflicts of Interest: The authors declare no conflict of interest.

References

1. Gustavsson, L.; Sathre, R. Variability in Energy and Carbon Dioxide Balances of Wood and Concrete Building Materials. *Build. Environ.* **2006**, *41*, 940–951. [CrossRef]
2. UNEP. *Buildings and Climate Change: Summary for Decision-Makers*; UNEP DTIE, Sustainable Consumption & Production Branch: Paris, France, 2009; ISBN 987-9-28073-064-7.
3. Buchanan, A.H.; Honey, B.G. Energy and Carbon Dioxide Implications of Building Construction. *Energy Build.* **1994**, *20*, 205–217. [CrossRef]
4. Sathre, R.; Gustavsson, L. Using Wood Products to Mitigate Climate Change: External Costs and Structural Change. *Appl. Energy* **2009**, *86*, 251–257. [CrossRef]
5. UNECE; FAO. *Circularity Concepts in Wood Construction*; United Nations Publication: Geneva, Switzerland, 2023; ISBN 978-9-21117-328-4.
6. Aicher, S.; Christian, Z.; Dill-Langer, G. Hardwood Glulams—Emerging Timber Products of Superior Mechanical Properties. In Proceedings of the World Conference on Timber Engineering (WCTE), Quebec City, QC, Canada, 10–14 August 2014.
7. Camú, C.T.; Aicher, S. A Stochastic Finite Element Model for Glulam Beams of Hardwoods. In Proceedings of the World Conference on Timber Engineering, Seoul, Republic of Korea, 20–23 August 2018.
8. De Rigo, D.; Caudullo, G. Fagus Sylvatica in Europe: Distribution, Habitat, Usage and Threats. In *European Atlas of Forest Tree Species*; Publication Office of the European Union: Luxembourg, 2016; pp. 94–95, ISBN 978-9-27617-291-8.
9. Antonucci, S.; Santopuoli, G.; Marchetti, M.; Tognetti, R.; Chiavetta, U.; Garfì, V. What Is Known About the Management of European Beech Forests Facing Climate Change? A Review. *Curr. For. Rep.* **2021**, *7*, 321–333. [CrossRef]
10. Gómez-Royuela, J.L.; Majano-Majano, A.; Lara-Bocanegra, A.J.; Reynolds, T.P.S. Determination of the Elastic Constants of Thermally Modified Beech by Ultrasound and Static Tests Coupled with 3D Digital Image Correlation. *Constr. Build. Mater.* **2021**, *302*, 124270. [CrossRef]
11. Ozyhar, T.; Hering, S.; Niemz, P. Viscoelastic Characterization of Wood: Time Dependence of the Orthotropic Compliance in Tension and Compression. *J. Rheol.* **2013**, *57*, 699–717. [CrossRef]
12. Ozyhar, T.; Hering, S.; Sanabria, S.J.; Niemz, P. Determining Moisture-Dependent Elastic Characteristics of Beech Wood by Means of Ultrasonic Waves. *Wood Sci. Technol.* **2013**, *47*, 329–341. [CrossRef]
13. Ozyhar, T.; Hering, S.; Niemz, P. Moisture-Dependent Orthotropic Tension-Compression Asymmetry of Wood. *Holzforschung* **2013**, *67*, 395–404. [CrossRef]
14. Ozyhar, T.; Hering, S.; Niemz, P. Moisture-Dependent Elastic and Strength Anisotropy of European Beech Wood in Tension. *J. Mater. Sci.* **2012**, *47*, 6141–6150. [CrossRef]

15. Widmann, R.; Fernandez-Cabo, J.L.; Steiger, R. Mechanical Properties of Thermally Modified Beech Timber for Structural Purposes. *Eur. J. Wood Wood Prod.* **2012**, *70*, 775–784. [CrossRef]
16. Franke, S.; Magnière, N. The Embedment Failure of European Beech Compared to Spruce Wood and Standards. In *Materials and Joints in Timber Structures*; Aicher, S., Reinhardt, H.-W., Garrecht, H., Eds.; RILEM Bookseries; Springer: Dordrecht, The Netherlands, 2014; pp. 221–229, ISBN 978-9-40077-810-8.
17. Gómez-Royuela, J.L.; Majano-Majano, A.; Lara-Bocanegra, A.J.; Xavier, J.; de Moura, M.F.S.F. Evaluation of R-Curves and Cohesive Law in Mode I of European Beech. *Theor. Appl. Fract. Mech.* **2022**, *118*, 103220. [CrossRef]
18. Gómez-Royuela, J.L.; Majano-Majano, A.; Lara-Bocanegra, A.J.; Xavier, J.; de Moura, M.F.S.F. Shear Traction-separation Laws of European Beech under Mode II Loading by 3D Digital Image Correlation. *Wood Sci. Technol.* **2022**, *56*, 1631–1655. [CrossRef]
19. Ehrhart, T.; Steiger, R.; Frangi, A. European Beech Glued Laminated Timber. *Bautechnik* **2021**, *98*, 104–114. [CrossRef]
20. Blass, H.; Frese, M.; Enders-Comberg, M. Beech LVL—High Strength Material for Engineered Timber Structures. In *Structures and Architecture: Beyond Their Limits*, 1st ed.; Cruz, P.J., Ed.; CRC Press: Boca Raton, FL, USA, 2016; pp. 102–109. ISBN 978-1-31573-076-9.
21. Lara-Bocanegra, A.J. Láminas Reticulares de Madera Deformadas Elásticamente. Del Material a La Construcción (In English: Elastic Timber Gridshells. From Material to Construction). Ph.D. Thesis, Universidad Politécnica de Madrid, Madrid, Spain, 2022. [CrossRef]
22. Caldeira, T. Caracterização Experimental e Numérica Do Comportamento Frágil de Ligações Com Cavilhas Em Estruturas de Madeira (In English: Experimental and Numerical Characterisation of the Brittle Behaviour of Dowel Connections in Timber Structures). Ph.D. Thesis, Universidade de Trás-os-Montes e Alto Douro, Vila Real, Portugal, 2011.
23. Franke, S.; Franke, B.; Heubuch, S.; Frangi, A.; Jockwer, R. *Anschlüsse in Buchenholz. Ermittlung von Grundlagen Zur Bemessung von Anschlüssen Für Die Marktimplementierung in Der Schweiz*; Berner Fachhochschule: Bern, Switzerland, 2019; ISBN 978-3-90687-806-5.
24. *EN 1995-1-1*; Eurocode 5: Design of Timber Structures. Part 1-1: General. Common Rules and Rules for Buildings. European Committee for Standardization: Brussels, Belgium, 2016.
25. Johansen, K.W. Theory of Timber Connections. *Int. Assoc. Bridge Struct. Eng.* **1949**, *9*, 249–262.
26. Van der Put, T.A.C.M. Tension Perpendicular to the Grain at Notches and Joints. In Proceedings of the International Council for Research and Innovation in Building and Construction (CIB-W18), Lisbon, Portugal, 10–14 September 1990.
27. Van der Put, T.A.C.M.; Leitjen, A.J.M. Evaluation of Perpendicular to Grain Failure of Beams Caused by Concentrated Loads of Joints. In Proceedings of the International Council for Research and Innovation in Building and Construction (CIB-W18), Delft, The Netherlands, 28–30 August 2000.
28. Jensen, J.L.; Quenneville, P. Splitting of Beams Loaded Perpendicular to Grain by Connections. Some Issues with EC5. In Proceedings of the International Council for Research and Innovation in Building and Construction (CIB-W18), Alghero, Italy, 29 August–1 September 2011.
29. Ballerini, M. A New Set of Experimental Tests on Beams Loaded Perpendicular to Grain by Dowel Type Joints. In Proceedings of the International Council for Research and Innovation in Building and Construction (CIB-W18), Graz, Austria, 23–25 August 1999.
30. Möhler, K.; Lautenschläger, R. *Großflächige Queranschlüsse Bei Brettschichtholz*; Forschungsbericht des Lehrstuhls für Ingenieurholzbau und Baukonstruktionen, Univ. Karlsruhe: Karlsruhe, Germany, 1978.
31. Möhler, K.; Siebert, W. *Ausbildung von Queranschlüssen Bei Angehängten Lasten an Brettschichtholzträger*; Forschungsbericht des Lehrstuhls für Ingenieurholzbau und Baukonstruktionen, Univ. Karlsruhe: Karlsruhe, Germany, 1980.
32. Ehlbeck, J.; Görlacher, R. *Tragverhalten von Queranschlüssen Mittels Stahlblechformteilen, Insbesondere Balkenschuhen, Im Holzbau*; Forschungsbericht der Versuchsanstalt für Stahl, Holz und Steine, Abt. Ingenieurholzbau, Univ. Karlsruhe: Karlsruhe, Germany, 1983.
33. Ballerini, M. Beams Transversally Loaded by Dowel-Type Joints: Influence on Splitting Strength of Beam Thickness and Dowel Size. In Proceedings of the International Council for Research and Innovation in Building and Construction (CIB-W18), CO, USA, 11–14 August 2003.
34. Schoenmakers, J.C.M. Fracture and Failure Mechanisms in Timber Loaded Perpendicular to the Grain by Mechanical Connections. Ph.D. Thesis, Technische Universiteit Eindhoven, Eindhoven, The Netherlands, 18 May 2010.
35. Jensen, J.L.; Girhammar, U.A.; Quenneville, P.; Källsner, B. Splitting of Beams Loaded Perpendicular to Grain by Connections: Simple Fracture Mechanics Models. In Proceedings of the World Conference on Timber Engineering (WCTE), Auckland, New Zealand, 15–19 July 2012.
36. Leijten, A.J.M. Splitting of Timber Beams Caused by Perpendicular to Grain Forces of Multiple Connections. *Eng. Struct.* **2018**, *171*, 10–14. [CrossRef]
37. Franke, S.; Franke, B.; Quenneville, P. Analysis of the Failure Behaviour of Multiple Dowel-Type Connections Loaded Perpendicular to the Grain in LVL. In Proceedings of the World Conference on Timber Engineering (WCTE), Auckland, New Zealand, 16–19 July 2012.
38. Hindman, D.P.; Finkenbinder, D.E.; Loferski, J.R.; Line, P. Strength of Sawn Lumber and Wood Composite Dowel Connections Loaded Perpendicular to Grain. I: NDS Design Equations. *J. Mater. Civ. Eng.* **2010**, *22*, 1217–1225. [CrossRef]
39. Patel, M.C.; Hindman, D.P. Comparison of Single- and Two-Bolted LVL Perpendicular-to-Grain Connections. I: National Design Specification for Wood Construction Equations. *J. Mater. Civ. Eng.* **2012**, *24*, 339–346. [CrossRef]
40. Majano-Majano, A.; Lara-Bocanegra, A.J.; Xavier, J.; Guaita, M. Splitting Capacity of Eucalyptus Globulus Beams Loaded Perpendicular to the Grain by Connections. *Mater. Struct.* **2022**, *55*, 147. [CrossRef]

41. Franke, B.; Quenneville, P. Analyses of the Failure Behaviour of Transversely Loaded Dowel Type Connections in Wood. In Proceedings of the World Conference on Timber Engineering (WCTE), Trentino, Italy, 20–24 June 2010.
42. Dourado, N.; Silva, F.G.A.; de Moura, M.F.S.F. Fracture Behavior of Wood-Steel Dowel Joints under Quasi-Static Loading. *Constr. Build. Mater.* **2018**, *176*, 14–23. [CrossRef]
43. Shi, D.; Huang, H.; Li, N.; Liu, Y.; Demartino, C. Bolted Steel to Laminated Timber and Glubam Connections: Axial Behavior and Finite-Element Modeling. *Int. J. Mech. Sci.* **2023**, *252*, 108364. [CrossRef]
44. Boström, L. Method of Determination of the Softening Behaviour of Wood and the Applicability of a Nonlinear Fracture Mechanics Model. Ph.D. Thesis, Lund University, Lund, Sweden, 1992.
45. Franke, S.; Franke, B.; Magnière, N. Load Carrying Capacity of Cracked Beams. In Proceedings of the World Conference on Timber Engineering (WCTE), Vienna, Austria, 22–25 August 2016.
46. Jensen, J.L.; Gustafsson, P.J.; Larsen, H.J. A Tensile Fracture Model for Joints with Rods or Dowels Loaded Perpendicular to Grain. In Proceedings of the International Council for Research and Innovation in Building and Construction (CIB-W18), Colorado, CO, USA, 11–14 August 2003.
47. Franke, B.; Quenneville, P. Design Approach for the Splitting Failure of Dowel-Type Connections Loaded Perpendicular to Grain. In Proceedings of the International Council for Research and Innovation in Building and Construction (CIB-W18), Vancouver, BC, Canada, 26–29 August 2013.
48. *DIN 4074-5:2008;* Strength Grading of Wood-Part 5: Sawn Hard Wood. DIN: Berlin, Germany, 2008. (In German)
49. *EN 1912:2012;* Structural Timber. Strength Classes. Assignment of Visual Grades and Species. European Committee for Standardization: Brussels, Belgium, 2012.
50. *EN 13183-1:2002;* Moisture Content of a Piece of Sawn Timber. Part 1: Determination by Oven Dry Method. European Committee for Standardization: Brussels, Belgium, 2002.
51. *EN 408:2010+A1:2012;* Timber Structures. Structural Timber and Gluedlaminated Timber. Determination of Some Physical and Mechanical Properties. European Committee for Standardization: Brussels, Belgium, 2012.
52. *EN 384:2016+A1;* Structural Timber. Determination of Characteristic Values of Mechanical Properties and Density. European Committee for Standardization: Brussels, Belgium, 2019.
53. *ABAQUS 2021 Documentation;* Dassault Systèmes Simulia Corp.: Johnston, RI, USA, 2021.
54. Gonçalves, J.P.M.; de Moura, M.F.S.F.; de Castro, P.M.S.T.; Marques, A.T. Interface Element Including Point-to-surface Constraints for Three-dimensional Problems with Damage Propagation. *Eng. Comput.* **2000**, *17*, 28–47. [CrossRef]
55. Franke, B.; Quenneville, P. Numerical Modeling of the Failure Behavior of Dowel Connections in Wood. *J. Eng. Mech.* **2011**, *137*, 186–195. [CrossRef]
56. McKenzie, W.M.; Karpovich, H. The Frictional Behaviour of Wood. *Wood Sci. Technol.* **1968**, *2*, 139–152. [CrossRef]
57. Durão, L.M.P.; de Moura, M.F.S.F.; Marques, A.T. Numerical Simulation of the Drilling Process on Carbon/Epoxy Composite Laminates. *Compos. Part A Appl. Sci. Manuf.* **2006**, *37*, 1325–1333. [CrossRef]
58. Silva, F.G.A.; de Moura, M.F.S.F.; Dourado, N.; Xavier, J.; Pereira, F.A.M.; Morais, J.J.L.; Dias, M.I.R. Mixed-Mode I+II Fracture Characterization of Human Cortical Bone Using the Single Leg Bending Test. *J. Mech. Behav. Biomed. Mater.* **2016**, *54*, 72–81. [CrossRef]
59. Davila, C.; Camanho, P.; de Moura, M.F. Mixed-Mode Decohesion Elements for Analyses of Progressive Delamination. In Proceedings of the 19th AIAA Applied Aerodynamics Conference, Reston, VA, USA, 16–19 April 2001.
60. Jockwer, R.; Dietsch, P. Review of Design Approaches and Test Results on Brittle Failure Modes of Connections Loaded at an Angle to the Grain. *Eng. Struct.* **2018**, *171*, 362–372. [CrossRef]
61. Jensen, J.L.; Quenneville, P.; Girhammar, U.A.; Källsner, B. Brittle Failures in Timber Beams Loaded Perpendicular to Grain by Connections. *J. Mater. Civ. Eng.* **2015**, *27*, 4015026. [CrossRef]
62. Leijten, A.J.M.; Jorissen, A.J.M. Splitting Strength of Beams Loaded by Connections Perpendicular to Grain, Model Validation. In Proceedings of the International Council for Research and Innovation in Building and Construction (CIB-W18), Venice, Italy, 22–24 August 2001.
63. Jensen, J.L. Splitting Strength of Beams Loaded by Connections. In Proceedings of the International Council for Research and Innovation in Building and Construction (CIB-W18), CO, USA, 11–14 August 2003.
64. Jensen, J.L. Splitting Strength of Beams Loaded Perpendicular to Grain by Dowel Joints. *J. Wood Sci.* **2005**, *51*, 480–485. [CrossRef]
65. Ballerini, M. A New Prediction Formula for the Splitting Strength of Beams Loaded by Dowel-Type Connections. In Proceedings of the International Council for Research and Innovation in Building and Construction (CIB-W18), Edinburgh, Scotland, 31 August–3 September 2004.
66. Habkirk, R.; Quenneville, J.H.P. Bolted Wood Connections Loaded Perpendicular-to-Grain: Effect on Wood Species. In Proceedings of the World Conference on Timber Engineering (WCTE), Portland, OR, USA, 6–10 August 2006.
67. Reshke, R.G. Bolted Timber Connections Loaded Perpendicular-to-Grain. Influence of Joint Configuration Parameters on Strength. Ph.D. Thesis, Royal Military College of Canada, Kingston, ON, Canada, 11 May 1999.
68. Yasumura, M. Criteria for Damage and Failure of Dowel-Type Joints Subjected To Force Perpendicular To the Grain. In Proceedings of the International Council for Research and Innovation in Building and Construction (CIB-W18), Venice, Italy, 22–24 August 2001.

69. Jensen, J.L.; Quenneville, P.; Girhammar, U.A.; Källsner, B. Beams Loaded Perpendicular to Grain by Connections—Combined Effect of Edge and End Distance. In Proceedings of the International Council for Research and Innovation in Building and Construction (CIB-W18): Meeting Forty-Five, Växjö, Sweden, 27–30 August 2012. CIB-W18/45-7-2.
70. Kasim, M.; Quenneville, P. Effect of Row Spacing on the Capacity of Bolted Timber Connections Loaded Perpendicular-to-Grain. In Proceedings of the International Council for Research and Innovation in Building and Construction (CIB-W18), Kyoto, Japan, 16–19 September 2002.

applied sciences

MDPI

Article

Investigating the Out-of-Plane Bending Stiffness Properties in Hybrid Species Diagonal-Cross-Laminated Timber Panels

Shaghayegh Kurzinski [1,*] and Paul L. Crovella [2]

[1] School of Engineering, Computing and Construction Management, Roger Williams University, Bristol, RI 02809, USA

[2] Department of Sustainable Resources Management, College of Environmental Science and Forestry, State University of New York, Syracuse, NY 13210, USA; plcrovella@esf.edu

* Correspondence: skurzinski@rwu.edu

Abstract: Since the introduction of Cross-laminated Timber (CLT) in Austria in the early 1990s, the adoption of this 90°-crosswise-laminated product has seen exponential growth worldwide. Compared to traditional laminated timber products (e.g., glulam), CLT provides improved dimensional stability but with reduced out-of-plane bending stiffness. To improve the bending stiffness, while maintaining relative dimensional stability, a modified orientation of the inner layers in a diagonal direction can be used. This novel product is Diagonal-Cross-laminated Timber (DCLT), a composite timber product, consisting of inner layers which are rotated at different angle-ply orientations between 0 and 90 degrees to the outer layers. To properly model the out-of-plane bending behavior of the DCLT, analytical models and finite element analysis (FEA) were used, and the results were validated by four-point bending tests performed on DCLT panels with angle-ply orientations of 10°, 20°, 40°, 70°, and a conventional CLT 90° panel. The results indicate that DCLT panels with angle-ply cross layers have a structural advantage in out-of-plane bending over traditional CLT (90°) panels. The apparent bending stiffness from DCLT 90° to DCLT ± 10° has an increase of 33%, 24%, and 35%, respectively, regarding the assessed methods of experimental, theoretical, and FEM modeling. Using these panels would allow for increased spans or load-carrying capacity for a given panel span-to-depth ratio. The development of DCLT and its introduction to the industry not only could enable the use of lower-quality timber that would not otherwise satisfy structural requirements for CLT but also could help reduce the fabrication cost of CLT due to utilizing lower amounts of fiber.

Keywords: Cross-laminated Timber (CLT); grain orientation; Hankinson's formula; Diagonal-Cross-laminated Timber (DCLT); bending stiffness

1. Introduction

Wood, a natural orthotropic material, has unique properties in different grain directions of the three mutually perpendicular axes: longitudinal, radial, and tangential (as shown in Figure 1). The longitudinal axis L is parallel to the grain, the radial axis R is normal to the growth rings (perpendicular to the grain in the radial direction), and the tangential axis T is perpendicular to the grain but tangent to the growth rings. The strength and stiffness properties of wood in the grain direction are much higher than the strength and stiffness properties perpendicular to the grain direction. In addition to the mechanical properties, given its hygroscopic nature, wood dimensional changes (i.e., the shrinkage and swelling) occur in each direction and vary with moisture content (MC%). This property of wood is referred to as dimensional stability. The dimensional stability affects how a final timber product will distort in service. The percent shrinkage and swelling of wood are greatest in the tangential direction, less in the radial direction, and least in the longitudinal direction; the latter changes are mostly negligible.

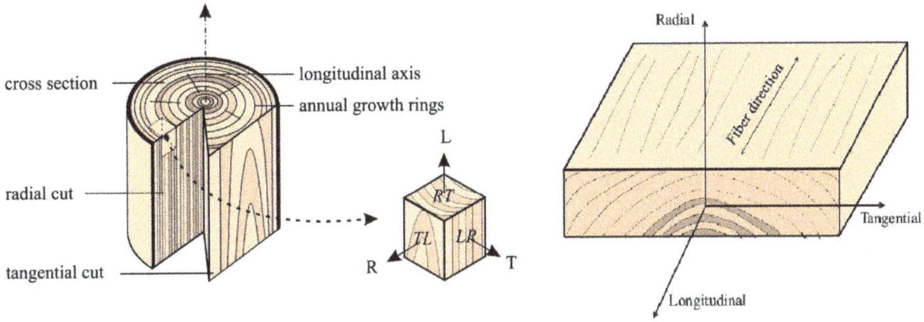

Figure 1. Three perpendicular axes of wood grain orientation.

The natural properties of wood have been the basis for innovations in mass timber products such as Glued-laminated Timber (GLT) and Cross-laminated Timber (CLT). GLT includes layers of timber boards glued together parallel in one major load-bearing direction (0°); CLT consists of crosswise orientated layers, where the layers are orientated perpendicular to each other, providing one major (0°) and one minor (90°) load-bearing direction for out-of-plane bending [1]. The cross layers of CLT help to minimize the overall shrinkage and swelling of the structural panel. Thus, CLT panels show higher dimensional stability compared to GLT beams. For bending out-of-plane, CLT's cross layers are responsible for the panel's bidirectional strength and work to distribute the load in the minor direction as well. However, the cross layers have lower stiffness and strength (in the major axis direction), leading to a greater panel deflection than with GLT. Recognizing these relative advantages and limitations of GLT and CLT, an optimized/modified orientation of the inner layers in a diagonal direction could be developed to improve the out-of-plane bending stiffness while maintaining structural and dimensional stability.

The product created with inner layers at varying angles is called Diagonal-Cross-laminated Timber (DCLT). DCLT can be seen as a standard counterpart to CLT, offering improved mechanical properties using the same panel thicknesses while using less fiber, resulting in lower fabrication and material cost. CLT is a balanced and symmetric laminate since the material properties, the thickness of its layers, and the orientation of its plies are symmetric about the midplane [2]. However, DCLT can be introduced as a balanced and antisymmetric laminate to improve the overall stiffness. This means that in DCLT panels, the plies with the same thickness are a mirror image of the geometrical midplane. Therefore, for each $+\alpha$ angle-ply laminate in the panel, there is an equally thick $-\alpha$ angle-ply laminate as well. Figure 2 illustrates the configurations of GLT, DCLT, and CLT to show how DCLT lay-up can be orientated compared to GLT and CLT products.

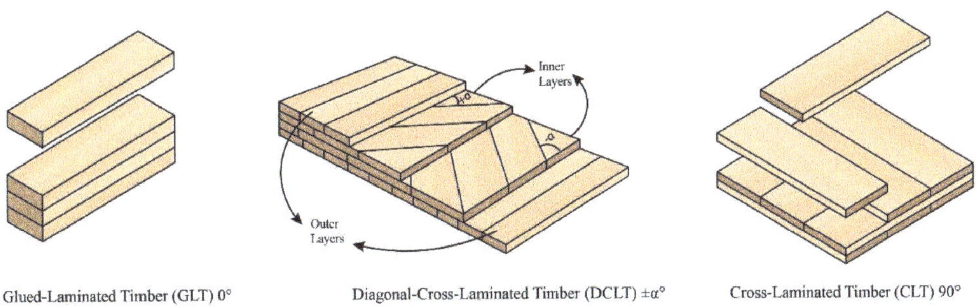

Figure 2. GLT, DCLT, and CLT lay-up configurations [3].

Several studies have been conducted to study the potential relationship between the grain orientation of DCLT's inner lamina and the panel's bending and shear stiffness properties. Buck et al. [4] investigated the difference between bending stiffness in conventional CLT panels and DCLT panels with inner layers of 45° grain orientation. They tested two groups of 20 five-layered panels, with orientations of 0°–90°–0°–90°–0° and 0°–45°–0°–45°–0° under four-point bending in the main load-carrying direction in a flatwise panel lay-up. Their evaluated results confirmed that the DCLT ± 45° bending strength and the apparent stiffness (EI_{app}) increased, respectively, by 35% and 15% over the conventional CLT panels with 90° layers. EI_{app} is defined as the apparent bending stiffness of the panel in the span direction including both bending and shear deformation, while the EI_{eff} of the panel refers to the effective bending stiffness in the span direction. In another study, these researchers [5] investigated the mechanics of the diagonally and conventionally layered CLT by a digital image correlation (DIC) technique and a destructive four-point bending test on samples with the conventional 90° and ±45° alternating inner layers. DIC evaluations of 100 CLT panel layers showed that a considerable part of the stiffness of conventional CLT is reduced by the shear resistance of transverse layers. The overall experimental results showed the shear strain in CLT ± 45° to be 60% lower than in CLT 90°. The reduced strain in the transverse layers contributes to a stiffer panel and therefore suggests the use of CLT panels with a ±45° layered configuration to be a more efficient material to use for timber engineering construction. Their results also showed that utilizing the lower-grade material for the transverse diagonal layers increased the yield values due to the increased stiffness properties of the overall product.

Other researchers have shown interest in studying both the shear and bending performance of DCLT panels with different angle-ply orientations. Bahmanzad et al. [6] performed research on three-ply DCLT panels fabricated from eastern hemlock. The inner layer had grain orientations of 30°, 45°, 60°, and CLT (90°) relative to the main load-bearing direction. Their results of the short-span three-point bending test showed higher shear and bending properties for the DCLT 30° panels, followed by successively lower values for the DCLT 45°, 60°, and CLT (90°) panels. The EI_{app} from their experimental short-span three-point bending results showed an increase of 174%, 112%, and 61%, respectively, for DCLT 30°, DCLT 45°, and DCLT 60°, compared to CLT 90°.

Also, Arnold et al. [2] investigated the load-bearing behavior and efficiency of DCLT under a biaxial bending configuration experimentally and theoretically. They tested seven replicas of CLT and DCLT ± 30° and ±45° panels with different thicknesses. Their evaluation included an evaluation of the increase in the torsional stiffness of DCLT panels compared to conventional CLT panels. They tested an applied parametric numerical model to evaluate the influence of the layer arrangement as well as the width and thickness of the laminations on the torsional stiffness. They also designed an analytical model for finding the torsional stiffness of the panels using two methods: the Kirchhoff–Love and Reissner–Mindlin equations. Their results showed a considerable increase in the torsional stiffness of DCLT panels compared to conventional Cross-laminated Timber panels. Their results confirmed that DCLT could improve biaxial bending stiffness. Therefore, it could be an alternative option to CLT for floor systems governed by serviceability limit states such as deflections or vibrations.

To determine wood properties at an angle to one of the principal axes, Hankinson's criterion is the most common method. Over time, researchers have proposed modifications to the base formula [7]; however, recent studies have confirmed the validity of fit to the experimental data for the original formula. For instance, Bahmanzad et al. [8] studied Hankinson's formula to determine its effectiveness in predicting shear moduli and strength based on grain orientation. The shear stiffness and strength results from the short-span bending tests on three-ply CLT and DCLT panels appeared to fit the data fairly well. Another study regarding the impact of grain orientation on the mechanical properties of DCLT was conducted by Franzoni et al. [9]. They developed a 3D FEM DCLT model with four configurations having the same total thickness but different spans and the mid-layers of the panels orienting from 0° to 90° with an increment of 10°. Their findings confirmed that the failure load and the deflection are functions of grain orientation. Their results state that the effect on bending stiffness of rotating the grain orientation becomes more significant for DCLT panels with shorter span-to-depth ratios.

This study continues the investigation on the effectiveness of orienting the inner layer of CLT products at another angle rather than the conventional 90° on the overall panel stiffness. This work focuses on the apparent bending stiffness of DCLT panels with various grain orientations in two asymmetric inner layers featuring a feasible lay-up for the future manufacturing of DCLT. In this study, three approaches were used: a theoretical approach, experimental testing, and FEM modeling. For this work, twelve four-ply DCLT $\pm\alpha$ panels and one four-ply CLT panel were fabricated from a local hardwood species, black locust (*Robinia pseudoacacia*), for the parallel (0°) top and bottom layers, and a local softwood species, eastern white pine (*Pinus strobus*), was used for the two diagonal ($\pm\alpha$°) inner layers. Before the experiment, an analytical model was developed using Hankinson's criterion with the shear analogy approach to predict the bending performance of each set of DCLT panels. The shear analogy method proposed by Kreuzinger [10], and used in the US CLT Handbook [11], gives the composite bending and shear stiffness (EI_{eff}, GA_{eff}, and EI_{app}) of the CLT panels using the values of the elastic and shear moduli parallel and perpendicular to the grain (E_0, E_{90}, G_{90}) of the species in each layer [12]. The bending stiffness properties of the DCLT and CLT panels from this theory were then compared to the experimental results from the four-point bending test. The finite element analysis also simulated the four-point bending test, and the results were compared with the results of the shear analogy method and experimental tests. This study provides methods of predicting the bending stiffness of DCLT and CLT panels with the available or measured elastic modulus parallel to the grain of the species used in the panel fabrication. These methods of prediction could help utilize and introduce non-standard species for the fabrication of these products.

2. Materials and Methodology

2.1. Theoretical Investigations

Initial investigations were conducted on predicting the bending stiffness of DCLT panels with different grain orientation angles and species. An analytical model was developed by applying Hankinson's formula in the shear analogy approach. Hankinson's formula (Equations (1) and (2)) was used to find the axial and shear stiffness values of the angle-ply inner layers, and the final panel stiffness (EI_{app}) was found using the shear analogy approach (Equations (3)–(5)),

$$E_\alpha = \frac{E_0 \times E_{90}}{E_0 \times \alpha + E_{90} \times \alpha} \tag{1}$$

$$G_\alpha = \frac{G_0 \times G_{90}}{G_0 \times \alpha + G_{90} \times \alpha} \tag{2}$$

where α is the angle of the grain orientation of the main axis, E_α is the modulus of elasticity in the major strength direction of the diagonal layer, and E_0 and E_{90} are the moduli of elasticity parallel and perpendicular to the grain. G_α is the shear modulus of the diagonal layer in the span direction, and G_0 and G_{90} are the shear moduli parallel and perpendicular to the grain,

$$EI_{eff} = \sum_{i=1}^{n} E_i \times b_i \times \frac{h_i^3}{12} + \sum_{i=1}^{n} E_i \times A_i \times z_i^2 \tag{3}$$

where i is the individual layer of the panel, EI_{eff} is the effective bending stiffness in the span direction, E_i is the modulus of elasticity of the individual layer that can be E_0 for outer layers and E_α for inner layers, b_i is the width of the effective cross-section, h_i is the height of the individual layer, A_i is the area of the effective cross-section, and z_i is the distance of the neutral axis of the individual layer from the neutral axis of the panel,

$$GA_{eff} = \frac{a^2}{\left[\left(\frac{h_i}{2\times G_i \times b}\right) + \left(\sum_{i=2}^{n-1} \frac{h_i}{G_i \times b_i}\right) + \left(\frac{h_n}{2\times G_n \times b}\right)\right]} \quad (4)$$

where, GA_{eff} is the effective shear stiffness in the span direction, a is the distance between the neutral axis of the outer layers, b_i is the width of the effective cross-section, h_i is the height of the individual layer, and G_i is the shear modulus of the individual layer that can be G_0 for outer layers and G_α for inner layers,

$$EI_{app} = \frac{EI_{eff}}{1 + \frac{K_s EI_{eff}}{GA_{eff} \times (l)^2}} \quad (5)$$

where, EI_{app} is the apparent bending stiffness in the span direction including both bending and shear deformation, EI_{eff} is the effective bending stiffness in the span direction using the shear analogy method, GA_{eff} is the effective shear stiffness in the span direction using the shear analogy method, K_s is a factor based on the ratio of deflection due to bending to deflection due to shear (12.96 for four-point loading [13]), and l is the length of the span.

Initially, lay-up alternatives for the hardwood and softwood species selection were assumed. Due to the higher rate of shrinkage/swelling of the hardwood species, black locust, the optimal lay-up was assumed to be a hybrid of both hardwood and softwood to increase the dimensional stability of the panels while still providing higher bending stiffness values. As Figure 3 shows, the stiffest hybrid lay-up was found to be the panel with hardwood in the outer layers and softwood in the inner layers.

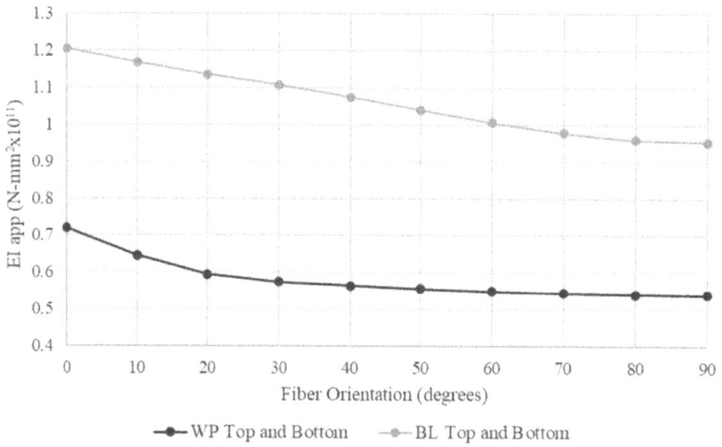

Figure 3. Apparent bending stiffness properties of two alternatives of hybrid DCLT panels with two species of white pine (WP) and black locust (BL).

2.2. DCLT Panel Preparation

A total of 1.8 m^3 (750 board feet) of ungraded, rough-sawn, random width, 31 mm thickness, 3 m length black locust boards were purchased from a sawmill in Newfield, NY. Also, a total of 24 NeLMA [14] finish grade, flat-sawn, eastern white pine boards with 22 mm thickness, 197 mm width, and 3.65 m length were purchased from a local lumber yard near Syracuse, NY. The eastern white pine boards had been kiln-dried and surfaced on four sides (S4S), but the black locust boards needed to be conditioned and surfaced (S4S) before fabrication. All the boards were then placed in an ambient temperature laboratory room to reach an equilibrium moisture content of approximately 12%. The average moisture content of the boards, determined with a Delmhorst (BD 2100) pin-type moisture meter, was measured as 11.8% at the time of fabricating the specimens. They were surfaced and planed to the desired thickness of 19 mm. Three replicates of each four-ply DCLT (±10°, ±20°, ±40°, and 70°) and one four-ply CLT (DCLT 90°) panel were fabricated to final dimensions of 76 mm thick, 30 cm wide, and 2.74 m long. Figure 4 illustrates the dimensions and the orientation of the panels. The width-to-thickness ratio of all the diagonal DCLT panels in this study followed the standard recommendations of 4:1 to limit the impact of rolling shear [15]. The white pine boards (angle plies) were cut to the desired size from straight grain and defect-free sections, eliminating the possible effects of knots on the results. However, because of the NeLMA finish grade's small knots and the difficult process of cutting the boards in different grain orientations, as shown in Figure 5a, minimal numbers of small knots were included in some of the boards. Before panel assembly, the longitudinal modulus of elasticity (MoE) of each black locust board (individual outer layer) and three eastern white pine sample boards was measured using the flatwise measured deflection of a simply supported board under a non-destructive center-point loading.

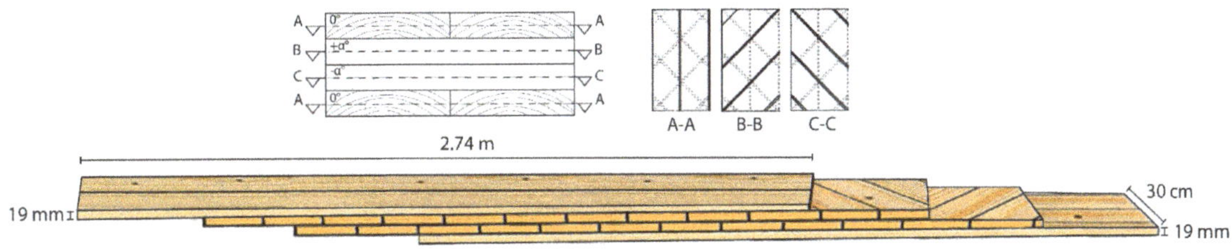

Figure 4. Dimensions and orientations of DCLT panels of the study.

Figure 5. From top left, (**a**) cutting of boards to desired grain orientation (representative sample at 10°), (**b**,**c**) DCLT panel's gluing process, and (**d**) DCLT panel's pressing process.

For panel fabrication, the boards were face-glued using Loctite HB X452 Purbond (Henkel Corporation, Bridgewater, NJ, USA) polyurethane adhesive with a recommended spread rate of 180 g/m^2 to glue the top and bottom of the lamina (face-glued) and without paying attention to the position of the board defects. The grain orientations of boards used in the inner layers were alternated in the second and third layers (Figure 4). The DCLT panels and the CLT panel clamping pressure was controlled by calibrated tightening torque set at 81.3 N.m (60 ft.lbf) per Equation (6) for torque-turn-tension control [16] to attain a minimum pressure of 410 kPa for 24 h, using a mechanical press and a pneumatic impact wrench.

$$T = K \times D \times F \tag{6}$$

where T is the required torque energy, K is the nut factor (depending on characteristics of the nut), D is the nominal diameter, and F is the desired force value. The nut factor is a variable determined by the nut geometry, the friction in the threads, and the friction between the nut, washer(s), and sealing surface. In this case, friction values were initially assumed for the well-oiled interfaces, and the actual clamping force was confirmed by placing pressure-sensitive indicating film (Fujifilm Prescale® Sensor Products Inc. Madison, NJ, USA) between the pressing platen and the CLT beam face.

The torque was applied until a specified "threshold" level was attained and an additional angle-of-turn was applied to finish the installation. Figure 5a–d show the panels' preparation process.

2.3. Four-Point Bending Test

Four-point load tests as outlined in ASTM D 198 [17] were performed on the panels using a Young universal testing machine outfitted with a 445 kN load cell with the crosshead speed of 0.001 mm/.min and two linear variable differential transducers (LVDTs). One LVDT was connected to the Young testing machine to measure the local displacement of load heads over a 0–150 mm range, and the other one was installed on the top center of the panel to measure the displacement of the top center of the panel relative to the load heads over a 0–75 mm range. The deflections from both LVDTs and loading data were recorded with LabView® Software (2020 Version) [18] on a dedicated personal computer. The total deflection was the sum of the two recorded deflections from LVDTs. The experimental setup is illustrated schematically in Figure 6 with the actual test set-up shown in Figure 7.

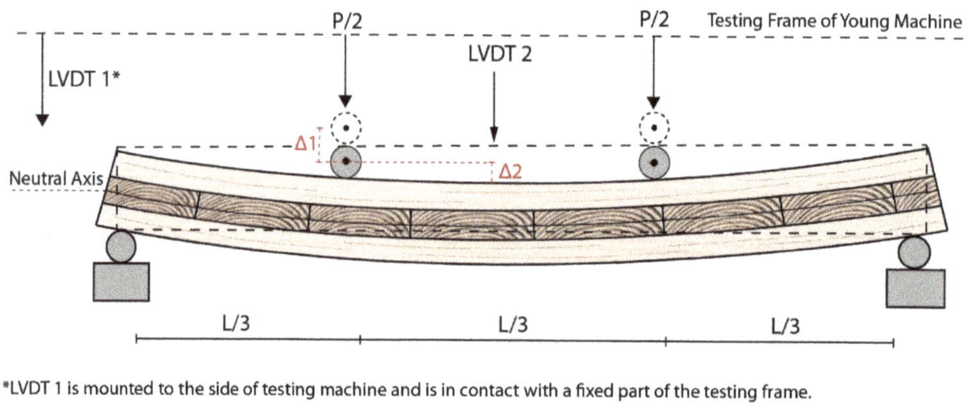

Figure 6. Schematic representation of testing arrangement and LVDTs' locations for DCLT bending specimens.

Figure 7. Actual testing arrangement for DCLT bending specimens.

The corresponding bending stiffness value from this test setup was acquired from Equation (7) ASTM D 198 [17]:

$$E_{app} = \frac{23 \times l^3}{108 \times b \times h^3} \times \frac{P}{\Delta} \quad (7)$$

where E_{app} is the apparent bending stiffness, P/Δ is the slope of the linear range of the load–deformation curve below the proportional limit with P and Δ accordingly being the maximum load and deflection at the proportional limit, l is the span length, and b and h are the width and the thickness of the specimen.

2.4. Finite Element Method

In order to further understand the bending performance of DCLT panels and the experimental results of the four-point test on the specimens, a numerical finite element (FE) model was developed using the research version of ANSYS Workbench 2022 [19] finite element software. The panels were modeled using two different linear elastic orthotropic element types, one for black locust in the top and bottom layers, and one for eastern white pine in the cross layers.

Nine independent elastic constants are required to define the mechanical response of an orthotropic material in the simulation. These constants are three elastic moduli (E), three Poisson's ratios (ν), and three shear moduli (G). For the material property values input in the simulations, the longitudinal elastic moduli values (E_L) for black locust and eastern white pine were obtained from the USDA Wood Handbook [20]. There are no available published values for the ratio of E_T/E_L, E_R/E_L, G_{TR}/E_L, G_{LR}/E_L, G_{TR}/E_L or the Poisson's ratios for the species used in this study.

The USDA Wood Handbook [20] provides elastic ratios with respect to the longitudinal elastic modulus for some softwood and hardwood species. However, there are no available published values or elastic ratios for the constants of E_R, G_{TL}, and E_T or the three required Poisson's ratios for eastern white pine. To predict these elastic constants using the available longitudinal (parallel to the grain) modulus of elasticity value, linear regressions (confidence level of 95%) were performed on each constant of the species in the USDA Wood Handbook [20] with available elastic ratios, regarding the longitudinal MoE as the predictor using Microsoft Excel [21]. The regressions showed R-squares ranging from 80% to 99% and a p-value less than the significance level (0.05) indicating a validation for prediction with a statistically significant correlation between the MoE and the response elastic constant [22].

The found ratios were repeated for another set of modeling considering the measured longitudinal MoE (from a simply supported bending test prior to panel fabrication) of black locust and eastern white pine boards as the longitudinal modulus of elasticity values. The final elastic constants for materials in FEM simulations are reported in Table 1.

Table 1. The ANSYS properties (black locust and eastern white pine) input for DCLT panels.

Species	Poisson's Ratio			Shear Modulus (MPa)			Young's Modulus (MPa)			Specific Gravity
	ν_{lt}	ν_{lr}	ν_{tr}	G_{lr}	G_{lt}	G_{rt}	E_t	E_r	E_l	SG
Black locust (Wood Handbook MoE)	0.286	0.220	0.196	1301	899	274	862	1697	14,134	0.69
Eastern white pine (Wood Handbook MoE)	0.233	0.223	0.220	624	592	76	467	766	8549	0.35
Black locust (Measured MoE)	0.286	0.220	0.196	1587	1096	334	1051	2070	17,236	0.69
Eastern white pine (Measured MoE)	0.233	0.223	0.220	548	520	67	411	673	7515	0.35

FEM ANSYS Workbench [19] simulations were performed under the "Static Structural" analysis system on one model of each DCLT $\pm$ 10°, DCLT $\pm$ 20°, DCLT $\pm$ 40°, DCLT $\pm$ 70°, and CLT specimens. The panels were modeled with the specifications for material, loading, and boundary conditions as shown in Figure 8 identical to the experimental four-point bending test to assure the accuracy of the comparison and validation.

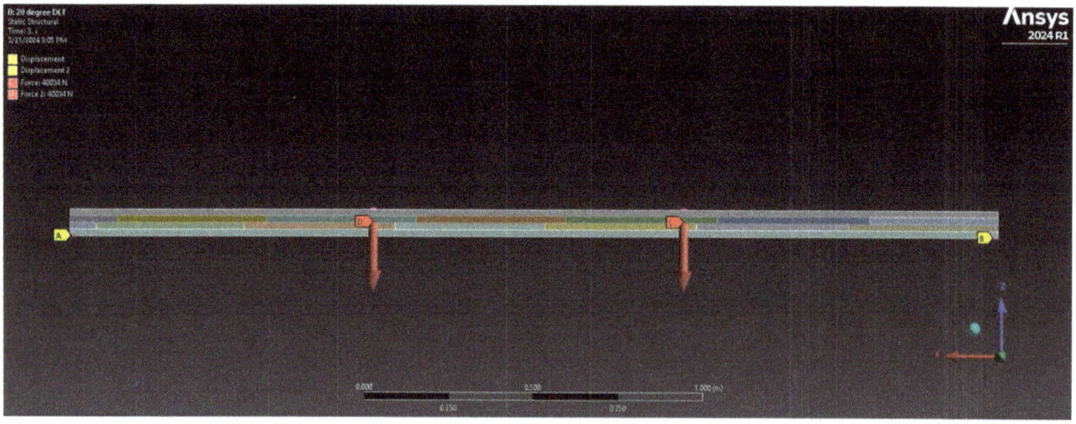

Figure 8. FEM modeling boundary condition for four-point bending test.

Mesh refinement as shown in Figure 9 was also performed under mechanical physics preference (program-controlled element order) by subdividing the mesh size to the rectangular elements with the size of 7 mm for each solid geometry. The mesh elements used for the sample specimens were higher order elements: SOLID186 element: 3D 20 nodes, TARGE170 3D target surface for the associated contact element of CONTA174: 3D 8 nodes surface-to-surface, and SURF154: 3D structural surface effect. Table 2 shows the details of the mesh sizing and the refinement quality of the model.

For each set of simulations, three coordinate systems were defined regarding the parallel (0°, global coordinate system) and two diagonal layers with $\pm\alpha°$ rotation of the main axis direction (x-y). The simulation's details and the details of different coordinate systems designed in the modeling are shown in Figure 10. After each simulation, load–deflection data were recorded to calculate the bending stiffness of each sample using Equation (7).

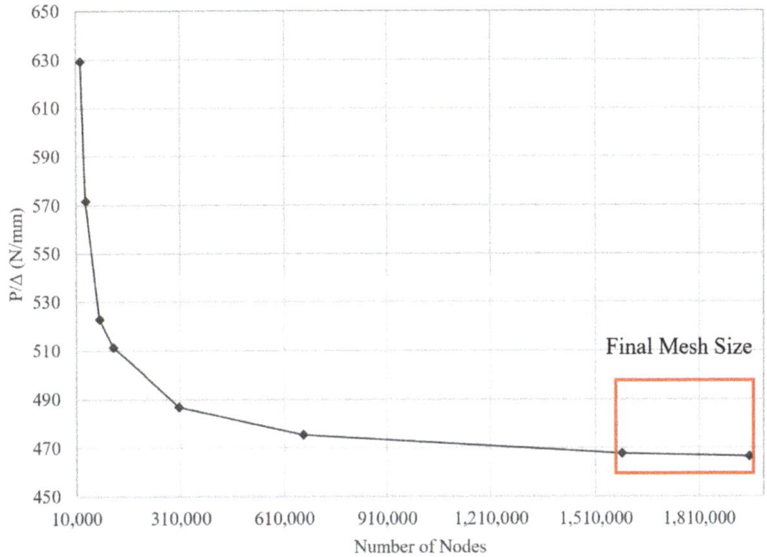

Figure 9. Mesh convergence for one DCLT representative sample.

Table 2. Mesh sizing and details for ANSYS modeling.

Group	Bounding Box Diagonal Size (cm)	Average Surface Area (cm^2)	Minimum Edge Length Size (cm)	Number of Nodes	Number of Elements
DCLT ± 10	276.93	803.67	0.96	2,111,626	413,142
DCLT ± 20	276.93	584.96	0.12	1,983,392	387,159
DCLT ± 40	277.08	383.61	0.43	1,466,501	283,758
DCLT ± 70	276.93	322.12	1.90	388,696	66,528
CLT	276.93	307.03	1.90	913,030	174,240

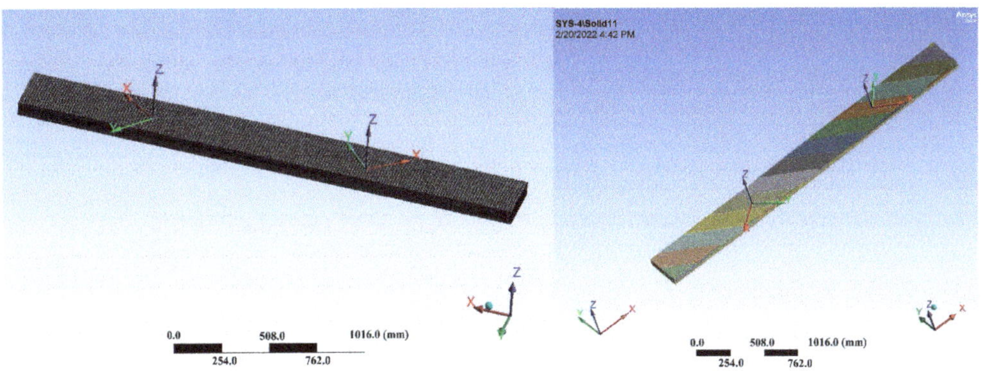

Figure 10. Representative samples of the finite element meshing and modeling showing details of relative mesh size and coordinate systems.

3. Results and Discussion

3.1. Hankinson's Theory Results

To evaluate the bending stiffness of the DCLT panels, the apparent bending stiffness has been calculated using the shear analogy approach. As mentioned in the Methodology section, Hankinson's equation has been used in the shear analogy approach to provide the stiffness properties of the angled inner layers of the panels.

Table 3 provides a summary of the bending stiffness values predicted by the shear analogy approach and the percent increase from each increment. Values for the theory use both the assumed MoE (i.e., USDA Wood Handbook [20]) and the measured MoE of the outer layers.

Table 3. Bending stiffness properties for different grain orientations obtained from Hankinson's theory.

EI_{App} (N-mm^2 × 10^{11})	**Grain Orientation**				
	10°	**20°**	**40°**	**70°**	**90°**
Handbook Values	1.45	1.39	1.31	1.21	1.18
MoE of the Layers	1.74	1.68	1.58	1.43	1.40
% Increase to CLT	23%	18%	12%	3%	--

3.2. Four-Point Bending Test Results

As shown in Figure 11, the slope of the load–deflection curves decreases, while the angle-ply orientation increases toward 90° or CLT. Therefore, the grain orientation of the inner layers is confirmed to affect the stiffness of the panels. Regarding the data from the testing shown in Table 4, the apparent bending stiffness of the panels is improved by 33% from CLT to DCLT ± 10°. This indicates that grain orientation had a statistically significant (p-value of 0.009) effect on the apparent bending stiffness of the DCLT panels. Table 4 also summarizes the difference percentages (%) between the bending stiffness properties of DCLT panels with different grain orientations in inner layers.

Table 4. Bending stiffness properties for different grain orientations obtained from experimental tests.

$EI_{App.}$ (N-mm^2 × 10^{11})	**Grain Orientation**				
	10°	**20°**	**40°**	**70°**	**90°**
Mean	2.28	2.14	1.95	1.76	1.71
% Increase to CLT	33%	25%	14%	3%	--
COV (%)	2.62	5.93	3.37	7.42	--
Min	2.23	2.00	1.90	1.67	1.71
Max	2.35	2.26	2.02	1.91	1.71

The beams were loaded to failure, and the first cracks started to form outside of the center third and closer to the ends due to rolling shear failure. This region provides the low secondary stresses originating from load heads or bearing supports, and with increasing load, the deformation propagates along the length of the panel toward the supports, where eventually the combination of the shear and bending stresses resulted in the final fracture.

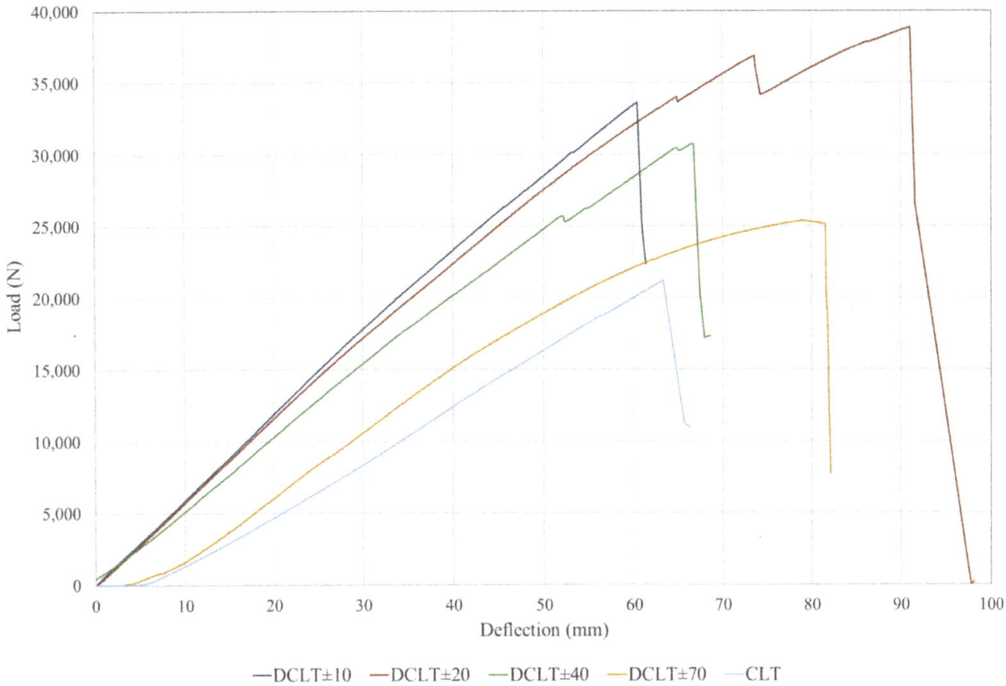

Figure 11. Load–deflection curves of DCLT ± α representative samples tested under four-point bending tests.

3.3. Finite Element Analysis

After each DCLT ± α sample was tested under the simulations, the deflection results from the four-point bending loading were recorded. As shown in Figure 12, the maximum deflection occurs in the center of the panel between the two loading areas. To demonstrate this process, the load–deflection curve for each sample model was recorded as shown in Figure 13, and slopes of the curves were used for measuring the bending stiffness of DCLT panels.

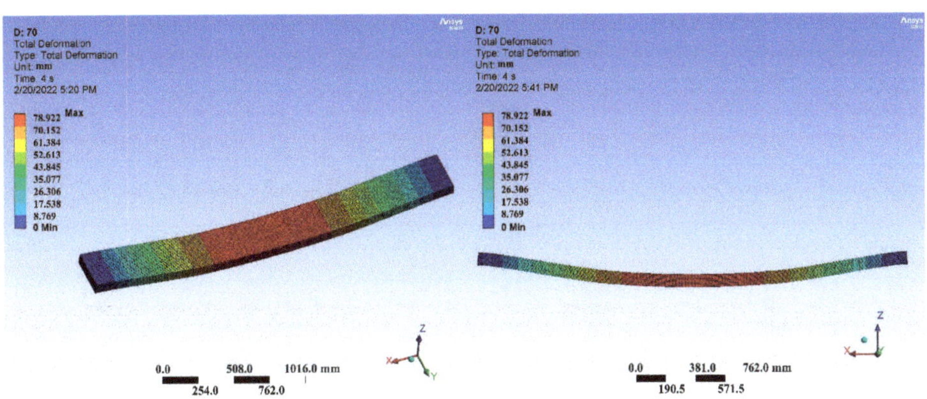

Figure 12. FEM test deflection results for a DCLT sample under applied load.

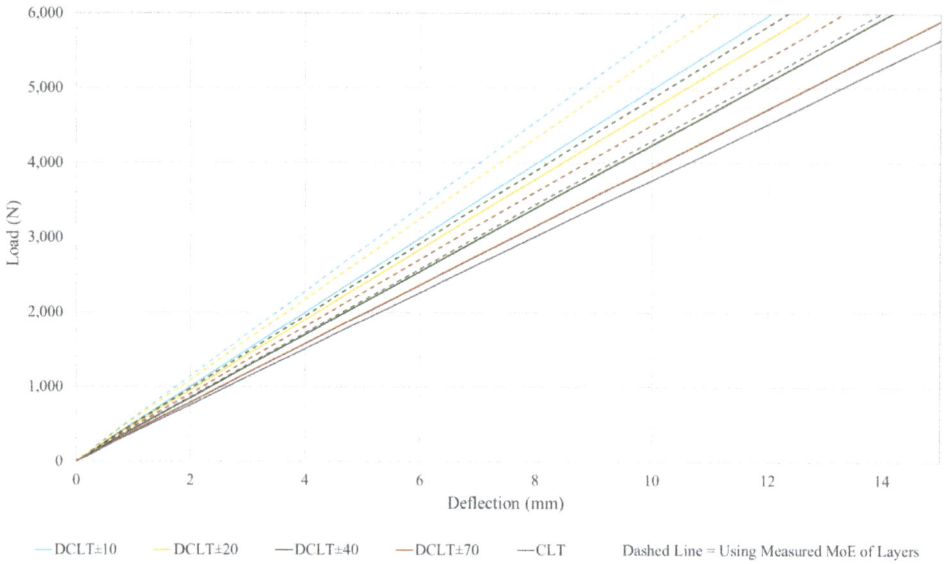

Figure 13. Load–deflection curves of DCLT $\pm$ α samples from FEM simulations.

The simulations were repeated one more time with the material properties of black locust and eastern white pine being considered with the measured MoE value instead of the published handbook values. As the results show, the simulated DCLT samples with the measured MoE have approximately 12–14% higher apparent bending stiffness than the results using the published E_L.

Regarding the data presented in Table 5, the FEM validates the experimental and theoretical results indicating that decreasing the angle of the grain orientation in the inner layers of the panels increases the bending stiffness. The finite element method data show that DCLT $\pm$ 10° has an improved apparent bending stiffness by approximately 35% (compared to 90° lay-up) when using either the USDA Wood Handbook [20] values or the measured MoE of the outer layers as the input for simulated models. Table 5 also summarizes the percent difference (%) between the bending stiffness properties of DCLT panels with different grain orientations in inner layers.

Table 5. Bending stiffness properties for different grain orientations obtained from FEM simulations.

EI_{App} ($N\text{-}mm^2 \times 10^{11}$)	Grain Orientation				
	10°	20°	40°	70°	90°
Handbook Values	1.94	1.85	1.65	1.53	1.46
MoE of the Layers	2.32	2.18	1.96	1.81	1.72
% Increase to CLT	33%	26%	13%	6%	--

3.4. DCLT's Bending Stiffness Comparison between the Methods

Considering all the assessed methods in this study, Figure 14 shows the final apparent bending stiffness results of each method for comparison. All the methods confirmed that decreasing the angle of the grain orientation in the inner layers of the panels increases the bending stiffness. This is due to the fact that by decreasing the inner lamina grain orientation, these layers will no longer be stressed under shear perpendicular to the grain but parallel to the grain (which is higher than shear perpendicular to the grain), which means the specimen then shows a higher bending stiffness. As the plots indicate, the highest values for all the grain orientations of DCLT panels and the CLT panel are from the results of FEM modeling using the measured MoE of the boards. This FEM result is very similar to the experimental results from four-point bending tests due to higher and more accurate elastic constant values as (or for) the material properties input. An obvious observation is that using the measured MoE of the layers provides more accurate models than using the reference MoE from the Wood Handbook when used in either numerical or theoretical modeling. The results from the theory (Hankinson's formula in shear analogy) under-predict the apparent bending stiffness The results from these experiments show approximately 50% higher values of apparent bending stiffness compared to the theoretical values from the Hankinson/shear analogy approach with the published USDA Wood Handbook reference values, which makes this method of prediction a conservative model requiring further investigations to provide a more accurate analytical model.

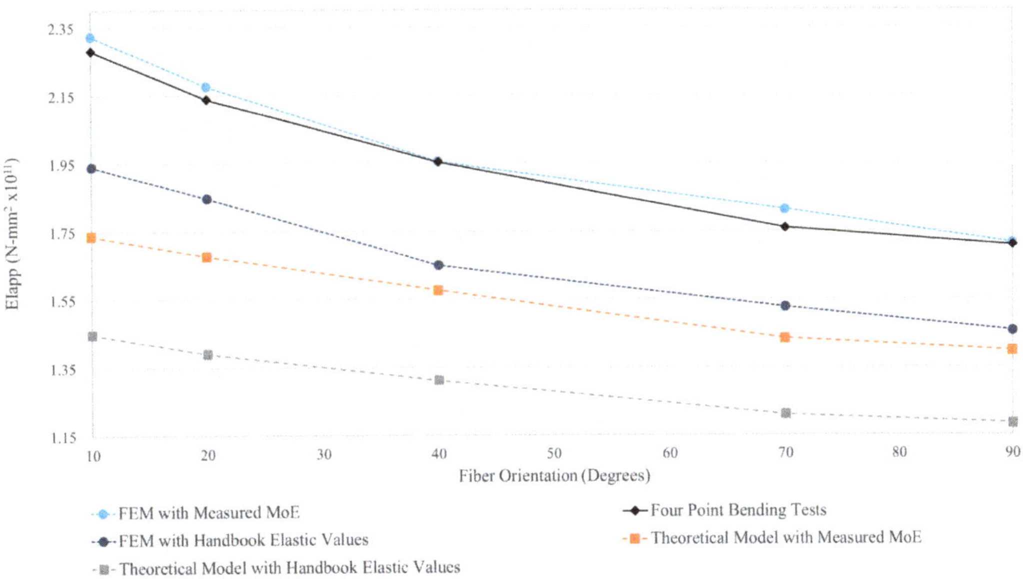

Figure 14. Apparent bending stiffness of DCLT/CLT panels predicted from four different methods.

This study is also in agreement with the previous literature on DCLT's structural advantages compared to conventional CLT. The current study shows an increase of 14% for the apparent bending stiffness from CLT to DCLT $\pm$ 40°, which is very similar to results from four-point bending tests on DCLT $\pm$ 45° panels conducted by Buck et al. [4], which showed an increase of 15% in the EI_{app} from CLT. However, the apparent bending stiffness increases from CLT to 30° (174%), 45° (112%), and 60° (61%) in Bahmanzad et al. [5]'s study, which was conducted using a short-span three-point bending test, which appears to be significantly different from what might be concluded to be the testing fixture and its application for measuring the shear stiffness rather than bending stiffness.

4. Conclusions

This study investigated the bending stiffness of four-ply Diagonal-Cross-laminated Timber (DCLT) panels consisting of five different grain orientations of 90° (CLT), 70°, 40°, 20°, and 10° in the cross layers. According to the four-point bending tests, the DCLT $\pm$ 10° panels were the stiffest followed successively by the DCLT $\pm$ 20°, $\pm$40°, $\pm$70°, and 90° (CLT) panels. These results support the modeled results acquired from Hankinson's input in shear analogy and FEM simulations. The results from the three methods agreed that the decrease in the angle of the inner layer grain reorientation would improve the bending stiffness properties of CLT panels. The apparent bending stiffness from DCLT 90° to DCLT ± 10° has an increase of 33%, 24%, and 35%, respectively, regarding the assessed methods of experimental, theoretical, and FEM modeling. This study indicates that DCLT panels with angle-ply cross layers have a structural advantage over traditional CLT (90°) panels. This study could be further expanded for investigation on other inner lamina orientations of DCLT products to propose an optimized grain orientation considering the structural and dimensional integrity of GLT and CLT. Using these panels would allow for increased spans or load-carrying capacity for a given panel span-to-depth ratio. The development of DCLT and its introduction to the industry could also enable the utilization of lower-quality timber that would not otherwise satisfy structural requirements for CLT. DCLT opens new possibilities for efficient material usage in massive timber construction. Additionally, this study validates methods of predicting the bending stiffness of DCLT and CLT panels with the available or measured elastic modulus parallel to the grain of the species used in the panel fabrication. This method of prediction could help utilize and introduce non-standard species for the fabrication of these products.

Author Contributions: Conceptualization, P.L.C.; Methodology, S.K.; Software, S.K.; Validation, P.L.C.; Formal analysis, S.K.; Investigation, S.K.; Resources, P.L.C.; Data curation, S.K.; Writing—original draft, S.K.; Writing—review & editing, P.L.C.; Visualization, S.K.; Supervision, P.L.C.; Project administration, P.L.C.; Funding acquisition, P.L.C. All authors have read and agreed to the published version of the manuscript.

Funding: This research was generously supported by a USDA Wood Innovation Grant, grant number: 20-DG-11094200-179.

Institutional Review Board Statement: Not applicable.

Informed Consent Statement: Not applicable.

Data Availability Statement: The data presented in this study are available in the article.

Conflicts of Interest: The authors declare no conflict of interest.

References

1. Bejtka, I. *Cross (CLT) and Diagonal (DLT) Laminated Timber as Innovative Material for Beam Elements;* Karlsruher Institut für Technologie (KIT) Scientific Publishing: Karlsruhe, Germany, 2011.
2. Arnold, M.; Dietsch, P.; Maderebner, R.; Winter, S. Diagonal laminated timber: Experimental; analytical, and numerical studies on the torsional stiffness. *Constr. Build. Mater.* **2022**, *322*, 126455. [CrossRef]
3. Kurzinski, S.; Crovella, P.; Smith, W. Evaluating the Effect of Inner Layer Grain Orientation on Dimensional Stability in Hybrid Species Cross-and Diagonal-Cross-laminated Timber (DCLT). *Mass Timber Constr. J.* **2023**, *6*, 11–16.
4. Buck, D.; Wang, X.A.; Hagman, O.; Gustafsson, A. Bending properties of Cross Laminated Timber (CLT) with a 45° alternating layer configuration. *BioResources* **2016**, *11*, 4633–4644. [CrossRef]
5. Buck, D.; Hagman, O. Mechanics of Diagonally Layered Cross-laminated Timber. In Proceedings of the World Conference on Timber Engineering WCTE 2018, Seoul, Republic of Korea, 20–23 August 2018.
6. Bahmanzad, A.; Clouston, P.L.; Arwade, S.R.; Schreyer, A.C. Shear Properties of Symmetric Angle-Ply Cross-Laminated Timber Panels. *J. Mater. Civ. Eng.* **2020**, *32*. [CrossRef]
7. Yoshito, M.; Hiroe, K.; Yukari, T.; Yasue, N. Relationship between the Strength and Grain Orientation of Wood: Examination and modification of the Hankinson's Formula. *Univ. Tokyo Grad. Sch. Agric. Life Sci. Exerc. Forest.* **1995**, *93*, 1–5.
8. Bahmanzad, A.; Clouston, P.L.; Arwade, S.R.; Schreyer, A.C. Shear Properties of Eastern Hemlock with Respect to Fiber Orientation for Use in Cross Laminated Timber. *J. Mater. Civ. Eng.* **2020**, *32*, 04020165. [CrossRef]
9. Franzoni, L.; Lebée, A.; Lyon, F.; Foret, G. Influence of orientation and number of layers on the elastic response and failure modes on CLT floors: Modeling and parameter studies. *Eur. J. Wood Wood Products.* **2016**, *74*, 671–684. [CrossRef]
10. Kreuzinger, H. Platten, Scheiban und Schalen. **1999**, *101*, 34–39. (In German)
11. Karacebeyli, E.; Douglas, B. *CLT Handbook—US Edition;* FPInnovations and Binational Softwood Lumber Council: Point-Claire, QC, Canada, 2013.
12. Kurzinski, S.; Crovella, P.L. Predicting the Strength and Serviceability Performance of Cross-Laminated Timber (CLT) Panels Fabricated with High-Density Hardwood. In Proceedings of the World Conference on Timber Engineering, WCTE 2021, Santiago, Chile, 9–12 August 2021.
13. American Wood Council's (AWC) Wood Design Standards Committee. *National Design Specification (NDS) for Wood Construction;* AWC: Leesburg, VA, USA, 2018.

14. American Lumber Standard Committee (ALSC). *Northeastern Lumber Manufacturers Association, Standard Grading Rules for Northeastern Lumber*; Northeastern Lumber Manufacturers Association: Cumberland Center, ME, USA, 2021.
15. *ANSI-APA/PRG 320*; Standard for Performance-Rated Cross-Laminated Timber. American National Standard Institute & APA—The Engineered Wood Association: New York, NY, USA, 2019.
16. Shoberg, R.S. *Engineering Fundamentals of Threaded Fastener Design and Analysis*; RS Technologies, a Division of PCB Load & Torque, Inc., 2000; pp. 1–39. Available online: http://www.hexagon.de/rs/engineeringfundamentals.pdf (accessed on 5 March 2024).
17. *ASTM D 198-15*; Standard Test Methods of Static Tests of Lumber in Structural Sizes. American Society for Testing and Materials Committee: West Conshohocken, PA, USA, 2015.
18. National Instruments Corporation. *Software Recording the Test*; LabVIEW 2020 Version: Austin, TX, USA, 2021.
19. ANSYS Inc. *ANSYS®Workbench*; ANSYS Inc.: Canonsburg, PA, USA, 2022.
20. Forest Products Laboratory. *USDA Wood Handbook—Wood as an Engineering Material*; General Technical Report FPL-GTR-190; Forest Products Laboratory: Madison, WI, USA, 2021. [CrossRef]
21. Microsoft Corporation. *Microsoft Excel*; Microsoft Corporation: Redmond, WA, USA, 2021.
22. Kurzinski, S.; Crovella, P.L. Theoretical and Experimental Investigation on Predicting Longitudinal and Tangential Elastic Constants and Ratios of Wood. In Proceedings of the World Conference on Timber Engineering, WCTE 2023, Oslo, Norway, 19–22 June 2023. [CrossRef]

applied sciences

MDPI

Article

Use of Carbon and Basalt Fibers with Adhesives to Improve Physical and Mechanical Properties of Laminated Veneer Lumber

Mario Núñez-Decap [1,2,*], Bárbara Sandoval-Valderrama [1], Camila Opazo-Carlsson [1], Boris Moya-Rojas [1], Marcela Vidal-Vega [1] and Alexander Opazo-Vega [1,2]

[1] Department of Civil and Environmental Engineering, Universidad del Bío-Bío, Avenida Collao 1202, Concepción 4051381, Chile

[2] Centro Nacional de Excelencia para la Industria de la Madera (CENAMAD), Pontificia Universidad Católica de Chile, Vicuña Mackenna 4860, Santiago 7820436, Chile

* Correspondence: mnunez@ubiobio.cl; Tel.: +56-41-311-1968

Abstract: Climate change is one of the main factors influencing the research of environmentally friendly materials. This is why the use of engineering fibers as a reinforcement technique in wood, in order to increase its mechanical properties, has recently been investigated. This research presents the results obtained from the use of carbon and basalt fiber fabrics as a reinforcement for microlaminated Radiata Pine wood panels at a laboratory scale using the adhesives epoxy resin and polyvinyl acetate. Tests were carried out in comparison to the control boards, relating the physical properties obtained in terms of thickness swelling by 48 h-water immersion with a decrease of 19% for the polyvinyl acetate and carbon fiber matrix reinforcement, about the mechanical properties evaluated, a better performance was obtained for the epoxy resin and carbon fiber matrix reinforcement and in terms of flexural stiffness and strength (in flatwise), tensile strength and Janka hardness, with an increment of 31%, 38%, 56% and 41%, respectively.

Keywords: laminated veneer lumber; basalt fiber; carbon fiber; epoxy resin; polyvinyl acetate

1. Introduction

In engineering, specifically in the field of construction, the aim is to use raw materials of the highest quality and at an affordable cost that allows for efficient construction. However, the materials commonly used in this sector (such as steel, concrete, and brick) emit more greenhouse gases than a design made of wood. Replacing these materials with wood could reduce global CO_2 emissions by 14–31% and global fossil fuel consumption by 12–19%, making a key contribution to the fight against climate change [1–3].

Solid wood is characterized by the presence of cracks, knots, growth stress, and limited length [4]. This is why wood composites were introduced, and among them, the creation of Laminated Veneer Lumber (LVL) arose from the need to upgrade solid wood's mechanical properties. LVL is an engineered wood-based product manufactured from veneers that are rotary peeled, dried, and laminated together with parallelly oriented grains under heat and pressure with a waterproof adhesive and whose veneer grain is usually oriented in a single direction, although cross-grained sections are also manufactured to offer tailored mechanical properties. Applications include beams, columns, headers, joists, rafters, scaffold planks, and truss chords [5–7]. LVL is a reconstituted dimensional timber that is usually twice the strength of dimensional timber of the same species manufactured from rotary peeled veneers of spruce, pine, or Douglas fir of 3 mm thickness; in general, their mechanical properties are more uniform compared with solid timber [5,8]. However, microlaminated boards produced from these tree species have low mechanical properties, and they attain good dimensional stability, good homogeneous mechanical properties, and great durability [4,7]. In order to generate an increase in LVL physical-mechanical and thermal properties and a high structural application, the addition of engineered fiber fabrics, such as carbon or basalt fiber, as reinforcement for LVL boards can be a great opportunity.

Nowadays, carbon fiber (CF) and basalt fiber (BF) are typical reinforcing materials in polymer composites, and recently, BF has been subjected to more investigation [9]. The advantages of carbon fiber (CF) compared to other reinforcement fibers are its low density, high tensile strength, stiffness, and chemical resistance [10]. Carbon fiber is a great option to use as reinforcement in structural terms, as well as in aerospace, automotive, and wind energy sectors, among others. Despite its great advantages, CF's main problem is a high production cost, which creates a considerable advantage for basalt fibers (BF), whose price is one-third that of carbon fiber. Among the benefits of basalt fiber are that it has high thermal resistance, high thermal insulation, high corrosion resistance and is environmentally friendly [10,11].

To generate the adhesion of the reinforcement fibers with wood, two types of adhesives will be studied. The first one is epoxy resin (ER), a thermosetting adhesive of high rigidity, resistance to chemicals, low shrinkage, and excellent thermo-mechanical characteristics [12]. The second one corresponds to a thermoplastic adhesive: Bicomponent Polyvinyl acetate (PVAc) type D4. Among its advantages are resistance to humidity under controlled environmental conditions, its price compared to epoxy resin, faster solidification, easier handling, cleanability, and lower environmental impact [13]. This last feature is related to the fact that, currently, most Fiber Reinforcement Polymers (FRPs) use epoxy or polyester resins as a matrix [14], but there are other adhesives that are less harsh to the environment than conventionally used resins, such as the PVAc adhesive that could substitute the polymeric matrix of FRP [13].

Since FRP materials have become popular, several authors have studied the reinforcement of different construction materials with engineering fibers, for example, in concrete mixtures [15], coral sand cement mortar [16], geopolymer concretes [17], geopolymers [18], oil well cement [19], solid wood [20,21], vintage wood [22], wood beams [23] and others, to improve their mechanical properties, restoring construction materials and finding environmentally friendly reinforcements.

Specifically in the case of wood-based panels, in 2023, Zhang et al. [24] proposed a hollow glulam beam design with stiffening plates in a hollow rectangular cross-section reinforced at the bottom with carbon and basalt reinforcement fibers. In 2022, Núñez-Decap et al. [11] reinforced plywood with carbon and basalt fiber combined with epoxy resin and polyvinyl acetate adhesives to enhance their physical and mechanical properties. During the same year, the authors Rescalvo et al. [25] also reinforced LVL with carbon and basalt fibers with polyurethane adhesive to enhance their mechanical properties under shear and compression stresses. In 2021, Gallego et al. [26] also reinforced LVL panels with carbon and basalt fibers to improve their bending properties, but in this case, the reinforcement fibers were placed between the last and the penultimate veneer of each side of the board. In 2021, Wdowiak-Postulak and Swit [27,28] studied the reinforcement of pine beams made from glued laminated timber with subsurface basalt fibers and basalt fiber reinforcement of bent heterogeneous glued laminated beams. In 2018, carbon fiber-reinforced LVL beams were tested to improve their bending properties by Globa et al. [29]. The effect of grain orientation on the carbon fiber reinforcement polymer (CFRP) to LVL bond has also been studied in order to analyze its behavior by Subhani et al. 2017 [30].

The main problem the present research tries to solve relates to the physical-mechanical low performance of wood veneers compared to the primary construction materials currently used, such as steel and concrete.

Given the foregoing, the main opportunity to be addressed in this research is to improve the physical-mechanical performance of LVL panels to structural applications through the incorporation of high-performance engineering fibers as reinforcement. Although PVAc adhesive does not have structural applications, we seek to compare it to epoxy resin to see if PVAc could be a non-toxic alternative to reinforcing LVL panels.

2. Materials and Methods

2.1. LVL Panels

The Laminated Veneer Lumber (LVL) panels were fabricated with five flawless veneers of *Pinus radiata* D. DON (with an average density of 456 kg/m^3 and an average moisture content of 6.90%) and bonded with Phenol Formaldehyde (PF) adhesive. The veneers' dimensions were 500 mm length × 500 mm width and a 2.2–3.6 mm thickness range. The difference in veneer thicknesses is due to raw material availability of common thicknesses used in Chile; however, all the samples were manufactured with the same configuration of veneer thicknesses (the thinner ones located towards the board surfaces, and the thicker ones were located in the center). The veneers used were donated by PRODIMA-LAB (Concepción, Chile).

2.2. Reinforcement Fibers

The engineering fibers used in the reinforcement of LVL panels were carbon fiber (CF) and basalt fiber (BF), both provided by Aura Industrial (Santiago, Chile). The fibers' properties, according to their data sheets, are listed in Table 1.

Table 1. Properties of the reinforcement fibers used by Aura Industrial data sheets [31].

Properties	Carbon Fiber	Basalt Fiber
Fabric	Bidirectional weaving	Bidirectional weaving
Thickness (mm)	0.28	0.18
Grammage (g/m^2)	200	200
Tensile Strength (MPa)	≥3500	≥1000

2.3. Adhesives

The fiber reinforcement polymer (FRP) panels FRP-LVL, were glued with polyvinyl acetate adhesive (PVAc-X016 D4 Franklin) supplied by SlipNaxos Chile S.A. (Santiago, Chile) and epoxy resin (ER) with its hardener (Aura 280 and its hardener Aradur 963) supplied by Aura Industrial and Comercial SpA (Santiago, Chile).

Two types of adhesives were used on each board because the manufacturer of the reinforcing fibers recommends using epoxy resin to adhere them to the wood veneer, and in search of a non-toxic replacement, it was decided to try PVAc.

2.4. Characterization of Adhesives

The adhesive viscosity was measured according to ASTM D1084-16R21 [32] with a Brookfield viscosimeter model LVF with a spindle N°6, 150 rpm to PF adhesive, and 200 rpm to PVAc adhesive.

The PF and PVAc adhesives' pH was measured according to ASTM E0070-19 [33] with a Hanna Instruments pH-meter model 211. The epoxy resin pH was measured with a pH-paper test. The electrical conductivity of the adhesives was measured with a multiparameter tester. And the adhesive's solid content was measured according to ASTM D1490-01R18 [34].

2.5. Reinforcement of LVL Panels

Three different types of LVL panels were configured, as shown in Figure 1. The first one corresponds to the control sample, composed of 5 veneers glued with phenol formaldehyde adhesive. The second one corresponds to the carbon fiber-reinforced LVL panel (CFRP-LVL), which have two layers of CF, one between the first and the second veneer and other one on the opposite surface (tension side), and the third panel corresponds to the basalt-reinforced LVL panel (BFRP-LVL), which have two layers of BF, one between the first and the second veneer and other one on the opposite surface (tension side). The purpose of the board being reinforced by only one surface is to allow the user to decide whether or not to leave the fiber exposed.

Figure 1. The control LVL panel, carbon fiber reinforced LVL, and basalt fiber reinforced LVL panel were configured as follows: (**a**) Control LVL panel; (**b**) LVL panel reinforced with carbon fiber; and (**c**) LVL panel reinforced with basalt fiber.

Both reinforcement fibers, CF and BF, were impregnated on both sides with ER and PVAc adhesives, forming different samples, while the rest of the veneers were glued with PF adhesive. The difference in the grammage used is due to the different solids content of the adhesives and also to the manufacturer's recommendations. (The experimental design is presented in Table 2).

Table 2. Design of experiment.

ID	N° LVL	Adhesive Grammage (g/m^2)
P	4	200
ER-CF	4	250
ER-BF	4	250
PVAc-CF	4	400
PVAc-BF	4	400

Once properly prepared, the panels were subjected to two pressing steps; the first step corresponded to high-temperature pressing, and the second step to cold pressing. The hot-pressing step consisted of subjecting the panels to a specific temperature of 130 °C, using a three-stage pressing cycle, as illustrated in Figure 2. The objective of the first cycle is to generate adhesion between the wood and the adhesive, the second cycle seeks to consolidate the adhesive, and the third cycle is to prevent the pressure decompensation from being abrupt and producing the breakage of the bonding chains (blowing).

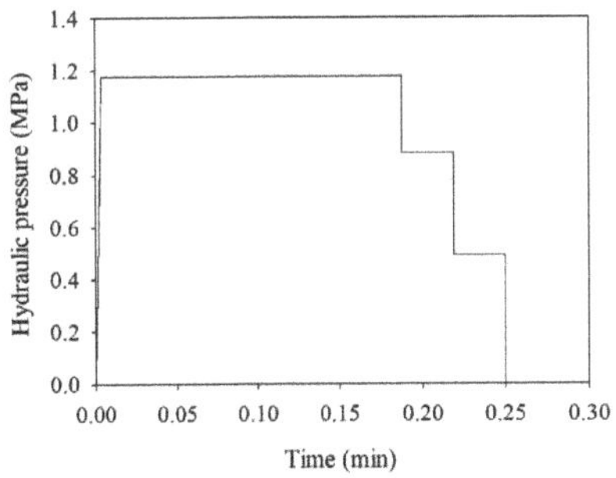

Figure 2. Three-stage pressing cycle for both hot and cold pressing.

After hot pressing, the panels were immediately subjected to the cold pressing stage at an ambient temperature of 20 °C, under the same pressure and time conditions of the first pressing step (Figure 2). The purpose of cold pressing is to accelerate the adhesive curing process.

After pressing, the LVL panels were conditioned for 7 days at 23 ± 3 °C and 65% relative humidity. Then, the LVL panels were formatted in 480 mm length × 480 mm width and a dimension range of 14.7–15.6 mm thickness and finally cut to the dimensions of each specimen according to European and American standards to determine their physical, mechanical, and thermal properties.

2.6. Physical Properties

Initially, density was assessed in a total of 40 specimens, with each specimen measuring 50 mm in length and 50 mm in width. The evaluation was conducted in accordance with the European standard UNE-EN 323 [35]. The Amersham plc AMCK6693 gamma-ray densimeter was employed to measure the density profile of the panel's specimens. The densimeter scanned directly through the sample's thickness with an incremental step of 0.1 mm. A total of 5 specimens from each LVL sample, measuring 50 mm in length and 50 mm in width, were subjected to density profiling.

To determine the moisture content, 20 specimens were assessed following the guidelines of the European standard UNE-EN 322 [36]. Furthermore, thickness swelling was evaluated in 40 specimens, each measuring 50 mm in length and 50 mm in width, by the conditions specified in the European standard EN 317 [37].

2.7. Mechanical Properties

2.7.1. Static Bending Tests

The bending properties of both the control and reinforced LVL panels were assessed through mechanical tests utilizing a four-point static bending setup. The positions tested included flatwise (FW) (oriented as shown in Figure 1) and edgewise (EW), following the specifications and methodology outlined in the European standard UNE-EN 408 [38]. For each LVL sample, a total of 10 specimens were subjected to testing.

In the flatwise position, the specimens measured 450 mm in length, 50 mm in width, and 15 mm in thickness. In the edgewise position, the specimens had dimensions of 450 mm in length, 20 mm in width, and 15 mm in thickness. The observed failures in the specimens were carefully noted and classified, adhering to the guidelines specified in the European standard EN 310 [39].

The bending properties were determined using an Instron 100-23 universal testing machine, which was equipped with BlueHill2 software for data analysis and recording.

2.7.2. Tensile Strength

A tensile strength test was conducted on both the control and reinforced LVL panels. A total of 12 specimens per sample were tested, following the dimensional requirements outlined in the American standards ASTM D3500—Method A [40].

2.7.3. Hardness Test

The Janka hardness test was carried out on both the control and reinforced boards. A total of 6 specimens per sample were used for the test, and they had dimensions of 76 mm in length and 52 mm in width, as specified by the American standard ASTM D1037-12 [41]. The hardness value per specimen corresponds to the average of 4 indentations, two on each side.

To measure the mechanical properties, a universal testing machine, specifically the Instron 100-23, was employed. The testing machine was equipped with BlueHill2 software, which facilitated the analysis and recording of the test results.

2.7.4. Thermal Properties

The thermal conductivity test was conducted on both the control and reinforced LVL boards. A total of 5 specimens were used for the test, and they had dimensions of 100 mm in length, 50 mm in width, and 15 mm in thickness, according to the specifications provided by the American standard ASTM D5334 [42].

The Decagon KD2 PRO equipment was utilized to perform the thermal conductivity measurements on the specimens.

2.7.5. Analysis of Data

After obtaining the results, an ANOVA was conducted to assess the differences between the averages. In this analysis, significance was accepted at a level of $p < 0.05$, indicating that the observed differences were considered statistically significant.

When the differences between the results of each tested specimen were found to be statistically significant, a multivariate analysis LSD test was employed. This test helps to identify specific pairwise differences between the results of the specimens. The Statgraphics Centurion 19 software was used to perform this analysis.

The outcomes of the data analysis, including the results of the ANOVA and the relevant tables indicating the statistically significant differences, are documented and marked accordingly.

3. Results and Discussion

3.1. Characterization of Adhesives

The characteristics of adhesives used in LVL production have an important role in the quality of the final product, which can be established based on the physical and mechanical properties of the boards [43].

The characterization of the adhesives used is presented in Table 3. According to the viscosity values of each adhesive, it can be seen that ER and PVAc viscosity values are very similar and around 75% lower than PF viscosity. These results are related to the penetration depth that each adhesive can achieve [44]; in this case, PF-wood penetration was more limited, which can also be related to a great bond quality.

Table 3. Characterization of adhesives.

ID	Viscosity (cP)	pH	Electrical Conductivity (mS)	Solid Content (%)
PF	4827	12.53	24.23	42.27
ER	1200	7.50	-	93.88
PVAc	1240	2.76	13.89	52.06

The pH has a significant influence on the glue-line formation as it affects the rate of adhesive curing [45]. Regarding the pH level results, PF and ER adhesives are in an alkaline condition, while PVAc is in an acidic condition. Anyways, the pH of the adhesives should not exceed the range of 2.5–11 because, beyond these limits, the resin causes degradation of the fibers of the wood, especially in the presence of moisture [46], a situation that could happen to veneers glued with PH adhesive. Normally, a low pH affects predominantly holocellulose, while a high pH affects mainly the lignin content of the wood. Therefore, a low pH may have a more negative effect on wood strength than a high pH.

PF adhesive is more conductor electrically than PVAc adhesive; this is probably because PF adhesive has more additives like fillers, which offer benefits like increased viscosity, better mechanical properties, increased electrical conductivity, achievement of higher thermal and dimensional stability, and others [47,48].

Regarding the content of solids, PF adhesive has the least solids in its composition, PVAc adhesive has almost half of the solids in its composition, while almost the entire composition of ER adhesive corresponds to solids; in other words, it has almost no liquid in its composition. When the panel is subjected to hot pressing, evaporation of the liquid component (cure), that is, the solidification of the adhesive, forms the glue line that is responsible for the bond between the substrates [49], so as the epoxy resin has a higher content of solids, in comparison to the other adhesives characterized, it may present a better bond quality.

3.2. Physical Properties

The FRP-LVL exhibited a higher density compared to the control panel, as indicated in Table 4. The density increase ranged from 6.23% for the ER-CF sample, with a density of 563 kg/m^3, to 10.38% for the PVAc-CF sample, with a density of 585 kg/m^3. However, these differences in density between the FRP-LVL samples were not found to be statistically significant.

Table 4. Results of physical properties.

ID	Density (kg/m^3)		Moisture Content (%)		Thickness (mm)		Thickness Swelling (%)			
							24 h		48 h	
	Mean	St. Dev.	Mean	St. Dev.	Mean	St. Dev.	Mean	St. Dev.	Mean	St. Dev.
P	530 ^{a}	27.29	8 ^{a}	0.64	14.33 ^{a}	0.31	1.26 ^{a}	1.81	5.13 ^{ab}	0.93
ER-CF	563 ^{ab}	43.13	10 ^{b}	0.82	14.97 ^{b}	0.44	0.74 ^{a}	1.22	4.43 ^{bc}	0.45
ER-BF	570 ^{b}	31.81	11 ^{c}	0.28	14.41 ^{a}	0.20	0.66 ^{a}	1.88	5.32 ^{a}	0.41
PVAc-CF	585 ^{b}	14.14	12 ^{d}	0.14	15.03 ^{b}	0.23	0.60 ^{a}	1.70	4.16 ^{c}	0.64
PVAc-BF	568 ^{b}	41.92	12 ^{cd}	0.40	14.20 ^{a}	0.24	0.00 ^{a}	0.00	4.83 ^{abc}	1.03

Column means are indicated with different letters (a, b, c, and d) to indicate homogeneous groups from the multiple-range test by Fisher's LSD method. This means that different letters are statistically different from each other at a 95% confidence level.

Furthermore, the density profile of the FRP-LVL samples revealed variations, particularly with a noticeable peak on the surface where the carbon and basalt fibers were positioned. This suggests that the distribution of fibers impacted the density profile of the FRP-LVL panels. Basalt fiber-reinforced samples revealed the highest density peaks, but in smaller thickness, while carbon fiber-reinforced samples revealed the lowest density peaks, but in larger thickness (Figure 3). This is consistent with the fiber's properties presented in Table 1 and with [50], who says that BF density is higher than CF density. The difference in thicknesses observed in Figure 3 is due to the thicknesses of the engineering fibers, glue lines, and small differences between the veneers themselves.

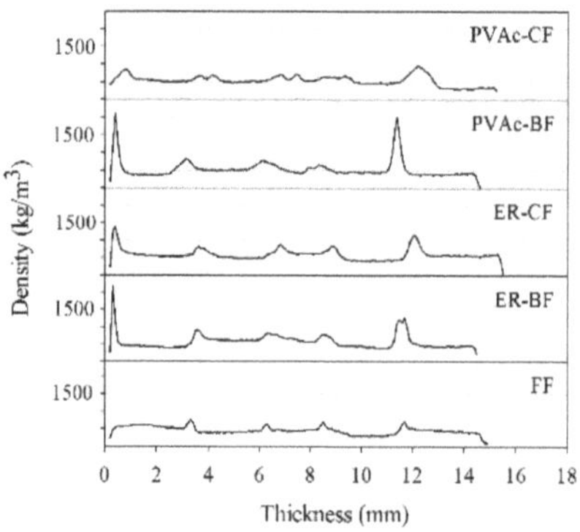

Figure 3. Density profile of the composite panels.

The density values of the FRP-LVL were found to be higher in comparison to LVL. This can be attributed to the incorporation of reinforcement fibers such as BF and CF, as well as the use of adhesives ER and PVAc, which have higher density values than wood. However, it was observed that there was no direct correlation between the type of reinforcement fiber or adhesive employed and the extent of density increase in the tested samples.

The moisture content (MC) values of the FRP-LVL samples were higher than the MC value of the control sample. Aside from the factors that are inherent to the wood itself, aspects related to the manufacturing processes of LVL influence its equilibrium MC [43]. All of the samples exhibited moisture content within the expected range for conditioning, ranging between 8% and 12%. This indicated that the samples were appropriately conditioned, and their moisture content levels were within the desired specifications.

The thickness swelling in 24 h of water immersion did not vary between the samples, but in 48 h, better dimensional stability was evidenced by the sample PVAc-CF (4.83%), which decreased by 18.91% compared to the control sample (5.13%). For the rest of the samples, there was no statistically significant difference, so in general, there was no difference in the dimensional stability between the FRP-LVL samples and the control sample, except for the sample PVAc-CF. The results of this property correspond to a percentage variation concerning the initial thickness (according to the standard indicated in the methodology).

3.3. Mechanical Properties

3.3.1. Strength and Stiffness Bending Properties

In general, the FRP-LVL samples glued with ER adhesive presented the highest results in stiffness (MOE) and strength (MOR).

In the flatwise position (Figure 4), the sample ER-CF (MOE: 8190 MPa; MOR: 86.27 MPa) presented an increase, about 30.53% and 37.61%, in MOE and MOR, respectively. For the rest of the samples, there were no statistically significant differences in comparison to the control sample.

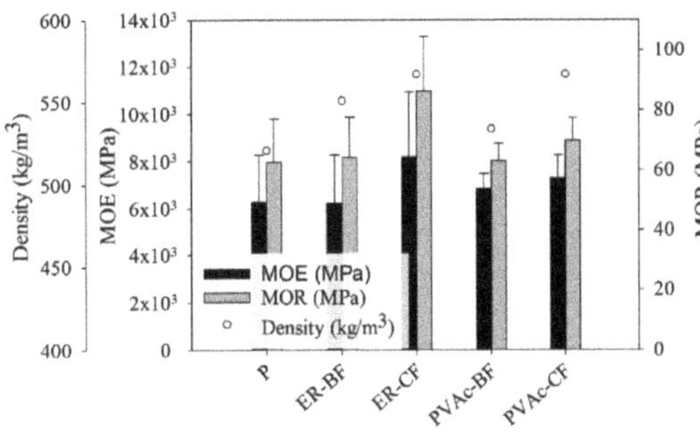

Figure 4. Results of strength (MOR) and stiffness (MOE) bending properties and density of each sample in the flatwise position (error bars indicate standard deviations).

In the edgewise position (Figure 5), there was an increase in MOE values, between ~14% and 35.93%, in the samples ER-CF; PVAc-CF (9626 MPa; 9754 MPa) and ER-BF (11,476 MPa), respectively. The MOR of the sample ER-BF (76.08 MPa) increased by 28.65%, while for the rest of the samples, in the MOR case, there were no statistically significant differences in comparison to the control sample.

In the investigations of Wang [51] and Rescalvo [25], CFRP-LVL samples presented higher values in bending properties compared to an unreinforced sample, which generally occurred in this investigation in MOE results. Furthermore, in the investigation of Wdowiak-Postulak [28], BFRP-glulam beam samples also presented higher values in bending properties compared to an unreinforced sample, which occurred in edgewise samples of this investigation.

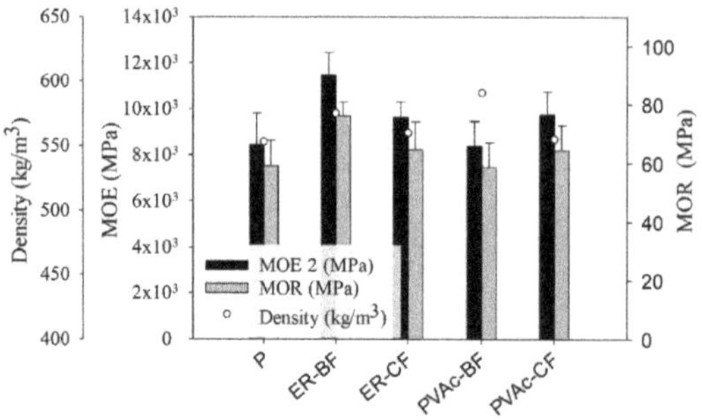

Figure 5. Results of strength (MOR) and stiffness (MOE) bending properties and density of each sample in the edgewise position (error bars indicate standard deviations).

The bending test results indicated that the samples glued with ER have a higher stiffness and strength compared to the samples glued with PVAc adhesive. This might be associated with the ten-times-higher MOE of ER than that of wood [52], but also the cure temperature needed for each adhesive, because while PVAc was hot pressed, it could have lost its mechanical properties by the water evaporation and its less solid content percentage. On the other hand, thermosetting polymers of ER make excellent structural adhesives because they undergo irreversible chemical change when cured [53], and thermoset adhesives generally are stronger than thermoplastic adhesives (PVAc) [49]. Anyways, good results of ER adhesive in the matrix are not consistent with the ecological perspective of this study.

The highest bending test stiffness (MOE) results were obtained by the EW position samples, while the highest strength (MOR) results were obtained by the flatwise position samples. This result can be attributed to veneer classification and board assembly that prioritized more resistant veneers for the external layers of LVL [43]. Furthermore, all the samples tested mainly had tensile failure. However, in several cases, shear failure was observed, especially in the flatwise position tested samples. The reason lies in the fact that the flexural strength contribution of the FRP exceeds the shear strength of the veneer itself, causing this type of failure [25]. Thus, the flatwise position tested samples did not reach higher values than the edgewise position in MOE results.

The bending test load-deflection curves of the flatwise and edgewise positions samples demonstrated that the FRP-LVL panels exhibited increased maximum load and maximum deflection when compared to the LVL panels (Figures 6 and 7); these results are consistent with the results of the authors Bakalarz and Kossakowski [54] who also obtained higher maximum load and deflection values for FRP-LVL compared to the control LVL.

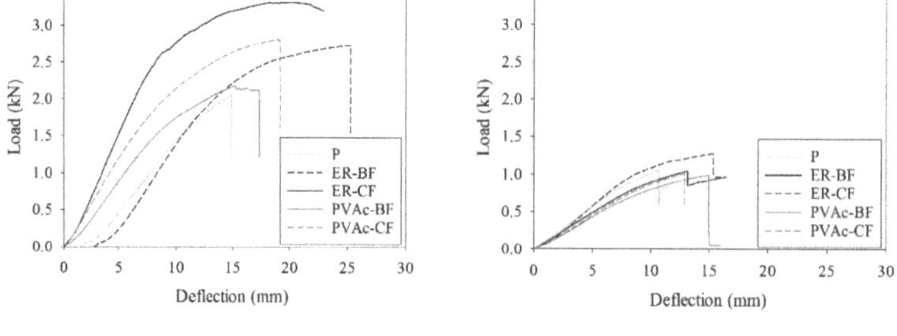

Figure 6. Load-deflection curves were obtained during the four-point bending test for the samples in the (**a**) flatwise position and in the (**b**) edgewise position. A characteristic curve for each sample was used as a representation of the behavior of each one.

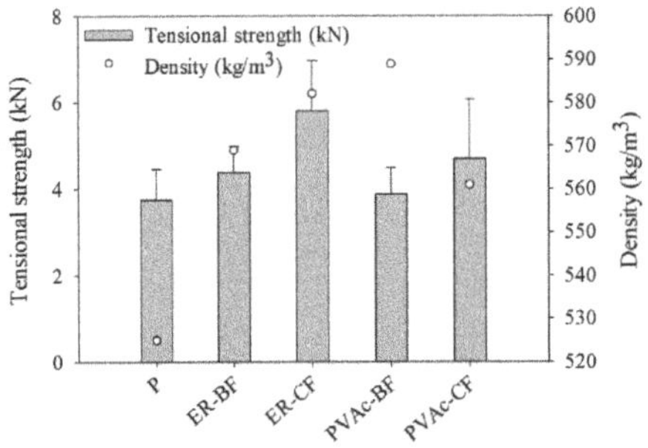

Figure 7. Results of tensile strength and density of each sample (error bars indicate standard deviations).

In both cases—flatwise and edgewise position—samples seem to present a more ductile performance than the control sample, especially samples glued with epoxy resin in the FW position. Results were also evidenced in the investigation of Rescalvo et al. [25] and probably obtained due to the high stiffness and strength properties of epoxy resin [12]. In the case of the EW position, there is not a clear patron.

About the reinforced fibers used, it can be noticed that samples reinforced with BF presented a higher deflection and a lower load than samples reinforced with CF, which evidenced the brittle property of CF and the high elongation property of BF [55]. The same results were reported by these authors.

3.3.2. Tensile Strength

The sample ER-CF (5.8 kN) increased by 54.30% compared to the control sample, while the rest of the samples did not present a statistically significant difference between them (Figure 7).

Figure 7 shows that samples reinforced with carbon fiber presented the highest values of tensile strength. In Table 1, the characteristics of each fiber used in this investigation are presented, and data sheets say that CF has a higher TS than BF. Furthermore, Wang et al. [51] explained in their investigation that the TS and MOE of CF are higher than for wood veneer, so using CF could improve the mechanical properties of LVL, and in this case, the tensile strength property. In addition, according to Dhand et al. [50], CF has a higher TS than BF.

3.3.3. Janka Hardness

The FRP-LVL increased the Janka Hardness results just in the samples reinforced with carbon fiber, between 31.52% and 41.43%, in the samples PVAc-CF (3.9 kN) and ER-CF (4.2 kN), respectively (Figure 8).

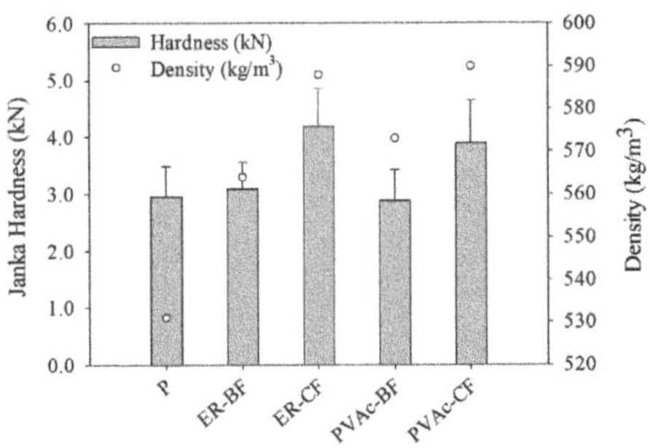

Figure 8. Results of Janka Hardness test and density of each sample (error bars indicate standard deviations).

Janka hardness is an important property in deciding the end use of FRP-LVL, especially when used as flooring components [56]. Carbon fiber-reinforced LVL showed a higher hardness than the rest of the samples, as in the study of Núñez-Decap et al. [11], probably because CF has the best mechanical properties in comparison to other kinds of reinforcement fibers.

3.4. Thermal Properties

In general, FRP-LVL maintained the thermal properties of LVL (Figure 9).

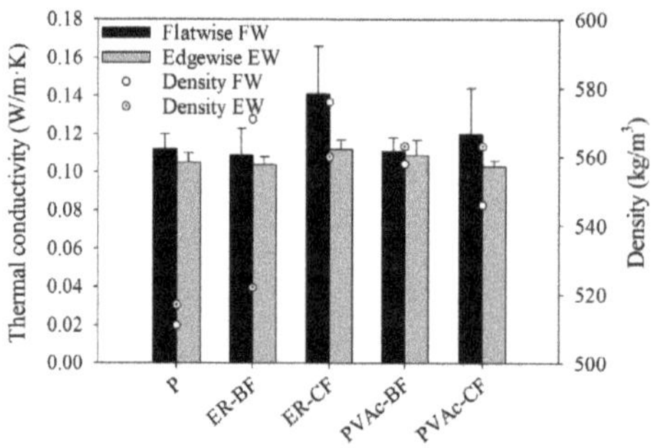

Figure 9. Results of the thermal conductivity test of each sample (error bars indicate standard deviations).

In the flatwise, the thermal conductivity of the sample ER-CF (0.14 W/m·K) increased by 25.78% compared to the control sample PF (0.11 W/m·K). This can be associated with the use of CF, as carbon-based materials, in general, present low density and high thermal conductivity [48]. On the contrary, one characteristic of the BF is its low thermal conductivity [57].

For the rest of the samples, in the flatwise and edgewise positions, there was no statistically significant difference in comparison to the control sample. Anyways, every sample in the flatwise and edgewise position presented thermal conductivity values between 0.10–0.14 W/m·K, which is the expected range, because the thermal conductivity values of panel boards, with a density of approximately 500 kg/m^3, should be around 0.13 W/m·K [58].

4. Conclusions

The tests performed evidenced that FRP-LVL modifies and, in general, improves LVL's physical and mechanical properties.

- In general, FRP-LVL increases their physical properties in 48 h water immersion when glued with PVAc adhesive. The sample PVAc-CF had the best performance, with a decrease in thickness swelling by 19%. These results introduce a good option for reinforced-LVL panels, as PVAc is a more environmentally friendly adhesive than epoxy resin and, in its case, evidenced good dimensional stability;
- FRP-LVL increase their flexural stiffness and strength properties when glued with ER adhesive; the samples ER-CF and ER-BF increased the MOE by 31% (FW), 36% (EW), and the MOR by 38% (FW) and 29% (EW), respectively;
- LVL panels, when reinforced with ER-CF, increase their tensile strength by 56% and their hardness by 41% in comparison to the control panel.

Author Contributions: Conceptualization, M.N.-D., B.S.-V. and C.O.-C.; methodology, M.N.-D., B.S.-V. and B.M.-R.; software, M.N.-D. and C.O.-C.; validation, M.N.-D., B.S.-V., C.O.-C. and M.V.-V.; formal analysis, M.N.-D., B.S.-V., C.O.-C., B.M.-R., M.V.-V. and A.O.-V.; investigation, M.N.-D., B.S.-V., C.O.-C., B.M.-R., M.V.-V. and A.O.-V.; resources, M.N.-D. and B.S.-V.; data curation, M.N.-D., B.S.-V. and C.O.-C.; writing—original draft preparation, M.N.-D., B.S.-V. and C.O.-C.; writing—review and editing, M.N.-D., B.S.-V. and C.O.-C.; visualization, M.N.-D., B.S.-V. and C.O.-C.; supervision, M.N.-D.; project administration, M.N.-D.; funding acquisition, M.N.-D. All authors have read and agreed to the published version of the manuscript.

Funding: This research received no external funding, and the APC was funded by Centro Nacional de Excelencia para la Industria de la Madera (CENAMAD).

Institutional Review Board Statement: Not applicable.

Informed Consent Statement: Not applicable.

Data Availability Statement: Data is unavailable due to privacy restrictions because this work is part of a current investigation project.

Acknowledgments: We extend our gratitude for the collaboration, equipment, and infrastructure provided by the Wood and Adhesive Engineering Products Laboratory (PRODIMA-LAB) at the Department of Civil and Environmental Engineering, Universidad del Bío-Bío. We also would like to thank the ANID BASAL FB210015.

Conflicts of Interest: The authors declare no conflict of interest.

References

1. CEI-Bois. La reducción de la huella de carbono y el impacto ambiental de edificios nuevos. In "Reportes ambientales para materiales constructivos". 1998–2001.
2. Lippke, B.; Wilson, J.; Perez-Garcia, J.; Bowyer, J.; Meil, J. CORRIM: Life-Cycle Environmental Performance of Renewable Building Materials. *For. Prod. J.* **2004**, *54*, 8–19.
3. IAAC. Why Er Need a New Generation of Wood Buildings and Professionals? By Michael Salka. Institute for advanced architecture of Catalonia. Available online: https://iaac.net/why-we-need-a-new-generation-of-wood-buildings-and-professionals/ (accessed on 27 August 2023).
4. Bal, B.C. Some physical and mechanical properties of reinforced laminated veneer lumber. *Constr. Build. Mater.* **2014**, *68*, 120–126. [CrossRef]
5. Ribeiro, A.S.; de Jesus, A.M.P.; Lima, A.M.; Lousada, J.L.C. Study of strengthening solutions for glued-laminated wood beams of maritime pine wood. *Constr. Build. Mater.* **2009**, *23*, 2738–2745. [CrossRef]
6. Çolak, S.; Çolakoğlu, G.; Aydin, I. Effects of logs steaming, veneer drying and aging on the mechanical properties of laminated veneer lumber (LVL). *Build. Environ.* **2007**, *42*, 93–98. [CrossRef]
7. Ramage, M.H.; Burridge, H.; Busse-Wicher, M.; Fereday, G.; Reynolds, T.; Shah, D.U.; Wu, G.; Yu, L.; Fleming, P.; Densley-Tingley, D.; et al. The wood from the trees: The use of timber in construction. *Renew. Sustain. Energy Rev.* **2017**, *68*, 333–359. [CrossRef]
8. Ross, P.; Downes, G.; Lawrence, A.; Technology, T. *Timber in Contemporary Architecture: A Designer's Guide*; TRADA Technology: London, UK, 2009.
9. Zhang, Y.; Yu, C.; Chu, P.K.; Lv, F.; Zhang, C.; Ji, J.; Zhang, R.; Wang, H. Mechanical and thermal properties of basalt fiber reinforced poly(butylene succinate) composites. *Mater. Chem. Phys.* **2012**, *133*, 845–849. [CrossRef]
10. Deans, M. *Inorganic and Composite Fibers*; Woodhead Publishing: Sawston, UK, 2018; p. iv. [CrossRef]
11. Núñez-Decap, M.; Barra-Rodríguez, Y.; Opazo-Carlsson, C.; Moya-Rojas, B.; Vidal-Vega, M.; Opazo-Vega, A. Use of Carbon and Basalt Fibers with Adhesives to Improve Physical and Mechanical Properties of Plywood. *Appl. Sci.* **2022**, *12*, 5114. [CrossRef]
12. Rafique, I.; Kausar, A.; Muhammad, B. Epoxy Resin Composite Reinforced with Carbon Fiber and Inorganic Filler: Overview on Preparation and Properties. *Polym. Plast. Technol. Eng.* **2016**, *55*, 1653–1672. [CrossRef]
13. Kramár, S.; Trcala, M.; Chitbanyong, K.; Král, P.; Puangsin, B. Basalt-Fiber-Reinforced Polyvinyl Acetate Resin: A Coating for Ductile Plywood Panels. *Materials* **2019**, *13*, 49. [CrossRef]
14. Weichand, P.; Gadow, R. Basalt fibre reinforced SiOC-matrix composites: Manufacturing technologies and characterisation. *J. Eur. Ceram. Soc.* **2015**, *35*, 4025–4030. [CrossRef]
15. Fořt, J.; Kočí, J.; Černý, R. Environmental Efficiency Aspects of Basalt Fibers Reinforcement in Concrete Mixtures. *Energies* **2021**, *14*, 7736. [CrossRef]
16. Qin, Y.; Xu, D.; Zhang, S.; Fan, X. Dynamic behavior of carbon nanotubes and basalt fiber reinforced coral sand cement mortar at high strain rates. *Constr. Build. Mater.* **2022**, *340*, 127396. [CrossRef]
17. Asil, M.B.; Ranjbar, M.M. Hybrid effect of carbon nanotubes and basalt fibers on mechanical, durability, and microstructure properties of lightweight geopolymer concretes. *Constr. Build. Mater.* **2022**, *357*, 129352. [CrossRef]

18. Łach, M.; Kluska, B.; Janus, D.; Kabat, D.; Pławecka, K.; Korniejenko, K.; Guigou, M.D.; Choińska, M. Effect of Fiber Reinforcement on the Compression and Flexural Strength of Fiber-Reinforced Geopolymers. *Appl. Sci.* **2021**, *11*, 10443. [CrossRef]
19. Zhang, W.; Zhang, Y.; Liu, B. Hybrid effect of basalt fiber and carbon fiber on the mechanical properties and microstructure of oil well cement. *Constr. Build. Mater.* **2021**, *286*, 122696. [CrossRef]
20. de la Rosa, P.; González, M.d.l.N.; Prieto, M.I.; Gómez, E. Compressive Behavior of Pieces of Wood Reinforced with Fabrics Composed of Carbon Fiber and Basalt Fiber. *Appl. Sci.* **2021**, *11*, 2460. [CrossRef]
21. Zhou, A.; Qin, R.; Chow, C.L.; Lau, D. Bond integrity of aramid, basalt and carbon fiber reinforced polymer bonded wood composites at elevated temperature. *Compos. Struct.* **2020**, *245*, 112342. [CrossRef]
22. Wdowiak-Postulak, A. Natural Fibre as Reinforcement for Vintage Wood. *Materials* **2020**, *13*, 4799. [CrossRef]
23. Wang, B.; Bachtiar, E.V.; Yan, L.; Kasal, B.; Fiore, V. Flax, Basalt, E-Glass FRP and Their Hybrid FRP Strengthened Wood Beams: An Experimental Study. *Polymers* **2019**, *11*, 1255. [CrossRef]
24. Zhang, X.; Zhang, Y.; Xie, X. Experimental and analytical investigation of the flexural behaviour of stiffened hollow glulam beams reinforced with fibre reinforced polymer. *Structures* **2023**, *50*, 810–822. [CrossRef]
25. Rescalvo, F.J.; Duriot, R.; Pot, G.; Gallego, A. Mechanical behaviour of poplar laminated veneer lumber with fiber reinforced polymer under shear and compression stresses. *Wood Mater. Sci. Eng.* **2022**, *18*, 893–901. [CrossRef]
26. Gallego, A.; Rescalvo, F.; Suarez, E. Madera microlaminada estructural de chopo reforzada con tejidos de carbono o basalto. *Alzada* **2020**, *120*, 32–43.
27. Wdowiak-Postulak, A.; Świt, G. Behavior of glulam beams strengthened in bending with BFRP fabrics. *Civ. Environ. Eng. Rep.* **2021**, *31*, 1–14. [CrossRef]
28. Wdowiak-Postulak, A. Basalt Fibre Reinforcement of Bent Heterogeneous Glued Laminated Beams. *Materials* **2021**, *14*, 51. [CrossRef] [PubMed]
29. Globa, A.; Subhani, M.; Moloney, C.; Al-Ameri, R. Carbon Fiber and Structural Timber Composites for Engineering and Construction. *J. Archit. Eng.* **2018**, *24*, 1–14. [CrossRef]
30. Subhani, M.; Globa, A.; Al-Ameri, R.; Moloney, J. Effect of grain orientation on the CFRP-to-LVL bond. *Compos. Part B Eng.* **2017**, *129*, 187–197. [CrossRef]
31. Núñez, M. Adhesivo de extractos de proteínas de levadura de riles cerveceros y panificadora, nanolignina y entrecruzante. *Chile Patent Appl.* **2022**, 202203793.
32. *ASTM D1084-16R21*; Standard Test Methods for Viscosity of Adhesives. ASTM International: West Conshohocken, PA, USA, 2021.
33. *ASTM E0070-19*; Standard Test Method for pH of Aqueous Solutions with the Glass Electrode. ASTM International: West Conshohocken, PA, USA, 2019.
34. *ASTM D1490-01R18*; Standard Test Method for Nonvolatile Content of Urea-Formaldehyde Resin Solutions. ASTM International: West Conshohocken, PA, USA, 2018.
35. *UNE-EN 323*; Wood-Based Panels—Determination of Density. Fernández de la Hoz: Madrid, Spain, 1994.
36. *UNE-EN 322*; Wood-Based Panels. Determination of Moisture Content; Fernández de la Hoz: Madrid, Spain, 1994.
37. *UNE-EN 317*; Particleboards and Fibreboards. Determination of Swelling in Thickness after Immersion in Water; Fernández de la Hoz: Madrid, Spain, 1994.
38. *UNE-EN 408*; Timber Structures. Structural Timber and Glued Laminated Timber. Determination of Some Physical and Mechanical Properties; Fernández de la Hoz: Madrid, Spain, 2012.
39. *UNE-EN 310*; Wood-Based Panels. Determination of Modulus of Elasticity in Bending and of Bending Strength; Fernández de la Hoz: Madrid, Spain, 1994.
40. *ASTM D3500-20*; Standard Test Methods for Wood Structural Panels in Tension. ASTM International: West Conshohocken, PA, USA, 2020; p. 6.
41. *ASTM D1037-12*; Standard Test Methods for Evaluating Properties of Wood-Base Fiber and Particle Panel Materials. ASTM International: West Conshohocken, PA, USA, 2012.
42. *ASTM D5334-22*; Standard Test Method for Determination of Thermal Conductivity of Soil and Rock by Thermal Needle Probe Procedure. ASTM International: West Conshohocken, PA, USA, 2000.
43. de Melo, R.R.; del Menezzi, C.H.S. Influence of Adhesive Type on the Properties of LVL Made from Paricá (Schizolobium amazonicum Huber ex. Ducke) Plantation Trees. *Drv. Ind.* **2015**, *66*, 205–212. [CrossRef]
44. Fu, Q.; Yan, L.; Ning, T.; Wang, B.; Kasal, B. Interfacial bond behavior between wood chip concrete and engineered timber glued by various adhesives. *Constr. Build. Mater.* **2020**, *238*, 117743. [CrossRef]
45. Yusoh, A.; Tahir, P.; Uyup, M.; Lee, S.; Husain, H.; Khaidzir, M. Effect of wood species, clamping pressure and glue spread rate on the bonding properties of cross-laminated timber (CLT) manufactured from tropical hardwoods. *Constr. Build. Mater.* **2020**, *273*, 121721. [CrossRef]
46. Wang, X.-M.; Casilla, R.; Zhang, Y.; Cooper, P.; Huang, Z. Effect of extreme pH on bond durability of selected structural wood adhesives. *Wood Fiber Sci.* **2016**, *48*, 245–259.
47. Hýsek, S.; Premysl, S.; Böhm, M.; Schönfelder, O.; Beran, R. Influence of Using Recycled Polyurethane Particles as a Filler on Properties of Polyurethane Adhesives for Gluing of Wood. *BioResources* **2018**, *13*, 2592–2601. [CrossRef]
48. Sanghvi, M.R.; Tambare, O.H.; More, A.P. Performance of various fillers in adhesives applications: A review. *Polym. Bull.* **2022**, *79*, 10491–10553. [CrossRef]

49. Lengowski, E.C.; Júnior, E.A.B.; Nishidate Kumode, M.M.; Carneiro, M.E.; Satyanarayana, G. Nanocellulose-Reinforced Adhesives for Wood-Based Panels. *Sustain. Polym. Compos. Nanocompos.* **2019**, *35*, 1001–1025. [CrossRef]
50. Dhand, V.; Mittal, G.; Rhee, K.Y.; Park, S.-J.; Hui, D. A short review on basalt fiber reinforced polymer composites. *Compos. Part B Eng.* **2015**, *73*, 166–180. [CrossRef]
51. Wang, J.; Guo, X.; Zhong, W.; Wang, H.; Cao, P. Evaluation of Mechanical Properties of Reinforced Poplar Laminated Veneer Lumber. *BioResources* **2015**, *10*, 7455–7465. [CrossRef]
52. Wei, P.; Wang, B.J.; Zhou, D.; Dai, C.; Wang, Q.; Huang, S. Mechanical Properties of Poplar Laminated Veneer Lumber Modified by Carbon Fiber Reinforced Polymer. *BioResources* **2013**, *8*, 4883–4898. [CrossRef]
53. Frihart, C.R.; Hunt, C.G. Adhesives with Wood Materials Bond Formation and Performance. In *Wood Handbook: Wood as an Engineering Material*; Forest Products Laboratory, United States Department of Agriculture Forest Service: Madison, WI, USA, 1910-2010.
54. Bakalarz, M.M.; Kossakowski, P.G. Strengthening of Full-Scale Laminated Veneer Lumber Beams with CFRP Sheets. *Materials* **2022**, *15*, 6526. [CrossRef]
55. Subagia, I.D.G.A.; Kim, Y.; Tijing, L.D.; Kim, C.S.; Shon, H.K. Effect of stacking sequence on the flexural properties of hybrid composites reinforced with carbon and basalt fibers. *Compos. Part B Eng.* **2014**, *58*, 251–258. [CrossRef]
56. de Souza, J.B.; de Azevedo, T.K.B.; Sousa, T.B.; da Silva, G.G.C.; Guimaraes, J.B.; Pimenta, A.S. Plywood bonding with an adhesive based on tannins from Acacia mangium Wild. bark from trees grown in Northeastern Brazil. *Rev. Bras. Cienc. Agrar. Agrar.* **2020**, *15*, 1–7. [CrossRef]
57. Yang, G.; Park, M.; Park, S.-J. Recent Progresses of fabrication and characterization of fibers-reinforced composites: A review. *Compos. Commun.* **2019**, *14*, 34–42. [CrossRef]
58. Studiengemeinschaft Holzeimbau e.V.; Finnish Woodworking Industries. *Laminated Veneer Lumber (LVL) Bulletin*, 3rd ed.; Studiengemeinschaft Holzeimbau e.V.: Wuppertal, Germany; Finnish Woodworking Industries: Helsinki, Finland, 2020; pp. 1–8. Available online: https://timberdevelopment.uk/resources/laminated-veneer-lumber-lvl-bulletin/ (accessed on 27 August 2023).

applied sciences

MDPI

Article

Effect of Primary Spruce Lamella Aging on the Bending Characteristics of Glulam Beams

Ladislav Reinprecht [1,*], Dávid Ciglian [1], Ján Iždinský [1] and Ján Sedliačik [2]

[1] Department of Wood Technology, Faculty of Wood Sciences and Technology, Technical University in Zvolen, T. G. Masaryka 24, 960 01 Zvolen, Slovakia
[2] Department of Furniture and Wood Products, Faculty of Wood Sciences and Technology, Technical University in Zvolen, T. G. Masaryka 24, 960 01 Zvolen, Slovakia
* Correspondence: reinprecht@tuzvo.sk

Abstract: Glued laminated (glulam) beams are used in the roofs, ceilings and walls of buildings as well as in bridges and towers. At present, with the limitation of tree harvesting, the production of glulam beams from recycled wood sources is implemented with the proviso that their mechanical properties and resistance to pests, fire and weathering will not be aggravated. This work deals with the primary effect of aging Norway spruce wood (*Picea abies* Karst. L.) lamellas on the moduli of rupture (MOR) and elasticity (MOE) in bending of three-layer glulam beams composed of sound and aged lamellas and polyurethane (PUR) glue. Three methods of lamella aging were used: (A) natural, lasting 60 years in the form of roof trusses with a greater or lesser degree of bio-attack by woodworm (*Anobium punctatum* De Geer); (B) artificial, caused by increased temperatures from 160 to 220 °C for 4 h; (C) artificial, caused by 2% water solutions of inorganic preservatives, namely, $CuSO_4 \times 5H_2O$, $ZnCl_2$, H_3BO_3 or $(NH_4)_2SO_4$, for 28 days. The lowest MOR values were determined for glulam beams in which all three lamellas or two surface lamellas had a greater degree of bio-attack (60.5 MPa, a decrease of 25.9%) or were exposed to primary aging at 220 °C (62.6 MPa, a decrease of 23.3%). On the contrary, the exposure of lamellas to 160 or 180 °C did not significantly influence the MOR of beams (76.0–82.7 MPa, an average decrease of 1.6%). The MOE of glulam beams ranged from 7540 to 10,432 MPa without an obvious influence of the method of lamella aging or their location in the beams. Linear correlations between the MOR or MOE of glulam beams and the shear strength (σ) of glued joints, if both composite types consisted of similarly aged lamellas, were only slightly significant or insignificant.

Keywords: Norway spruce wood; lamellas; natural and artificial aging; polyurethane glue; glulam; modulus of rupture; modulus of elasticity; shear strength

1. Introduction

Glued laminated (glulam) beams and similar wooden composites, e.g., cross-laminated timber (CLT), overlayed materials, laminated wood–nonwood composites and multiwood composites (COM-PLY), are commonly used in structural engineering, furniture, sport, musical and decoration products [1]. In glulam, the defects in the sawn timber are cut out, and lamellas with a thickness of ≤50 mm are bonded together [2]. Glulam beams can be made from coniferous woods (Douglas fir, larch, Norway spruce, hemlock, Southern pine, etc.), broadleaved woods (azobé, beech, doussie, keruing, padouk, red maple, resak, etc.), as well as densified low-density woods (batai, poplar, etc.) [2–5], preferentially using thermosetting glues such as polyurethane (PUR), resorcinol–formaldehyde (RF), phenol–resorcinol–formaldehyde (PRF), melamine–formaldehyde (MF) or urea–melamine–formaldehyde (UMF). In Europe, the Norway spruce is the most common wood species used for glulam [2,5,6].

The final quality of glulam beams is influenced by many factors: (1) the wood species, whether it is used in its natural state or after chemical, thermal, mechanical or biological modifications, and what are its typical structural characteristics and properties, e.g., the density and porosity, the type and amount of extractives, the pH value, roughness, free surface energy, moisture content at bonding, tension and compression strengths, elasticity, and resistance to pests, fire and weathering; (2) the glue type and its physical–chemical characteristics, e.g., the molecular weight, viscosity, surface tension, chemical reactivity, resistance to water and total durability; and (3) the technology of gluing, e.g., the spread rate of glue and its open time and the pressing diagram used [7–14]. The applicability of structural glulam beams depends mainly on their mechanical properties and resistance to biological damage, fire and weathering. Today, when new sources of wood are limited, preparation of glulam beams from older wood products previously aged in various environments is actual.

During the demolition and reconstruction of buildings, a large amount of construction waste is generated, including wood waste. This usually ends up in municipal waste landfills, while only a small part of it is reused in the form of recycling [15,16] or for energy production [17]. In general, to increase the efficiency of the use of waste materials, a method of cascading resources was developed [18]. This is a method in which the efficiency of the use of a material resource is increased through its gradual or repeated use for several applications in individual consecutive cascading steps. By applying cascading in practice, the efficiency of the use of wood raw materials can also be increased, and thus, trees and wood stocks in the forest can be saved. The theoretical concept of wood cascading was first described by Fraanje [19]. Subsequently, the studies by Höglmeier et al. [20,21], Lubke et al. [22] and other researchers emphasized the environmental and economic advantages of this method for wood recycling in practice. Currently, despite all of the favorable aspects of the cascading method, wood waste is primarily used for the production of particleboards [23], cement chipboards [24] and possibly other types of agglomerated materials [25]. However, during their production, it is necessary to first chip the waste wood material, which does not fulfill the basic principle of cascading. In order to maximize efficiency, it is necessary to achieve the greatest possible number of cascading steps, and in the first steps, wood particles from the waste material, even larger than wood chips, should be used. So far, relatively few workplaces have dealt with such research, and most of the research has focused on the production and testing of the properties of CLT panels from recycled solid wood [26,27].

Glued laminated (glulam) wood can be characterized as a construction material that is created by gluing individual wood lamellas under prescribed technological conditions. The lamellas are oriented so that the longitudinal wood fibers coincide with the longitudinal direction of the final wooden element [28]. Currently, glulam wood is used in various construction applications, from load-bearing elements in one- and multi-story residential and non-residential buildings, through arched trusses covering buildings with large spans [29], to load-bearing elements of bridge structures [30–32]. In principle, glulam can be produced from any wood species, as long as suitable types of glues are developed and tested for the given species of wood. In European countries, glulam beams are made mainly from coniferous wood species, especially from spruce wood [31,33]. The definitions of and requirements for wood lamellas and glulam products are prescribed by the standard EN 14080:2013 [6].

The theoretical analysis of the efficiency of solid wood recycling for the production of glulam beams and other glued composite materials was solved in the study by Risse et al. [34]. The results of this study indicate that the recycling of solid wood for preparing new glulam beams can be environmentally and economically significant. However, relatively few researchers have dealt with this issue so far, and thus, it is not sufficiently reviewed. For example, Syrovátko [35] tested glulam beams made of lamellas prepared from old wooden elements of the roof structure, and he discovered that the mechanical properties of these beams were comparable to those of beams made of new lumber. The work by Mali [36] showed that the mechanical properties of glulam beams from old waste wood decreased by 4.7 to 10.5% compared to glulam beams based on current healthy wood, while the rate of decrease in their strength depended on the proportion and position of old and new wood in individual types of glulam beams. Based on this knowledge, it can be concluded that old waste wood obtained from the demolition of buildings may not always lose its original quality during its use for its initial purpose, and its mechanical properties are often at the level of new wood. These findings were confirmed by the review studies by Cavalli et al. [37] and Kránitz et al. [38], in which the experimental results of several researchers dealing with changes in the properties of old wood were compared. The knowledge described in these review studies showed that the age of old wood does not have a dominant effect on changes in its mechanical properties. This is self-evident as long as the old wood is not simultaneously degraded (1) by biological agents, i.e., especially due to rotting caused by wood-decaying fungi and larval galleries caused by wood-damaging insects, and (2) by abiotic factors, i.e., especially by thermal attacks during fires at ~300 °C, in the vicinity of hot chimneys above 100 °C or specifically in roofs during long-term elevated summer temperatures of 50–70 °C, and specifically by chemical attacks due to aggressive substances such as inorganic biocides and flame retardants.

However, in practice, older wood elements in the roofs, ceilings or log-walls are, in several cases, damaged by brown or white rot, insect galleries and high temperatures or contain chemical preservatives that served to increase their durability during use [39]. Therefore, in the cascading method of wood source utilization for glulam or other supporting composite elements, the correct choice of wood material is very important. Decreases in the strength of wood deteriorated by decaying fungi, insects, high temperatures or aggressive chemicals are usually evidently higher as decrease of its density [40]. Bio-attacked wood quickly loses its stiffness, as well as its tension and bending strength [41,42]. Wood exposed to high temperatures above 160 °C loses its strength quickly and its stiffness slowly [43], it is less hygroscopic–more hydrophobic, and its surfaces are less wettable by polar liquid glues [43–45]. Bastani et al. [46] found that the processing time needed for the glue to be absorbed into thermally modified wood is longer due to the slower penetration rate. Changes in the pH value of the thermally modified wood surface might retard or accelerate the curing of glues, depending on their type [47]. Finally, the mechanical properties of glulam from thermally modified wood lamellas are usually lowered [48]. In many cases, chemical preservatives—many of which are aggressive chemical substances or such substances can be formed over time in a wet environment—negatively affect the wettability of wood lamellas with polar glues [12] and deteriorate the lignin–saccharide matrix of wood cells in lamellas [39]. So, such chemicals can worsen the adhesion of glues to wood lamellas and/or decrease the strength of wood lamellas and finally degrade the mechanical properties of glulam beams [49]. For example, in the study by Piao et al. [50], who dealt with the use of wood material from old pine telecommunication poles in the production of glulam beams, the strength of glulam prepared from aged wood was reduced due to the presence of the biocidal preservative CCA (chromated copper arsenate). However, today, in several European countries, old wood containing CCA, creosote, Cu-azole, pentachlorophenol or other more toxic contaminants of the fourth healthy-environmental class cannot be used for new panels including glulam [51].

The basic aim of this work was to examine the influence of the primary aging of spruce lamellas—natural aging under the action of wood-destroying insects, as well as artificial aging due to high temperatures or inorganic wood preservatives commonly used in the past—on the bending characteristics of glulam beams. Simultaneously, the relationship between the bending characteristics of glulam beams prepared from aged and sound spruce lamellas and the shear strength of glued joints between these lamellas was evaluated.

2. Materials and Methods

2.1. Spruce Wood Lamellas

For the experiment were used prisms of Norway spruce (*Picea abies* Karst. L.) wood with the dimensions of 2000–3000 mm (length) × 140–300 mm (width) × 24–30 mm (thickness). The prisms were cut (1) from several naturally aged 60-year-old bio-attacked roof trusses of a reconstructed house, No. 320 in village Brodzany, Slovakia, and (2) from one sound 96-year-old tree trunk obtained from the National Forest Centrum in Zvolen, Slovakia. Then, the prisms were conditioned for 1 year outside under shelter at Technical University in Zvolen. From the prism were cut wide pre-lamellas of type I with the dimensions of 580 mm × 80 mm × 10 mm (longitudinal × radial × tangential), which were used for three-layer glulam beams (Sections 2.3 and 2.4), and lamellas of type II with the dimensions of 80 mm × 20 mm × 5 mm (longitudinal × radial × tangential), which were used for the double-bonded sets (Section 2.5). Both of the top surfaces of lamellas were gradually milled in a milling machine (D630-EL; Robland, Belgium) and subsequently machine-sanded with 120-grit sandpaper.

2.2. Aging of Lamellas

The natural aging of lamellas was implemented directly in the 60-year-old roof trusses from which they were prepared. The trusses were subjected to non-homogeneous bio-attacks by larval galleries of woodworm—the common furniture beetle (*Anobium punctatum* De Geer) with a diameter up to 1.5–2 mm. The degree of the bio-attack of prepared lamellas was determined visually, i.e., a greater degree (GD) for lamellas from the outer zone, usually the sap zone, of truss beams, with 50 or more larval galleries per m^2, and a lesser degree (LD) for lamellas prepared from the inner zone of truss beams, with no or less than 50 larval galleries per m^2 (Figure 1).

(**a**) (**b**)

Figure 1. Sixty-year-old roof trusses bio-attacked by larval galleries of *Anobium punctatum* (**a**), and scheme of prism used for the lamella preparation with a greater degree (GD) and lesser degree (LD) of bio-attack (**b**).

For the artificial aging, lamellas were cut from a 96-year tree trunk. They did not have sap zones, knots, fungal rots or insect galleries. The thermal aging of lamellas took place at atmospheric pressure and temperatures of 160, 180, 200 or 220 °C for 4 h in a drying oven (Memmert UFE 500; Schwabach, Germany). The chemical aging of lamellas was performed using a double-coating treatment (2 × 120 ± 10 g/m^2, using a 24 h break) with 2% water solutions of inorganic wood preservatives traditionally used in the past (copper sulfate $CuSO_4 \times 5H_2O$, zinc chloride $ZnCl_2$, boric acid H_3BO_3 and ammonium sulfate $(NH_4)_2SO_4$)), followed by their 28-day exposure at 20 °C ± 2 °C and 65% ± 5% RH.

Aged and sound spruce lamellas for glulam beams were conditioned at 20 °C ± 2 °C and 65% ± 5% RH, with achieving their moisture content (MC) of 9% ± 3%. For example, lamellas aged at 220 °C had an MC of ~6%, while the MC of the sound and naturally aged lamellas was ~12%.

The density of spruce lamellas in their conditioned state ranged in a narrow interval from 418 to 467 kg·m^{-3}, i.e., 459 kg·m^{-3} for sound lamellas; 433 and 418 kg·m^{-3} for naturally aged lamellas from the inner lesser (LD) and outer greater (GD) bio-attacked zones of old trusses; 451 to 426 kg·m^{-3} for lamellas thermally aged at 160 to 220 °C; and 448 to 467 kg·m^{-3} for chemically aged lamellas. So, these small differences in the densities of the lamellas should not affect the mechanical properties of glued wood elements.

2.3. PUR Glue and Gluing of Lamellas

The one-component PUR glue Kestopur 1030 (Kiilto, Tampere, Finland) is characterized by the following technical parameters—viscosity of 7000 mPa·s, density of 1200 kg/m^3, open time of 30 min and pressing time of 90–120 min. PUR glue was applied using coating technology on the interconnected surfaces of spruce lamellas at the recommended rate of 180 g/m^2 ± 10 g/m^2. The three-layer glulam beams (Figure 2), as well as the double-bonded sets (see figure in [12]), were prepared by pressing spruce lamellas coated with PUR glue. Pressing was performed in steel bolts at a pressure of 1.2 MPa, a temperature of 20 °C ± 2 °C and an RH of 65% ± 5% for 2 h (Figure 2a). Before testing the quality of the prepared glulam beams and double-bonded sets (see Sections 2.4 and 2.5), the 7-day conditioning of these beams to set equilibrium MCs of 9% ± 3% was performed under the same air conditions.

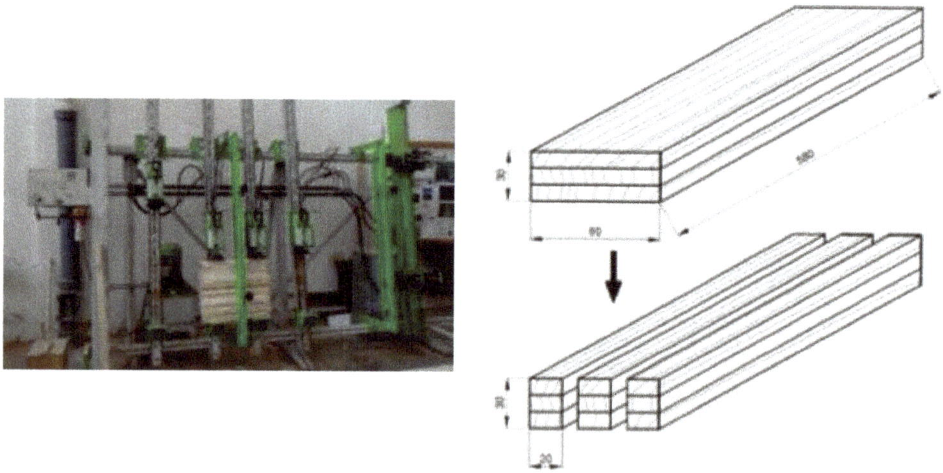

(**a**) (**b**)

Figure 2. Pressing of wide pre-lamellas (580 mm × 80 mm × 10 mm) coated with PUR glue into three-layer glulam composites (580 mm × 80 mm × 30 mm) (**a**); preparation of the three-layer glulam beams (580 mm × 20 mm × 30 mm) from the composite (**b**).

2.4. Bending Characteristics—MOR and MOE

The conditioned three-layer glulam beams, with the dimensions of 580 mm × 20 mm × 30 mm, were subjected to the 3-point bending test in a TiraTest 2200 (VEB TIW Rauenstein, Germany) device with a maximum force of 10 kN. The distance of the supports was 540 mm, which is 18 times the beam height. The values of MOR and MOE in bending were determined according to the standards EN 408:2010+A1:2012 [52] and EN 310:1993 [53], using Equations (1) and (2):

$$\sigma_{MOR} = \frac{3 \times F_{max} \times l_0}{2 \times b \times h^2} \text{ (Mpa)} \quad (1)$$

$$\sigma_{MOE} = \frac{\Delta F \times l_0{}^3}{4 \times \Delta y \times b \times h^3} \text{ (Mpa)} \quad (2)$$

where *MOR* is the modulus of rupture in bending (Mpa), *MOE* is the modulus of elasticity in bending (Mpa), F_{max} is the force corresponding to the breaking strength (N), l_0 is the distance between the supports (mm), b is the width of the glulam beam (mm), h is the height of the glulam beam (mm), ΔF is the load increment, and Δy is the deflection increment.

The 1st (reference) group of glulam beams consisted of three sound lamellas, the 2nd group consisted of one aged lamella in the middle layer and two sound lamellas in the surface layers, the 3rd group consisted of two aged lamellas in the surface layers and one sound lamella in the middle layer, and the 4th group consisted of three aged lamellas (Figures 2 and 3).

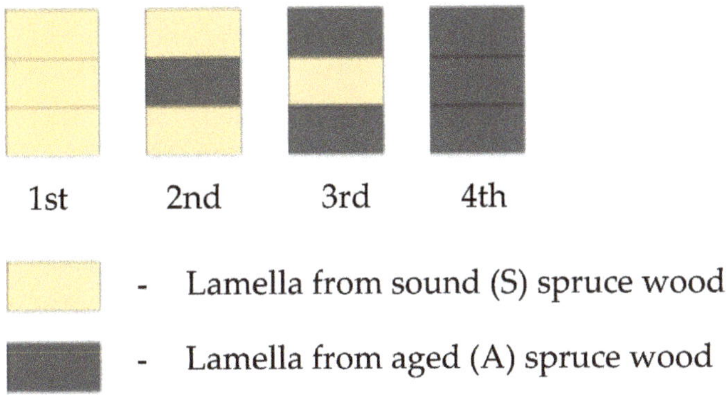

Figure 3. Glulam beams in the 1st, 2nd, 3rd and 4th groups. Note: Glulam beams in the 1st group consist of three sound "S" lamellas; those in the 2nd group consist of two sound "S" lamellas on surfaces and one aged "A" lamella in middle; those in the 3rd group consist of one sound "S" lamella in middle and two aged "A" lamellas on surfaces; and those in the 4th group consist of three aged "A" lamellas.

2.5. Shear Strength

The shear strength (σ) of glued joints in the conditioned double-bonded sets, which consisted of two lamellas of 80 mm × 20 mm × 5 mm with a mutual contact area of 10 mm × 20 mm, was determined according to the standard EN 205:2016 [54] using a LabTech 4.050 (LaborTech s.r.o., Opava, Czech Republic) device. The 1st (reference) double set consisted of two sound lamellas, the 2nd double set consisted of one aged lamella and one sound lamella, and the 3rd double set consisted of two aged lamellas. Double sets from naturally aged lamellas were prepared and tested in this work, while those from artificially aged lamellas were prepared and tested previously in the works of Vidholdová et al. [55] and Ciglian and Reinprecht [12].

2.6. Statistical Analyses

The evaluations of the measured data (arithmetic means and standard deviations) and the processing of linear correlations between the bending characteristics (MOR or MOE) of glulam beams and the shear strength of glued joints in double-bonded sets were performed using Microsoft Excel (Version 2010; Microsoft Corporation, Redmond, WA, USA). The statistical software STATISTICA 12 (StatSoft Inc., Tulsa, OK, USA) was used to evaluate Duncan's tests to measure specific differences between pairs of means (with indexes of significance: a—very significant decrease at the 99.9% level; b—significant decrease at the 99% level; c—less significant decrease at the 95% level; and d—insignificant change (decrease or increase) at the <95% level) of the bending characteristics of glulam beam replicates containing various combinations of aged and sound lamellas (2nd–4th groups) in comparison to reference replicates containing sound lamellas (1st group).

3. Results and Discussion

3.1. Bending Characteristics of Glulam Beams Affected by Aging of Lamellas

3.1.1. Influence of Lamella Aging Methods and Their Locations in Beams

For the 1st–4th groups of three-layer glulam beams, consisting of spruce lamellas of type I (Section 2.1) and PUR glue (Figures 2 and 3), the bending characteristics MOR and MOE were determined (Table 1 and Figure 4).

Table 1. The bending characteristics MOR and MOE of the three-layer glulam beams in the 1st, 2nd, 3rd and 4th groups (Figure 3) prepared from Norway spruce lamellas of type I and PUR glue (Figure 2).

Glulam Beam	MOR (MPa)	MOE (MPa)
1st group (S-S-S)		
Sound	81.63 (2.66)	10,432 (742)
2nd group (S-A-S)		
GD bio-attack	72.78 (5.38) b	10,864 (693) d
LD bio-attack	74.45 (0.82) c	9956 (1298) d
T-160 °C	81.74 (1.63) d	9566 (332) d
T-180 °C	82.65 (1.93) d	11,574 (391) d
T-200 °C	77.53 (2.65) d	9540 (1360) d
T-220 °C	68.29 (4.75) a	10,564 (1470) d
$CuSO_4 \cdot 5H_2O$	75.29 (5.69) c	9701 (598) d
$ZnCl_2$	70.78 (5.60) a	9806 (674) d
H_3BO_3	80.03 (3.05) d	10,905 (1055) d
$(NH_4)_2SO_4$	73.08 (7.18) b	8856 (384) b
3rd group (A-S-A)		
GD bio-attack	60.98 (3.21) a	8292 (913) a
LD bio-attack	67.37 (3.74) a	9614 (426) d
T-160 °C	80.85 (3.50) d	9145 (1456) c
T-180 °C	79.88 (6.54) d	11,554 (499) d
T-200 °C	70.38 (6.50) a	10,443 (484) d
T-220 °C	62.60 (4.63) a	9518 (731) d
$CuSO_4 \cdot 5H_2O$	71.09 (4.58) a	10,591 (704) d
$ZnCl_2$	68.92 (1.36) a	9151 (414) c
H_3BO_3	71.20 (1.63) a	8415 (517) a
$(NH_4)_2SO_4$	70.82 (5.38) a	9052 (269) c
4th group (A-A-A)		
GD bio-attack	60.02 (9.86) a	7540 (1032) a
LD bio-attack	74.99 (7.18) c	9738 (670) d
T-160 °C	80.97 (7.51) d	9841 (860) d
T-180 °C	76.03 (6.19) d	9826 (1436) d
T-200 °C	69.29 (7.38) a	10,155 (543) d
T-220 °C	62.56 (2.17) a	8508 (811) a
$CuSO_4 \cdot 5H_2O$	69.20 (4.20) a	9686 (396) d
$ZnCl_2$	70.32 (3.55) a	10,542 (1689) d
H_3BO_3	75.45 (2.54) a	10,825 (920) d
$(NH_4)_2SO_4$	69.87 (1.54) a	9012 (1399) c

Notes: Mean values are from 6 measurements. Standard deviations are in parentheses. Duncan's tests were used to evaluate the bending characteristics of the 2nd, 3rd or 4th groups of glulam beams consisting of 1, 2 or 3 aged "A" lamellas, relative to the 1st (reference) group of glulam beams consisting of 3 sound "S" lamellas, with indexes of significance: a—very significant decrease at 99.9% level; b—significant decrease at 99% level; c—less significant decrease at 95% level; and d—insignificant change (decrease or increase) at <95% level.

The MOR of glulam beams, in many cases, significantly depended—as determined by Duncan's tests—both on the method of lamella aging and their locations in the beams (Table 1). The highest MORs were determined for glulam beams in the first (reference) group consisting only of sound spruce lamellas (81.63 MPa) and those in the second group, in which the middle lamella was initially aged at lower temperatures of 160 or 180 °C (81.74 or 82.65 MPa). However, the MOR in the second group of glulam beams was evidently smaller at the 99.9% or 99% significance level if the middle lamellas were made from naturally aged prisms with a greater degree (GD) of bio-attack (72.78 MPa; a decrease of 10.8%) or if the middle lamellas were artificially aged at a maximum temperature of 220 °C or in the presence of $ZnCl_2$ and $(NH_4)_2SO_4$ (68.29–73.08 MPa; decreases of 16.3–10.5%).

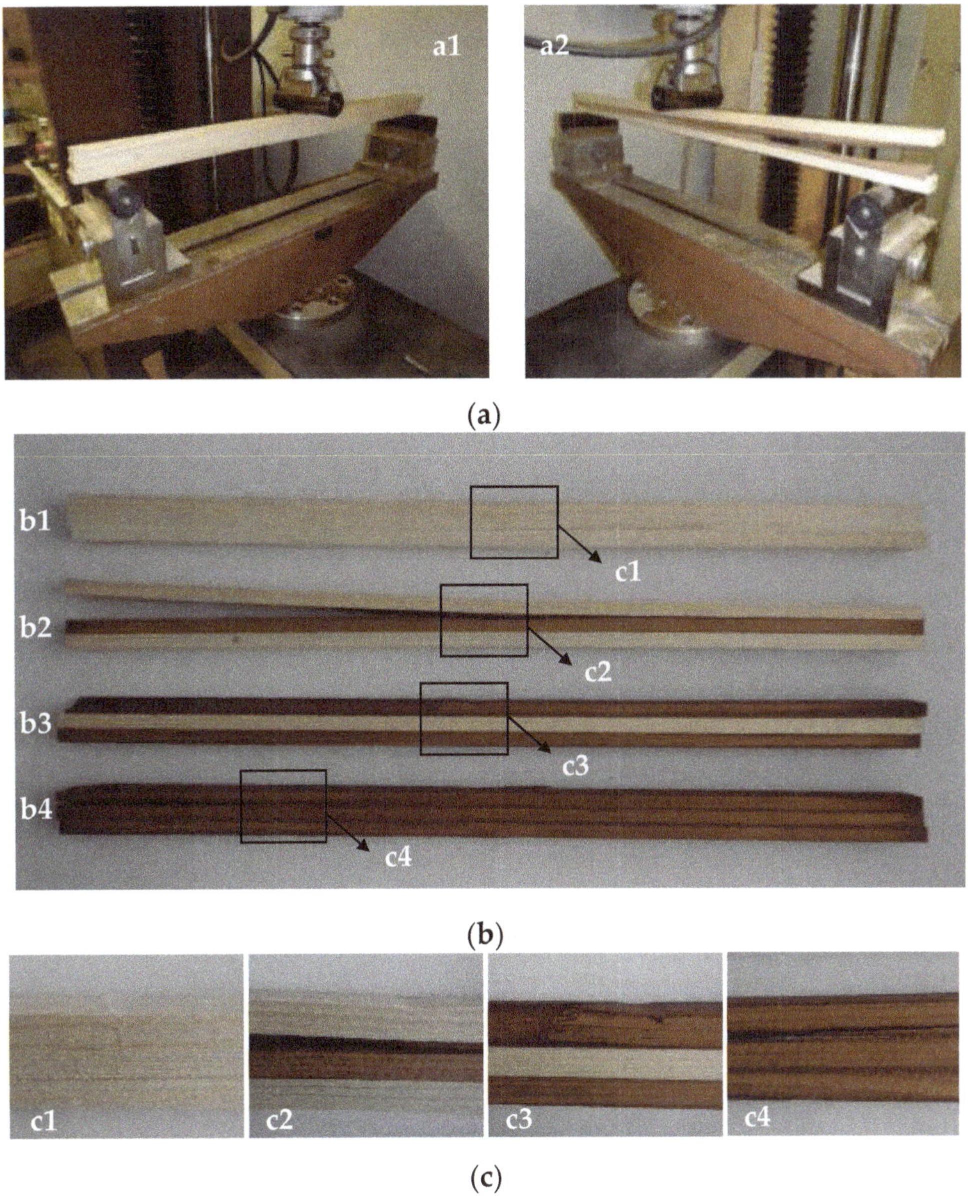

Figure 4. Examples of the 3-point bending test of three-layer glulam beams ((**a**): **a1** = SSS; **a2** = SAS, if A aged at T-200 °C), showing typical damages in the bond joints and/or wood lamellas ((**b**): **b1** = SSS; **b2** = SAS; **b3** = ASA; **b4** = AAA, if A always aged at T-220 °C) and details of these damages ((**c**): **c1** to **c4**).

In accordance with the theoretical assumption, the lowest MOR were obtained for glulam beams of the third group and fourth group, in which aged lamellas were localized in the surface "tension and compression" zones. For example, the MOR decreased by 25.9–23.3% if such lamellas were prepared from the outer greater bio-attacked zone of naturally aged prisms (~60.5 MPa), or if these lamellas were artificially aged at a maximum temperature of 220 °C (~62.6 MPa). The determination of similar values of the MOR for the third group (A-S-A) and the fourth group (A-A-A) of glulam beams created from lamellas aged (A) artificially at 220 °C or aged naturally with a greater attack by *A. punctatum* is in good accordance with the theory that for the bending strength of beams, including glulam beams, the most important zones are their tension and compression zones.

The MOE of glulam beams depended less, compared to the MOR, on the method of spruce lamella aging and their locations in glulam beams. This was determined by Duncan's tests (Table 1). Quite high MOEs, approximately from 9000 to 11,000 MPa, were obtained for various glulam beam types, i.e., not only in the first and second groups but, in some cases, also from the third and fourth groups. Apparently lower MOEs, from 8000 and 9000 MPa, were observed for some glulam beams in which aged lamellas were localized (1) in the middle "neutral" zone of the second group—aged by $(NH_4)_2SO_4$; (2) in the surface "tension and compression" zones of the third group—with a GD of bio-attack or aged by H_3BO_3; (3) in all three zones of the fourth group—aged at 220 °C. The lowest MOE, only 7540 MPa, was obtained for glulam beams in the fourth group consisting of three lamellas with a greater degree (GD) of bio-attack by *A. punctatum*.

Generally, a decrease in the bending properties of glulam beams made from aged lamellas can be explained (1) by the creation of poorly glued joints, as well as (2) by the presence of deteriorated wood in beams having lower strength and stiffness. Poorly glued joints are created when using wood with a more hydrophobic surface or a surface weakened by bio-, thermal and/or chemical attacks. Worse bending properties of wood damaged by larval galleries were found by Reinprecht and Pánek [42]. Temperatures from 160 °C to 220 °C cause only a minor change ("increase or decrease") in the MOE but a large decrease in MOR from 10% to 50% with a rise in temperature [56–58]. The mechanical properties of wood decrease after its attack by acids, bases and other aggressive chemicals [40]. For example, the negative effect of wood impregnation with compounds of trivalent boron on the strength of glued joints and the strength of wooden composites significantly depends on the type of wood and the type of polymer used as the glue or coating [59,60]. Keskin and Mutlu [61] researched the effect of impregnating beech, oak and pine woods with boron preservatives on wood strength, and they found that wood containing sodium tetraborate ($Na_2B_4O_7 \cdot 10H_2O$) had increased bending strength, whereas many mechanical properties of wood decreased in presence of H_3BO_3 and other trivalent boron types.

3.1.2. Influence of Temperatures during Lamella Aging

With the increase in the temperature applied during the initial aging of spruce lamellas from 160 to 220 °C, the MOR of the prepared glulam beams gradually decreased. This was proved by the individual linear correlations in the second group ($MOR = 84.63 - |0.040 \times T|$; $R^2 = 0.24$), third group ($MOR = 85.97 - |0.070 \times T|$; $R^2 = 0.38$) and fourth group ($MOR = 85.90 - |0.076 \times T|$; $R^2 = 0.34$) of beams, as well as by the summary linear correlation evaluating all three groups of glulam beams together ($MOR = 90.39 + |0.087 \times T|$; $R^2 = 0.29$) (Figure 5a). On the contrary, the MOE values of glulam beams were not apparently influenced by the initial thermal aging of spruce lamellas, as was confirmed based on the small coefficients of determination: R^2 from 0.0003 to 0.19 for the individual linear correlations of the 2nd–4th groups and $R^2 = 0.015$ for the summary linear correlation (Figure 5b).

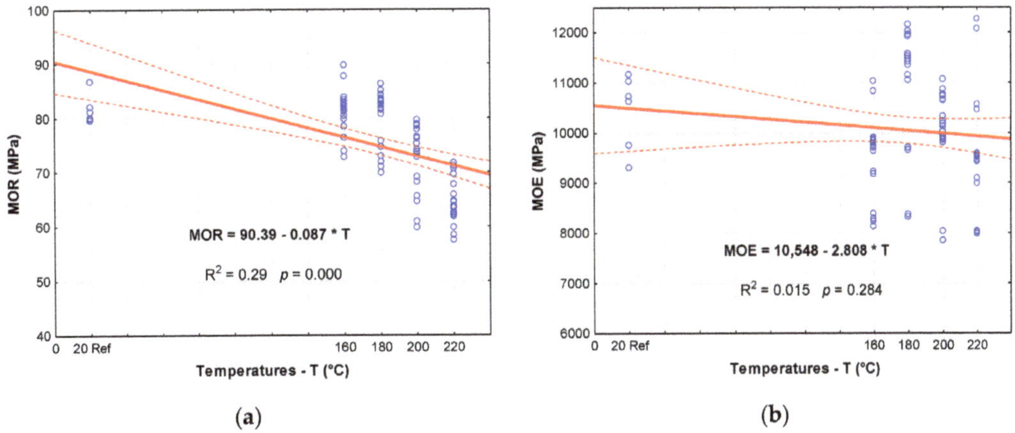

Figure 5. Summary linear correlations between the bending characteristic MOR (**a**) or MOE (**b**) of the three-layer glulam beams belonging to the 2nd, 3rd and 4th groups and the temperature during the artificial aging of spruce lamellas.

3.2. Shear Strength of Glued Joints Affected by Aging of Lamellas

The shear strengths (σ) of glued joints in the first, second and third double-bonded sets prepared from naturally aged lamellas of type II (Section 2.1) and PUR glue (Sections 2.3 and 2.5) are presented in Table 2.

Table 2. The shear strengths (σ) of glued joints in the double-bonded sets consisting of Norway spruce lamellas of type II and PUR glue.

Double-Bonded Set	Shear Strength—σ (MPa)
1st set (S-S)	
Sound	10.82 (1.39)
2nd set (S-A)	
GD bio-attack	10.85 (1.24) d
LD bio-attack	11.02 (0.86) d
3rd set (A-A)	
GD bio-attack	9.90 (1.30) d
LD bio-attack	10.94 (1.49) d

Notes: Mean values are from 10 measurements. Standard deviations are in parentheses. Duncan's tests were used to evaluate the shear strength of the 2nd or 3rd double-bonded set consisting of 1 or 2 aged "A" lamellas, relative to the 1st (reference) double-bonded set consisting of 2 sound "S" lamellas, with indexes of significance: a—very significant decrease at 99.9% level; b—significant decrease at 99% level; c—less significant decrease at 95% level; d—insignificant change (decrease or increase) at <95% level.

According to Duncan's test for the shear strength, no significant differences occurred between the first set prepared from two sound lamellas (σ = 10.82 MPa) and the second or third sets prepared from one or two naturally aged lamellas (σ from 9.90 to 11.02 MPa). This unexpected result was potentially caused by PUR glue possibly seeping into *A. punctatum* larval holes when, after glue curing, quasi-pin joints between the jointed lamellas can be created with their orientation perpendicular or at a defined angle to shear forces, and thanks to which the shear strength of glulam beams from the naturally aged lamellas did not worsen. For this reason, the correlation analyses of MOR or MOE = f (σ) in Section 3.3 had to be performed without the sets of naturally aged lamellas and the groups of glulam beams made from them.

The shear strengths (σ) of glued joints of the first, second and third double-bonded sets prepared from artificially aged spruce lamellas were published and discussed by Vidholdová et al. [55], who used thermally aged lamellas, and by Ciglian and Reinprecht [12], who used chemically aged lamellas. In this work, the achieved shear strength values at Section 3.3 were applied in the correlation analyses of MOR or MOE = f (σ). The strength of the glued joints apparently decreased at the level of significance of 95% and as high as 99.9% when one lamella of the double set was aged at a temperature of $\geq$200 °C or in the presence of $ZnCl_2$ and H_3BO_3, or when both lamellas of the double set were aged at a temperature of $\geq$160 °C or in the presence of one of the four used inorganic preservatives.

3.3. MOR and MOE in Bending of Glulam Beams Versus Shear Strength of Glued Joints

Comparing the bending characteristics of glulam beams with the shear strength (σ) of glued joints, certain significant links were observed only for the MOR; however, none were observed for the MOE (Figure 6).

Generally, the artificial aging processes of spruce lamellas manifested in glulam beams as an apparent loss of MOR; however, they underwent less significant changes in their MOE (Table 1). This observation is in accordance with a greater negative effect of high temperatures above 160 °C or several aggressive chemicals on the wood strength than on the wood deflection or other deformation types [43]. As the bending characteristics of glulam beams depend not only on the used wooden materials-lamellas (species, quality, amount and location in beam, etc.) but also on the used glue (type, quality, amount, etc.), the mechanical properties of glulam beams significantly depend on the "wood-glue" interfaces. Their quality is reflected in the adhesion strength, which, in practice, is usually determined by the shear strength.

The shear strength (σ) of sets created from spruce lamellas and PUR glue was negatively influenced by the primary artificial aging of lamellas at high temperatures of $\geq$200 °C [55], as well as by the presence of inorganic preservatives, namely, $CuSO_4 \times 5H_2O$, $ZnCl_2$, H_3BO_3 and $(NH_4)_2SO_4$ [12].

In wood treated at high temperatures, there in polysaccharides and lignin are created new substances containing fewer polar -OH groups [43]. Such wood is more hydrophobic, has worse wettability and, finally, also forms weaker adhesion to polar glues. According to Taghiyari et al. [62], a reduction in the shear strength of glued joints created from thermally modified wood can be attributed to (1) a reduction in polar groups in the cell walls of wood due to the degradation of amorphous polysaccharides by heat treatment, resulting in fewer sites available for bonding; (2) the increased stiffness of cell walls after heat treatment, which results in a reduction in internal surfaces for the chemical bonding or mechanical interlocking of adhesives; and (3) a reduction in the wettability of wood, which may retard the proper penetration of glues. The formation of micro-cracks and checks, as well as the decrease in wood density due to heat treatment at temperatures above 180 °C [63], also might contribute to a decline in the shear strength of bonded wood.

A decrease in the shear strength of glued joints for wood impregnated with chemicals can be explained by many factors, e.g., by the reduced penetration of glue into lumens and intercellular spaces in wood filled with chemicals, followed by reduced adhesion, as documented by Atar et al. [64] and Özçifçi and Okcu [65].

Generally, due to the action of high temperatures or aggressive chemicals with a low or high pH value, the oxidative character of the lignin–saccharide matrix of wood can possibly deteriorate, the individual wood cells can be defibrated and, finally, a decrease in the mechanical properties of wood occurs [40,66]. A decrease in adhesion strength also occurs owing to the presence of new improper, especially hydrophobic, substances on the surfaces of thermally aged wood, as well as owing to the lower penetration of glue into cell lumens filled with preservatives or other chemical substances [67–69]. Subsequently, a decrease in the mechanical properties of wood and changes in the wood–glue phase interface can negatively affect the strength of glued joints [48,65].

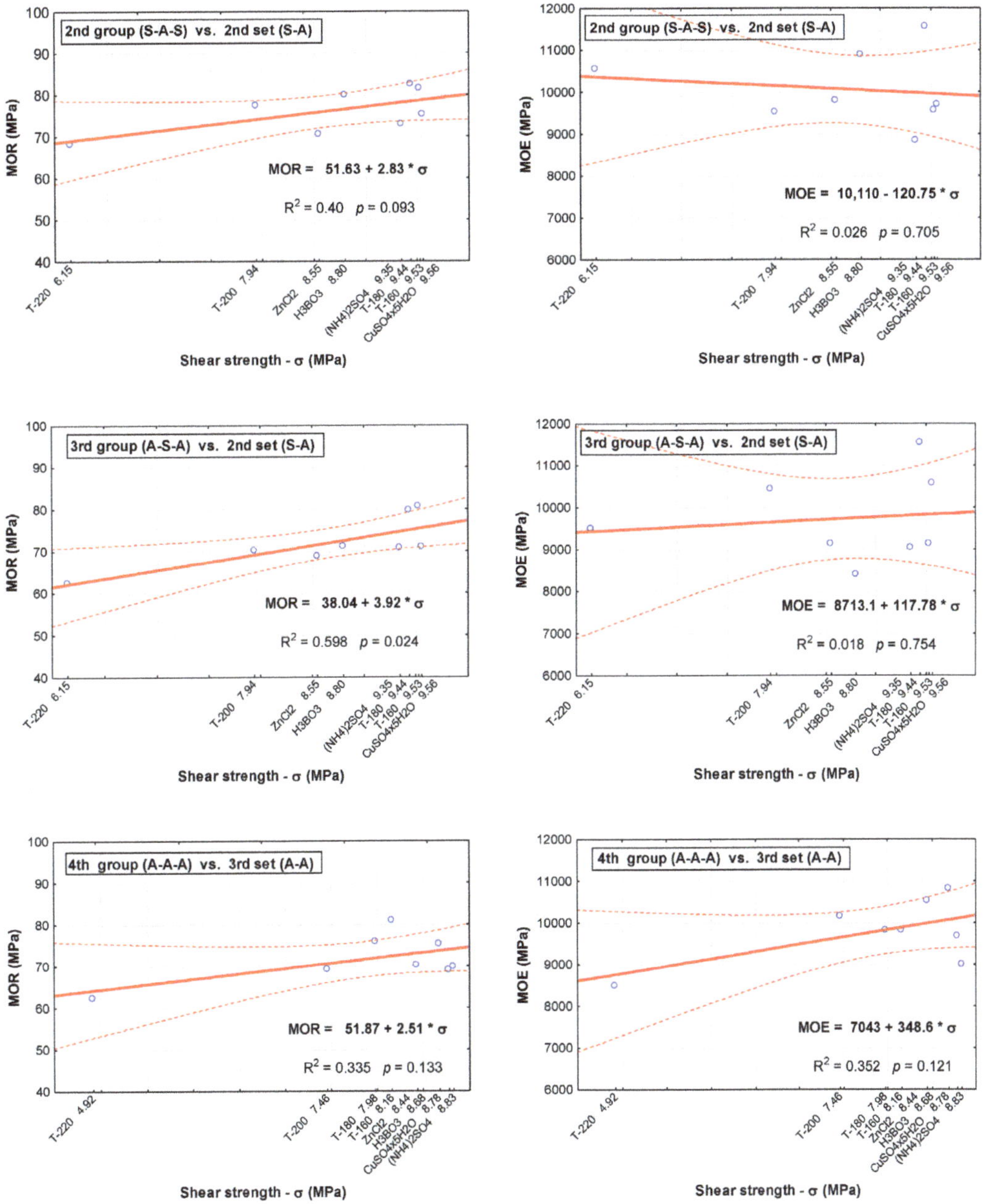

(a) (b)

Figure 6. The linear correlations between the mean shear strengths (σ) of glued joints of the 2nd or 3rd sets and the mean bending characteristics MOR (**a**) or MOE (**b**) of the three-layer glulam beams of the 2nd, 3rd and 4th groups.

From the above-mentioned views, it is possible to raise the question of why only slightly significant relationships were found between the MOR and σ (Figure 6a), or even none between the MOE and σ (Figure 6b). This question can be explained by other types of forces acting during bending, in addition to shear forces in the adhesive joint, e.g., by the action of tension and compression forces localized directly in the wood cells of surface lamellas, as well as by the combination of shear and non-shear loadings of the adhesive layers. At the same time, it is necessary to point out that while during the thermal aging of spruce lamellas, the quite homogeneous degradation of wood anatomical elements in their outer and inner zones could take place [70], on the contrary, during the chemical aging of spruce lamellas, the deeper penetration of inorganic chemicals into the refractory spruce wood is limited [71], and therefore, the degradation of the anatomical elements could occur only in the surface parts of spruce lamellas exposed to bonding, i.e., without a negative effect on the strength of the lamellas [12].

4. Conclusions

- The type of natural and artificial aging of spruce lamellas and the location of aged lamellas in three-layer glulam beams affected the MOR more than the MOE.
- The largest negative effect on the MOR of glulam beams was exerted by the primary thermal aging of lamellas at a maximum temperature of 220 °C.
- The smallest MOR was acquired by glulam beams in which the location of aged lamellas was in their surface "pressure and tension" zones, in accordance with bending theory.
- The linear correlation analyses confirmed certain relationships between the MOR of glulam beams and the shear strength (σ) of bonded joints also containing aged lamellas. However, in this experiment, these relationships were significant only in one case, i.e., if glulam beams had a sound lamella in the middle (A-S-A) and the bonded joints consisted of a sound lamella and aged lamella (S-A).
- In general, it can be stated that lamellas from recycled wood sources with damages caused by a higher number of insect galleries or by temperatures above 180 °C or the presence of certain inorganic wood preservatives should not be used for glulam beams designed for load-bearing structures.
- In practice, individual lamellas from aged woods should be selected and then, in the glulam beams, positioned according to their defect type, degree and range to ensure minimal decreases in the MOR and MOE. Before gluing, the lamellas should be dried to a suitable moisture content (usually 6–15%), and the surface of each lamella should be accurately machined to ensure that the thickness of the glue layer will be even throughout and that no dirt, grease and clumps of chemical substances will be present there.

Author Contributions: Conceptualization, L.R. and D.C.; methodology, L.R.; software, L.R., D.C. and J.I.; validation, L.R., D.C. and J.I.; formal analysis, L.R., D.C. and J.I.; investigation, L.R. and D.C.; resources, L.R., D.C., J.I. and J.S.; data curation, L.R., D.C. and J.I.; writing—original draft preparation, L.R. and D.C.; writing—review and editing, L.R. and J.I.; visualization, L.R. and D.C.; supervision, L.R.; project administration, L.R. and J.I.; funding acquisition, L.R. All authors have read and agreed to the published version of the manuscript.

Funding: This work was supported by the Slovak Research and Development Agency under contract No. APVV-17-0583.

Institutional Review Board Statement: Not applicable.

Informed Consent Statement: Not applicable.

Data Availability Statement: Not applicable.

Acknowledgments: This work was supported by the Slovak Research and Development Agency under contract No. APPV-17-0583, No. APVV-18-0378 and No. APVV-21-0049.

Conflicts of Interest: The authors declare no conflict of interest.

References

1. Stoeckel, F.; Konnerth, J.; Gindl-Altmutter, W. Mechanical properties of adhesives for bonding wood-a review. *Int. J. Adhes. Adhes.* **2013**, *45*, 32–41. [CrossRef]
2. Mohamad, W.H.W.; Razlan, M.A.; Ahmad, Z. Bending strength properties of glued laminated timber from selected Malaysian hardwood timber. *Int. J. Civ. Environ. Eng.* **2011**, *11*, 7–12.
3. Bourreau, D.; Aimene, Y.; Beauchêne, J.; Thibaut, B. Feasibility of glued laminated timber beams with tropical hardwoods. *Eur. J. Wood Wood Prod.* **2013**, *71*, 653–662. [CrossRef]
4. Liew, K.C.; Tan, Y.F.; Albert, C.M.; Raman, V. Cross-laminated timber and glulam from low-density Paraserianthes falcataria: A look into densification and shear strength. *Forests* **2022**, *13*, 1540. [CrossRef]
5. Lissouck, R.O.; Mvondo, R.R.N.; Ateba, F.R.; Pommier, R. Investigation of green-glued laminated timber from the Congo Basin: Durability, mechanical strength and variability trends of the bondlines. *South. For.* **2022**, *84*, 1–17. [CrossRef]
6. *EN 14080*; Timber Structures. Glued Laminated Timber and Glue Solid Timber-Requirements. European Committee for Standardization: Brussels, Belgium, 2013.
7. Kamke, F.A.; Lee, J.N. Adhesive penetration in wood-a review. *Wood Fiber Sci.* **2007**, *39*, 205–220.
8. Konnerth, J.; Kluge, M.; Schweizer, G.; Miljković, M.; Gindl-Altmutter, W. Survey of selected adhesive bonding properties of nine European softwood and hardwood species. *Eur. J. Wood Wood Prod.* **2016**, *74*, 809–819. [CrossRef]
9. Bockel, S.; Mayer, I.; Konnerth, J.; Niemz, P.; Swaboda, C.; Beyer, M.; Harling, S.; Weiland, G.; Bieri, N.; Pichelin, F. Influence of wood extractives on two-component polyurethane adhesive for structural hardwood bonding. *J. Adhes.* **2018**, *94*, 829–845. [CrossRef]
10. Hänsel, A.; Sandak, J.; Sandak, A.; Mai, J.; Niemz, P. Selected previous findings on the factors influencing the gluing quality of solid wood products in timber construction and possible developments: A review. *Wood Mater. Sci. Eng.* **2021**, *17*, 1925963. [CrossRef]
11. Hu, L.; Qin, L.; Wu, D.; Xu, H.; Yang, Z. Radial variation in bonding performance of preservative-treated wood within Pinus elliottii trees and its relationships with wood density and adhesive penetration. *BioResources* **2021**, *16*, 2204–2214. [CrossRef]
12. Ciglian, D.; Reinprecht, L. The effect of inorganic preservatives in the Norway spruce wood on its wettability and adhesion with PUR glue. *Appl. Sci.* **2022**, *12*, 5642. [CrossRef]
13. Ding, L.; Han, X.; Jiang, S. Impregnation of poplar wood with multifunctional composite modifier and induction of in-situ polymerization by heating. *J. Wood Chem. Technol.* **2021**, *45*, 220–228. [CrossRef]
14. Han, X.; Wu, W.; Wang, J.; Tian, Z.; Jiang, S. Hydrogen-Bonding-Aided Fabrication of Wood Derived Cellulose Scaffold/Aramid Nanofiber into High-Performance Bulk Material. *Materials* **2021**, *14*, 5444. [CrossRef] [PubMed]
15. Iždinský, J.; Vidholdová, Z.; Reinprecht, L. Particleboards from recycled wood. *Forests* **2020**, *11*, 1166. [CrossRef]
16. Iždinský, J.; Reinprecht, L.; Vidholdová, Z. Particleboards from recycled pallets. *Forests* **2021**, *12*, 1597. [CrossRef]
17. Goverse, T.; Hekkert, M.P.; Groenewegen, P.; Worrell, E.; Smits, R.E. Wood innovation in the residential construction sector; opportunities and constraints. *Resour. Conserv. Recycl.* **2001**, *34*, 53–74. [CrossRef]
18. Besserer, A.; Troilo, S.; Girods, P.; Rogaume, Y.; Brosse, N. Cascading recycling of wood waste: A review. *Polymers* **2021**, *13*, 1752. [CrossRef]
19. Fraanje, P.J. Cascading of pine wood. *Resour. Conserv. Recycl.* **1997**, *19*, 21–28. [CrossRef]
20. Höglmeier, K.; Weber-Blaschke, G.; Richter, K. Utilization of recovered wood in cascades versus utilization of primary wood-a comparison with life cycle assessment using system expansion. *Int. J. Life Cycle Assess.* **2014**, *19*, 1755–1766. [CrossRef]
21. Höglmeier, K.; Steubing, B.; Weber-Blaschke, G.; Richter, K. LCA-based optimization of wood utilization under special consideration of a cascading use of wood. *J. Environ. Manag.* **2015**, *152*, 158–170. [CrossRef]
22. Lubke, H.; Ihnát, V.; Kuňa, V.; Balberčák, J. A multi-stage cascade use of wood composite boards. *Wood Res.* **2020**, *65*, 843–854. [CrossRef]
23. Yang, T.H.; Lin, C.J.; Wang, S.Y.; Tsai, M.J. Characteristics of particleboard made from recycled wood-waste chips impregnated with phenol formaldehyde resin. *Build. Environ.* **2007**, *42*, 189–195. [CrossRef]
24. Wang, L.; Chen, S.S.; Tsang, D.C.; Poon, C.S.; Shih, K. Value-added recycling of construction waste wood into noise and thermal insulating cement-bonded particleboards. *Constr. Build. Mater.* **2016**, *125*, 316–325. [CrossRef]
25. Akinyemi, B.A.; Okonkwo, C.E.; Alhassan, E.A.; Ajiboye, M. Durability and strength properties of particle boards from polystyrene-wood wastes. *J. Mater. Cycles Waste Manag.* **2019**, *21*, 1541–1549. [CrossRef]
26. Arbelaez, R.; Schimleck, L.; Sinha, A. Salvaged lumber for structural mass timber panels: Manufacturing and testing. *Wood Fiber Sci.* **2020**, *52*, 178–190. [CrossRef]
27. Llana, D.F.; González-Alegre, V.; Portela, M.; Íñiguez-González, G. Cross laminated timber (CLT) manufactured with European oak recovered from demolition: Structural properties and non-destructive evaluation. *Constr. Build. Mater.* **2022**, *339*, 127635. [CrossRef]
28. Serrano, E. Mechanical performance and modelling of glulam. In *Timber Engineering*; Thelandersson, S., Larsen, H.J., Eds.; John Wiley & Sons, Ltd.: Chichester, UK, 2003; pp. 67–80.
29. Kotwica, E.; Krzosek, S. Glulam, LVL and X-lam-engineering wood products in the buildings of the future. *Ann. Wars. Univ. Life Sci. -SGGW For. Wood Technol.* **2018**, *104*, 256–260.

30. Ritter, M.A.; Williamson, T.G. State of the art report-Glulam timber bridge design in the US. In Proceedings of the International Council for Building Research Studies and Documentation. Working Commission W18-Timber Structures, Sydney, Australia, 7–8 July 1994; CIB-W18/27-l; Timber Scientific Publishing: Karlsruhe, Germany, 1995; 10p.
31. Malo, K.A.; Angst, A. Lepené lamelové drevo. (Glued laminated wood.). In *Handbook 1-Dřevěné konstrukce. Leonardo da Vinci Pilot Project-Educational Materials for Designing and Testing of Timber Structures CZ/06/B/F/PP/168007*; Chapter 5; TU Graz: Graz, Austria, 2008; pp. 46–59.
32. Ong, C.B. Glue-laminated timber (Glulam). In *Wood Composites*; Elsevier, Ltd.: Amsterdam, The Netherlands, 2015; pp. 123–140.
33. Thelandersson, S.; Larsen, H.J. *Timber Engineering*; John Wiley & Sons, Ltd.: Chichester, UK, 2003; 429p.
34. Risse, M.; Weber-Blaschke, G.; Richter, K. Eco-efficiency analysis of recycling recovered solid wood from construction into laminated timber products. *Sci. Total Environ.* **2019**, *661*, 107–119. [CrossRef]
35. Syrovátko, J. *Pružnostní a Pevnostní Charakteristiky Lepeného Nosníku Využívajícího Recyklovaného Dřeva (Elasticity and Strength Characteristics of Glued Beams Using Wood Recyclates)*; Mendelova univerzita v Brně, Lesnická a dřevařská fakulta: Brno, Czech Republic, 2019; 63p.
36. Mali, A.B. Using Salvaged Lumber as a Feedstock for Manufacturing Structural Glulam. Doctor's dissertation, Michigan State University, East Lansing, MI, USA, 2022; 166p.
37. Cavalli, A.; Cibecchini, D.; Togni, M.; Sousa, H.S. A review on the mechanical properties of aged wood and salvaged timber. *Constr. Build. Mater.* **2016**, *114*, 681–687. [CrossRef]
38. Kránitz, K.; Sonderegger, W.; Bues, C.T.; Niemz, P. Effects of aging on wood: A literature review. *Wood Science and Technology* **2016**, *50*, 7–22. [CrossRef]
39. Unger, A.; Schniewind, A.P.; Unger, W. Conservation of Wood Artifacts. Springer-Verlag: Berlin, Germany, 2001; 578p.
40. Reinprecht, L. *Strength of Deteriorated Wood in Relation to its Structure*; Monography-VPA 2/1992; Technical University of Zvolen: Zvolen, Slovakia, 1992; 76p, ISBN 80-228-0195-X.
41. Wilcox, W.W. Detection of early stages of wood decay with ultrasonic pulse velocity. *For. Prod. J.* **1988**, *38*, 68–73.
42. Reinprecht, L.; Pánek, M. Ultrasonic technique for evaluation of biodefects in wood: Part 2-In-situ and in-vitro analyses of old beams using ultrasonic and bending tests. *Int. Wood Prod. J.* **2013**, *4*, 22–29. [CrossRef]
43. Hill, C.A. Wood Modification: Chemical, Thermal and Other Processes. John Wiley & Sons., Ltd.: Chichester, UK,, 2006; 260p.
44. Reinprecht, L.; Repák, M. The impact of paraffin-thermal modification of beech wood on its biological, physical and mechanical properties. *Forests* **2019**, *10*, 1102. [CrossRef]
45. Baar, J.; Brabec, M.; Slávik, R.; Čermák, P. Effect of hemp oil impregnation and thermal modification on European beech wood properties. *Eur. J. Wood Wood Prod.* **2020**, *79*, 161–175. [CrossRef]
46. Bastani, A.; Adamopoulos, S.; Militz, H. Effect of open assembly time and equilibrium moisture content on the penetration of polyurethane adhesive into thermally modified wood. *J. Adhes.* **2016**, *93*, 575–583. [CrossRef]
47. Cai, C.; Antikainen, J.; Luostarinen, K.; Mononen, K.; Heräjärvi, H. Wetting-induced changes on the surface of thermally modified Scots pine and Norway spruce wood. *Wood Sci. Technol.* **2018**, *52*, 1181–1193. [CrossRef]
48. Pulngern, T.; Udtaranakron, T.; Chanto, K. Physical and mechanical behaviours of thermally modified rubberwood glulam beam under sustained and cyclic loading. *Wood Fiber Sci.* **2020**, *52*, 298–312. [CrossRef]
49. Felton, C.C.; De Groot, R.C. The recycling potential of preservative-treated wood. *For. Prod. J.* **1996**, *46*, 37–46.
50. Piao, C.; Monlezun, C.J.; Shupe, T.F. Glueline bonding performance of decommissioned CCA-treated wood. Part I: Without retreatment. *For. Prod. J.* **2009**, *59*, 36–42.
51. Deroubaix, G. *Waste Plan of the Strategic Committee of the Wood Sector*; Stakes, Observations and Orientations: Le Havre, France, 2017.
52. *EN 408*; Timber Structures. Structural Timber and Glued Laminated Timber. Determination of Some Physical and Mechanical properties. European Committee for Standardization: Brussels, Belgium,, 2012; (2010+A1:2012).
53. *EN 310*; Wood-based Panels. Determination of Modulus of Elasticity in Bending and of Bending Strength. European Committee for Standardization: Brussels, Belgium, 1993.
54. *EN 205*; Adhesives. Wood Adhesives for Non-Structural Applications. Determination of Tensile Shear Strength of Lap Joints. European Committee for Standardization: Brussels, Belgium, 2016.
55. Vidholdová, Z.; Ciglian, D.; Reinprecht, L. Bonding of the thermally modified Norway spruce wood with the PUR and PVAc adhesives. *Acta Fac. Xylologiae Zvolen* **2021**, *63*, 63–73.
56. Bekhta, P.; Niemz, P. Effect of high temperature on the change in color, dimensional stability and mechanical properties of spruce wood. *Holzforschung* **2003**, *57*, 539–546. [CrossRef]
57. Boonstra, M.; van Acker, J.; Tjeerdsma, B.; Kegel, E. Strength properties of thermally modified softwoods and its relation to polymeric structural wood constituents. *Ann. For. Sci.* **2007**, *64*, 679–690. [CrossRef]
58. Bytner, O.; Drożdżek, M.; Laskowska, A.; Zawadzki, J. Influence of thermal modification in nitrogen atmosphere on the selected mechanical properties of Black poplar wood (*Populus nigra* L.). *Materials* **2022**, *15*, 7949. [CrossRef] [PubMed]
59. Özçifçi, A. Impacts of impregnation with boron compounds on the bonding strength of wood materials. *Constr. Build. Mater.* **2008**, *22*, 541–545. [CrossRef]
60. Reinprecht, L.; Tiňo, R.; Šomšák, M. The impact of fungicides, plasma, UV-additives and weathering on the adhesion strength of acrylic and alkyd coatings to the Norway spruce wood. *Coatings* **2020**, *10*, 1111. [CrossRef]

61. Keskin, H.; Mutlu, E. The bending strength properties of some wood types impregnated with fire retardant chemicals. In Proceedings of the 4th International Symposium on Innovative Technologies in Engineering and Science (ISITES2016), Alanya/Antalya, Turkey, 3–5 November 2016; pp. 1074–1083.
62. Taghiyari, H.R.; Esmailpour, A.; Adamopoulos, S.; Zereshki, K.; Hosseinpourpia, R. Shear strength of heat-treated solid wood bonded with polyvinyl-acetate reinforced by nanowollastonite. *Wood Res.* **2020**, *65*, 183–194. [CrossRef]
63. Andromachi, M.; Ekaterini, R. Adhesive bond performance of heat-treated fir wood (Abies Borrissiregis). *Wood Res.* **2018**, *63*, 909–916.
64. Atar, M.; Keskin, H.; Yavuzcan, H.G. Impacts of impregnation with boron compounds on the bonding strength of some wood materials. In *Bonding of Modified Wood, Proceedings of the 5th COST E34 International Workshop, Bled, Slovenia, 6 September 2007*; Biotechnical Faculty, University of Ljubljana: Ljubljana, Slovenia, 2007; pp. 61–69.
65. Özcifci, A.; Okçu, O. The influence of the impregnating chemicals on the bonding strength of impregnated wood materials. *J. Appl. Polym. Sci.* **2008**, *107*, 2871–2876. [CrossRef]
66. Lai, Y.Z. Chemical degradation. In *Wood and Cellulosic Chemistry*; Dekker: New York, NY, USA, 2000; pp. 443–512.
67. Lorenz, L.F.; Frihart, C. Adhesive bonding of wood treated with ACQ and copper azole preservatives. *For. Prod. J.* **2006**, *56*, 90–93.
68. Ozdemir, T.; Temiz, A.; Aydin, I. Effect of wood preservatives on surface properties of coated wood. *Adv. Mater. Sci. Eng.* **2015**, 631835. [CrossRef]
69. Alade, A.A.; Naghizadeh, Z.; Wessels, C.B.; Tyhoda, L. A review of the effects of wood preservative impregnation on adhesive bonding and joint performance. *J. Adhes. Sci. Technol.* **2021**, *36*, 1–25. [CrossRef]
70. Mamoňová, M.; Ciglian, D.; Reinprecht, L. SEM analysis of glued joints of thermally modified wood bonded with PUR and PVAc glues. *Materials* **2022**, *15*, 6440. [CrossRef] [PubMed]
71. Lehringer, C.H.; Richter, K.; Schwarze, F.W.M.R.; Militz, H. A review on promising approaches for liquid permeability improvement in softwoods. *Wood Fiber Sci.* **2009**, *41*, 373–385.

applied sciences

MDPI

Article

The Differences in Shape Stability for Hornbeam (*Carpinus betulus* L.) Lumber with and without Spiral Grain

Peter Vilkovský *, Hugo Miroslav Uličný, Ivan Klement and Tatiana Vilkovská

Department of Wood Technology, Faculty of Wood Sciences and Technology, Technical University in Zvolen, T. G. Masaryka 24, 96001 Zvolen, Slovakia; xulicny@tuzvo.sk (H.M.U.); klement@tuzvo.sk (I.K.); tatiana.vilkovska@tuzvo.sk (T.V.)

* Correspondence: xvilkovsky@tuzvo.sk

Abstract: Spiral grains are one of the most important tree growth defects that can remarkably affect the shape stability of lumber. This growth defect is observed by an oblique arrangement of axially oriented cell elements in the wood. This research focused on identifying and measuring selected shape changes caused by the spiral grains in the log. Hornbeam logs with spiral grain and with no spiral grain aged 60 and 52 years were selected to detect shape changes. Following the research and the norms of STN EN 1309-3 (49 1013), DIN (4074-5), and ASTM D245, it can be stated that spiral grain is an essential factor affecting the shape stability of hornbeam lumber. In the case of lumber with spiral grain, the spiral grain increased by up to 8° within three months. The warp of both types (bow, crook) increased exponentially throughout the study, peaking between the third and fourth week with an increase in warp bow of 26 mm and 17 mm in warp crook.

Keywords: hornbeam; spiral grain; shape changes; lumber; bow; crook

1. Introduction

The accuracy of the log dimensions, the material dimensional stability, stiffness, strength, resistance to rot, density, or even the appearance of the material are some of the requirements that the customer may place on timber. One of the most common reasons customers avoid using wood, e.g., in timber structures, is lower dimensional stability. Visible changes in shape, leading to unacceptable difficulties in wood processing, are the most significant adverse effects of timber construction. Since spiral grain in wood causes issues in timber structures, there are reasons why wood is often not preferred for use in structures and is replaced by other materials, such as concrete with steel. Moreover, the negative impact of the spiral grain on the quality of timber and its products can be observed, as the fibers copy the curvature but not the saw cut. Such a log has lower strength properties than a log with no spiral grain. When processing logs, the size of the spiral grain must be considered. This fact is essential in the case of a warp twist because it negatively affects the tensile properties of wood. Logs with extreme spiral grain are unacceptable as the material is further processed. Depending on the purpose of use, the spiral grain plays an essential role in wood utilization; therefore, it was also included in the classification criteria [1]. The changes in the mechanical properties (Hickory, Ash, Elm, Beech, Maple) due to the spiral grain were also investigated in [2]; it was found that the grain deviation angle (from 0° to 15°) from the rectilinear pattern has a significant impact on the values of mechanical properties. For all species (Hickory, Ash, Elm, Beech, Maple), they investigated an absolute decrease in value exceeding 50% of the initial value; this decrease was even over 60% for the two examined wood species. These findings have direct implications for the construction industry, highlighting the need for careful consideration of spiral grain in timber selection and utilization. Similar results were discovered in [3]. The authors investigated the effect of the global slope of grain on the bending strength of scaffold boards. Results showed that the magnitude of the slope of grain has a very marked effect on its bending strength (boards were 24% weaker in comparison with boards made of equivalent timber quality). The impact of spiral grain formation on timber quality is not just a theoretical concept. It is a practical concern that is reflected in the standards, which clearly state the permissible curvature values when sorting timber [4]. For instance, according to the American ASTM standards, left-handed logs and subsequent products may not be suitable for certain applications, such as ceiling beams, rafters, purlins, etc. However, material with left-handed fibers can still be used in timber structures for load-bearing walls, highlighting the nuanced understanding and application of this knowledge in the industry.

The formation of spiral grains is controlled genetically. However, environmental factors also have a noticeable influence on spiral grain—genetic predisposition, older age, position, type of cell formation, rotation of the earth, or the gravitational effects of the moon [5–8]. Most of the data from previous studies indicate that tortuosity formation originates in the cambium and may be associated with cell division in this region. This process can be regulated according to studies of the plant hormones [9]. Three methods of cell division are thought to cause spiral grain. The angle of rotation of the spiral grain depends on the transverse divisions of the vertically oriented cambial cells. The cumulative effect of transverse divisions accompanied by poor growth has received much attention for over a century [10]. Poor growth alone with no transverse division, an essential factor in spiral grain, means that the fibers in the cambial layer slide past each other, changing their relative orientation, i.e., the two ends of the cells move in opposite directions, resulting in a slight rotation of the cells. Therefore, the angle between the cell and the pith changes. The third process, which is determined to have a significant effect on the formation of spiral grain, is the imperfect division and differentiation of cells that accompany such divisions, i.e., the division is considered incomplete if the partially newly formed cell wall does not extend from one end of the cell to the other. However, there is still no evidence that these theories are more likely or, conversely, less likely to be correct than others [11].

In several articles, the authors confirm that standing on rocky and one-sided terrain produced significantly more trees with spiral grain than those on open stands with irregular spacing between trees. Border trees in stands are often dominant and have wider annual rings and relatively large living crowns compared to average trees in the same stand. According to several studies, the trees at the border of the forest have more pronounced spiral grains than trees inside the forest [12].

The shape instability of the hornbeam tree (*Carpinus betulus* L.) could be better for woodworking. Its sensitivity to moisture and low durability prevent its use outdoors [13]. This is one of the reasons why it is mainly used as firewood. Shape stability is an important property that significantly affects the final product based on changes in its mechanical properties. In the Slovak Republic, timber structures can only be built from Spruce wood (*Picea abies* (L.) H. Karst.) or certified timber, but why not use hornbeam timber? There is a significant amount of hornbeam, and in the future a combination with Beech (*Fagus sylvatica* L.) could be used, as these two trees occur in the same forest stands, which could save funds for mining and construction material for construction, e.g., in timber construction, where an upward construction trend is still expected.

Shape changes caused by changes in humidity during drying or operation are the main problem of wood materials intended for timber structures. Changes in moisture content are accompanied by significant shrinkage or swelling of the material. The stability of timber is often connected with the cylindrical geometry caused by the growth rings, the wood material's orthotropic nature, and the wood fiber's tendency to grow spirally around the trunk; this is often considered the most damaging deformation of timber [14]. Forsberg and Warenjö [15] found that the angle of the spiral grain on the surface of the trunk and the slope of the angle of the fibers of the curve are strongly correlated with the rotation angle. Ekevad [16] described the basic mechanical understanding of the causes of shape changes associated with the tortuosity of fibers. Ormarsson [17] states that the spiral grain angle is the angle between the pith and the spiral grain direction. This angle often shows a radial deviation in round wood. Therefore, the spiral grain angle is one of the most critical parameters in shape changes linked to warps. It also affects warp bow but almost does not affect transverse curvatures [18]. Shape changes in dried lumber can depend on several factors. One is the original position of the lumber before the log's breakdown: several authors [18–21] addressed this issue by focusing on the original position of the log by subsequent comparison of shape stability between produced radial and tangential lumber. This study showed a significant redistribution of stress when the logs were cut, resulting in a distinctly visible warp bow, especially on the side lumber. The warps created during the processing of the logs and the remaining growth stresses affect the shape changes after drying [20].

The main goal of the presented research was to analyze and compare the shape stability of the lumber that contained or did not contain spiral grain during natural drying of the wood hornbeam (*Carpinus betulus* L.).

2. Materials and Methods

The logs of Hornbeam (*Carpinus betulus* L.) were selected in the district of Ziar nad Hronom (Slovakia), in the cadastral territory of the municipality of Hronska Dubrava, with an average altitude of 457 m above sea level (ASL). The logs were yielded by the University Forest Enterprise in Zvolen. The geological substratum in this location consists mainly of andesites and pyroclastics. The exact sample location is shown in the map marked in green (Figure 1). The forests in this area are dominated by deciduous trees, mainly Beech (*Fagus sylvatica* L.) and Hornbeam (*Carpinus betulus* L.).

Figure 1. Marking of the sample location (green dot) using a map from the company FATMAP (https://fatmap.com/). GPS coordinates: trunk with no spiral grain—N 48°34′57.9″ E 18°59′43.6″; with spiral grain—N 48°34′58.5″ E 18°59′41.0″.

It was necessary to select two types of trees for the research. The first trees were without spiral grain, and the second type of trees with spiral grain (Figure 2). The diameters of the logs without spiral grain averaged about 38 cm, and the logs with spiral grain averaged about 35 cm. It is also possible to see in Figure 2 the amount of lumber that was used for research. For logs without spiral grain, this was 17 pieces of lumber; for logs with spiral grain, it was 14. Five selected samples were measured of the warp crook and warp bow and changes in the spiral grain during three months of air drying (Figure 2).

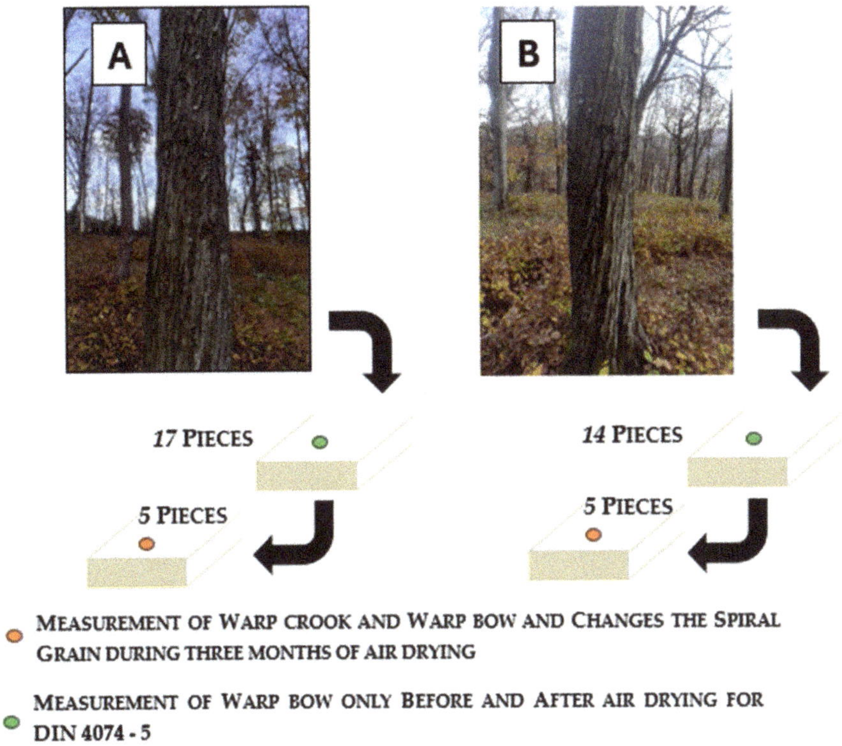

Figure 2. Showed trees without spiral grain (**A**) and with a spiral grain (**B**).

2.1. Preparation of Hornbeam Lumber

A band saw (Mebor 1200, MEBOR d.o.o., Železniki, Slovenia) was used to process the logs. The through-and-through sawing pattern used in the research can be seen in Figure 3. The selection of lumber in a log with or without spiral grain was influenced by visual sorting. Before sawing, the spiral grain was measured on the logs with the spiral grain following the standard [22].

The radial and tangential lumber was obtained by a through and through sawing pattern. Subsequently, it was marked and prepared for edging. While processing logs on the band saw, lumber was marked with numbers so that we knew in which order the logs were processed. Subsequently, to this sample marking, we added the order marking from the main lumber after edging, plus whether the lumber is from the trunk with no spiral grain or with spiral grain. The last data of the marking were the type of lumber, and whether it is tangential or radial, e.g., 6-3/N/R means sixth row (vertically), third in a row (horizontally), N lumber with no spiral grain, R radial lumber. Using an edger saw, the sideboards were removed.

The lumber was thoroughly sorted into radial and tangential. Pieces of lumber with thickness (t) = 0.025 m, width (w) = 0.095 m, and length (l) = 2 m were selected from each log. The selected pieces were placed in the stack. One type of stack is for lumber with no spiral grain, and the other is for lumber with spiral grain. Both stacks (Figure 4) were built on four concrete bases with a height of 0.435 m and 1.80 m apart on a reinforced surface. The lumber in the drying was stacked in the order: tangential–radial–tangential–radial–tangential.

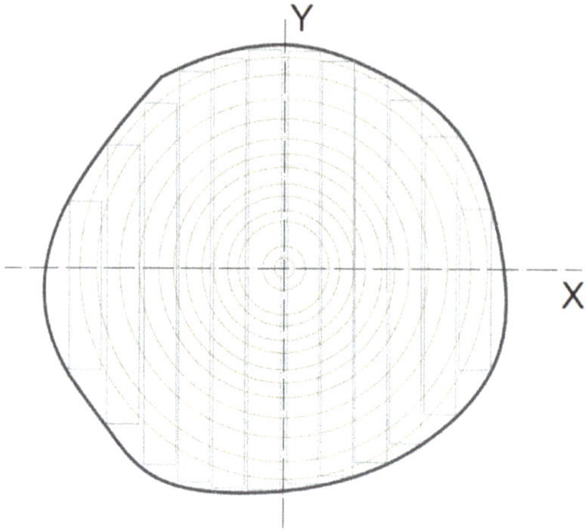

Figure 3. The through and through sawing pattern used in the production of the hornbeam lumber.

Figure 4. Drying stack for lumber with (pattern of blue) and without (pattern of green) spiral grain.

2.2. Conditions of Air Drying

The air drying was conducted from 10 November 2022 to 14 February 2023. The air-drying conditions were recorded using the device Data Logger Comet S3631 (COMET System, Rožnově pod Radhoštěm, Czech Republic). Hourly intervals were set for the research due to the high accuracy needed in the measured data. The air temperature was recorded in °C and the relative humidity in %.

Moisture loss was detected using a sample coated on both sides with a silicone layer and put in the middle of a stack. The silicone layer was painted to prevent rapid drying of the cross-section of the sample. The reason for applying silicone was the smaller sample compared to the other samples.

2.3. Types of Measurement on the Lumber

The fiber angle of deviation from the log longitudinal axis with spiral grain was measured on the lumber. The angles were measured before and after air drying. The measurements were according to the standard (Figure 5) [22].

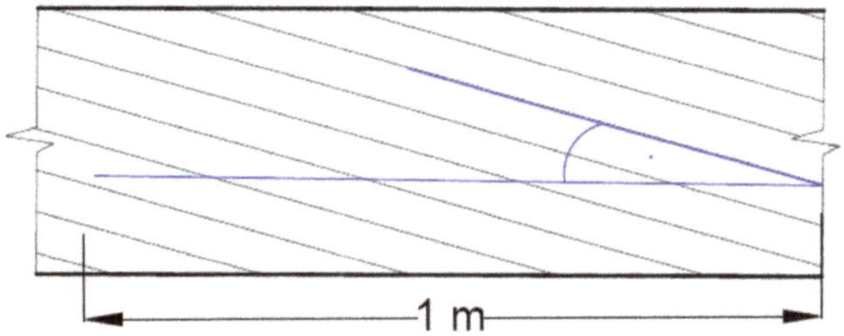

Figure 5. Measurement of the angle of deviation of the fibers from the longitudinal axis.

2.4. Measured Change of Shape by STN EN 1309-3 (49 1013)

As a part of the research focused on the shape stability of lumber, two types of warps were measured: warp bow and warp crook (Figure 6). The measurement was carried out using a digital caliper in the laboratories of the Technical University in Zvolen according to the standard [22].

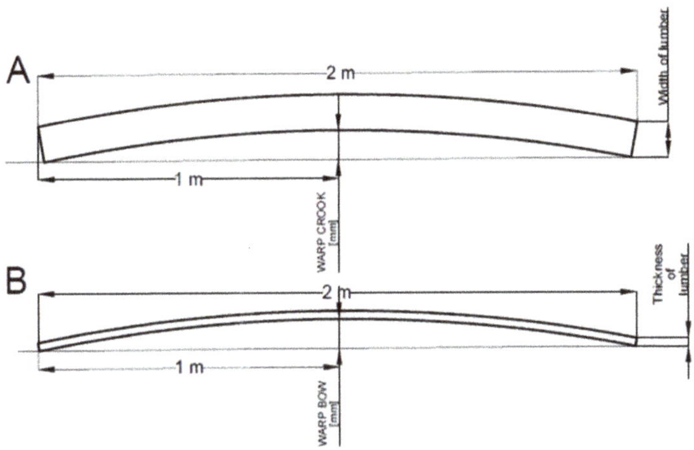

Figure 6. Measured change of shape by [22]: Warp crook (**A**), Warp bow (**B**).

2.5. Classification of Lumber According to Standards

The studies of two types of standards, [23] visually graded hardwood and [24] Standard Practice for Establishing Structural Grades and Related Allowable Properties for Visually Graded Lumber, were used and compared with our measured data.

2.6. Evaluation of Shape Stability by Statistical Methods

All values of warps were also processed at the level of inductive statistics. The analysis of variance ANOVA method was chosen for the above statistical evaluations, which was depicted using a Box 2D graph using a significance level ($p < 0.05$).

3. Results and Discussion

Air drying conditions were recorded, and temperatures ranged from 9 °C in November to −15 °C in February. Relative humidity ranged from 90% in November 2022 to 60% in February (Figure 7).

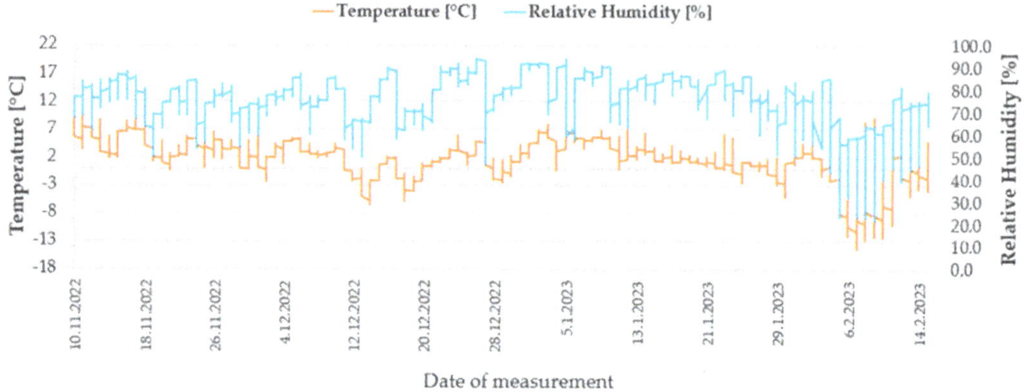

Figure 7. The conditions for air drying during the 3 months of research.

3.1. *Changes of Spiral Grain*

Wood is an orthotropic material with significant differences in volumetric shrinkage between the longitudinal, radial, and tangential directions during drying.

After the air drying, we first evaluated changes of spiral grain on the surface of the lumber with and without spiral grain (Figure 8). Changes on the surface were detected in the case of lumber with spiral grain, and most on lumber 4-1/To/R, where the spiral grain changed from 5° to 13°. An increase in the deviation of fibers from the longitudinal axis of the sample also occurred in the case of lumber with no spiral grain marked 2-1/N/T, from 0° to 3°. This change could mainly be due to a significant shift in shape, which we detected for lumber with spiral grain in combination with loss of moisture content during air drying. This statement confirmed studies [17,18,20]. Ormarsson [17], based on numerical simulations to determine deformations that develop in lumber during changes in moisture, determined that the most common reasons for formation shape change are spiral grain, reaction wood, location of lumber in the log, etc. Similar results were observed in the research of Ormarsson et al. [18], which discovered that the spiral grain angle and the lumber location within the log were the main contributors to the twist deformation.

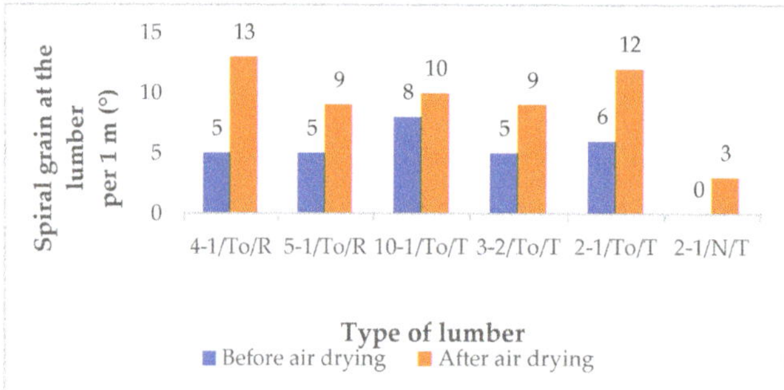

Figure 8. Spiral grain occurring in lumber.

3.2. *The Effect of Spiral Grain on Strength Properties*

Following the ASTM standard [24] indicating a probable decrease in the strength properties of lumber based on spiral grain, the estimated percentage reductions in bending strength were assigned to the measured data (Table 1). Table 1 shows that strength reduction in lumber with spiral grain should occur in all lumber, which was determined by changes in spiral grain after air drying, at about 60%. The estimated percentage reductions in bending strength is directly connected with an increase in the grain angle, as confirmed in several research studies [8,10,12]. Authors [2,3] investigated the magnitude of the spiral grain, which has a very marked effect on decreased values for mechanical properties. Results in [2] showed that species such as Beech or Hickory decreased Modulus of Rupture (MOR) by about 44% for Hickory and 55% for Beech, and Modulus of Elasticity (MOE) for Hickory by 47% and for beech by 52%. Similar bending strength decreased in [3] by about 24%, compared to boards of equivalent timber quality. In the case of lumber without spiral grain, the reduction in strength could probably occur only in lumber 2-1/N/T, where there was a change in angle of the grain from 0° to 3° due to shape changes after drying.

Table 1. Estimated reduction in strength according to ASTM D245 [22].

Spiral Grain in Degrees (°)		Strength Reduction in Bending (Experimental)	Strength Reduction in Bending (Numerical Simulation)
		(%)	
4-1/To/R	13	60	50
2-1/To/T	12	60	50
10-1/To/T	10	60	50
5-1/To/R	9	60	50
3-2/To/T	9	60	50
2-1/N/T	3	0	10

3.3. Warp Bow

Figure 9 shows a warped bow during the air drying of lumber containing spiral grain and lumber without spiral grain for three months. In Figure 9A, more considerable shape changes in the observed warps were achieved in the case of lumber containing spiral grain, where the maximum values reached more than 25 mm, compared to lumber without spiral grain, where the maximum values reached only 7 mm (Figure 9B). The most significant warp in a warped bow occurred in the case of lumber 2-1/To/T. This lumber was placed closer to the girth of the log. According to the statistical data processed using the program STATISTICA 13.3, no significant difference was confirmed. This means that the initial position of the lumber in the log did not affect this type of shape change (warp bow). In general, however, several authors claim [18–20,25] that shape changes for warp bow and warp cup depend on the original position in the log.

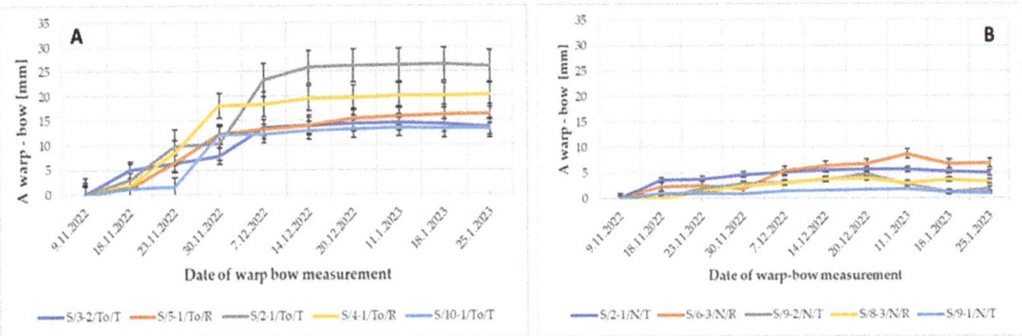

Figure 9. Warp bow (**A**) with spiral grain; (**B**) without spiral grain.

3.4. Warp Crook

Figure 10 shows a warp crook after three months of air drying the lumber containing spiral grain (A), and vice versa without spiral grain (B). It was confirmed that the warp crook was more extensive in lumber containing spiral grain (18 mm) than without spiral grain (9 mm). The highest values were measured for lumber outside the center (pith) of the log with a spiral grain (2-1/To/T, 4-1/To/T) and even without a spiral grain (2-1/N/T, 9-1/N/T). Results showed that, similar to those for the first shape change (warp bow), it was confirmed to affect the original position in the log on the size measured for shape changes. The results of [17,19,20] showed clearly that shape changes in the lumber are strongly influenced by the original location of the separate pieces within the log and how they are oriented in the products. Their results also showed that material parameters, such as the shrinkage coefficients and the spiral grain angle, considerably influence the warping [20].

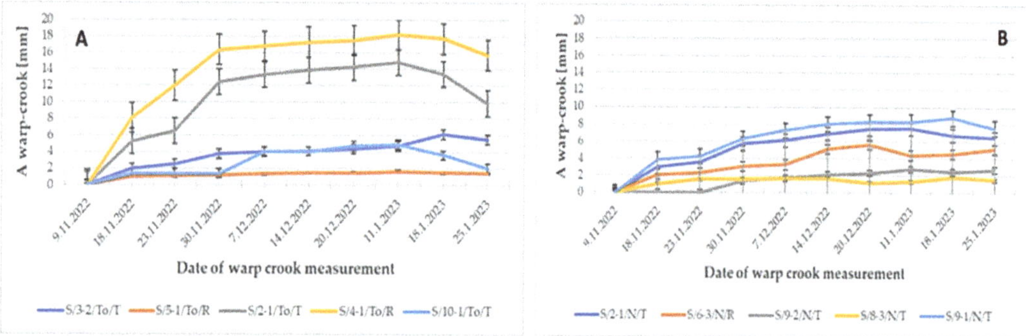

Figure 10. Warp crook (**A**) with spiral grain; (**B**) without spiral grain.

3.5. Visual Evaluation of Shape Changes

After three months of air drying, shape changes in lumber were visually assessed. All lumber from the cage without spiral grain and with spiral grain were disassembled and visually evaluated. Remarkable differences were observed already on the first inspection for comparison (Figure 11 with Figure 12). Remarkable shape changes in the lumber (Figure 11) were caused by a strong spiral grain for lumber (5–8° per 1 m of the length) as well as strong spiral grain for log (0.156 m per 1 m of the length).

Figure 11. Lumber with spiral grain after the three-month research.

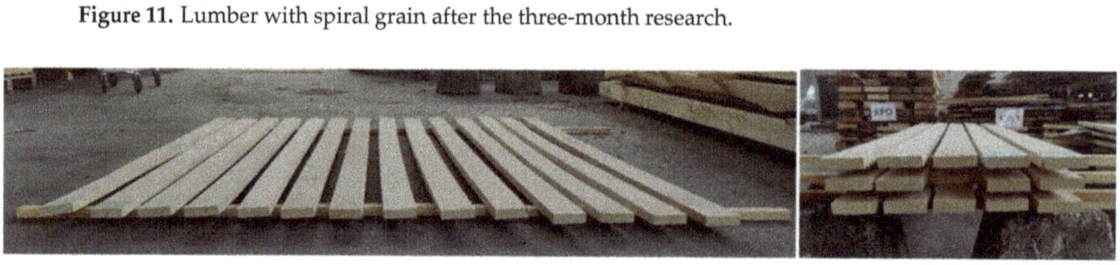

Figure 12. Lumber without spiral grain after the three-month research.

Shape changes occur during desorption or with water sorption by wood, around FSB = 30%. The size of the shape changes is not constant; they are different in the cross-section of the lumber. The shape size changes depending on the wood density and the fiber orientation, as also mentioned by [26].

3.6. Classification of Lumber Strength

Lumber was classified into strength classes based on warp bow values following [23]. Lumber was divided into three strength classes (LS $_7$, LS $_{10}$, LS $_{13}$), or it was classified as reclaimed wood because it did not meet even the lowest requirements for classifying lumber in the strength classes LS $_7$, LS $_{10}$, and LS $_{13}$. In the case of lumber with spiral grain, only one sample was included in strength class LS $_7$ (Table 2). On the contrary, for lumber with no spiral grain, 12 pieces were classified (Table 3).

Table 2. Classification of lumber with spiral grain according to the standard (DIN 4074-5) [23].

Classification of Lumber with Spiral Grain According to the Standard (DIN 4074-5)					
Sample	**Bow (mm)-O**	**LS $_7$**	**LS $_{10}$**	**LS $_{13}$**	**Reclaimed Wood**
3-2/To/T	21.49				O
5-1/To/R	18.93				O
2-1/To/T	23.23				O
4-1/To/R	21.94				O
10-1/To/T	17.35				O
7-1/To/R	36.92				O
6-1/To/R	9.68	O			
6-2/To/T	32.59				O
8-2/To/T	26.63				O
3-1/To/T	13.65				O
9-1/To/T	16.26				O
7-2/To/T	26.2				O
8-1/To/R	20.16				O
5-1/To/R	13.34				O

Table 3. Classification of lumber without spiral grain according to the standard (DIN 4074-5) [23].

Classification of Lumber with No Spiral Grain According to the Standard (DIN 4074-5)					
Sample	**Bow (mm)-O**	**LS $_7$**	**LS $_{10}$**	**LS $_{13}$**	**Reclaimed Wood**
1-1/N/R	12.74				O
8-2/N/T	13.14				O
8-1/N/R	10.55	O			
10-2/N/T	10.51	O			
7-1/N/R	11,92	O			
3-3/N/R-T	6.71		O		
2-2/N/T	7.89		O		
11-1/N/T	16.63				O
3-2/N/R	8.68	O			
10-1/N/T	12.98				O
5-2/N/T	17.11				O
6-1/N/R	6.86		O		
2-1/N/T	4.97			O	
6-3/N/R	6.89		O		
9-2/N/T	1.81			O	
8-3/N/T	3.32			O	
9-1/N/T	1			O	

3.7. Statistical Evaluation—Warp Bow

Statistical evaluation based on the Box 2D plot at the significance level ($p < 0.05$) confirmed a significant difference between lumber with and with no spiral grain in warp bow. The resulting p-value for the given warp reached $p = 0.00001$ (Figure 13).

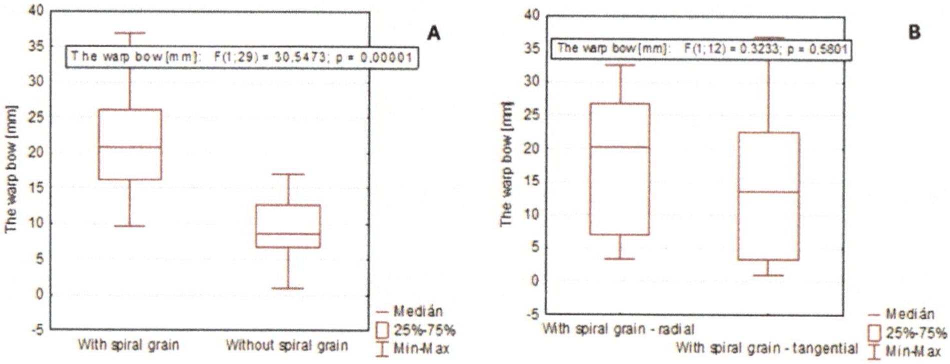

Figure 13. Statistical evaluation. Warp bow (**A**) difference between the log with and without spiral grain; (**B**) difference between tangential and radial lumber.

When comparing tangential and radial lumber, no significant difference was confirmed, which is also indicated by the resulting value of $p < 0.5801$, i.e., the warp does not depend on the initial position of the log, Figure 13B.

3.8. Statistical Evaluation—Warp Crook

In the case of this type of warp, i.e., a warp crook, there is a significant difference between the two types of logs with and with no spiral grain. The final *p*-value for warp crook reached $p = 0.0436$, Figure 14A.

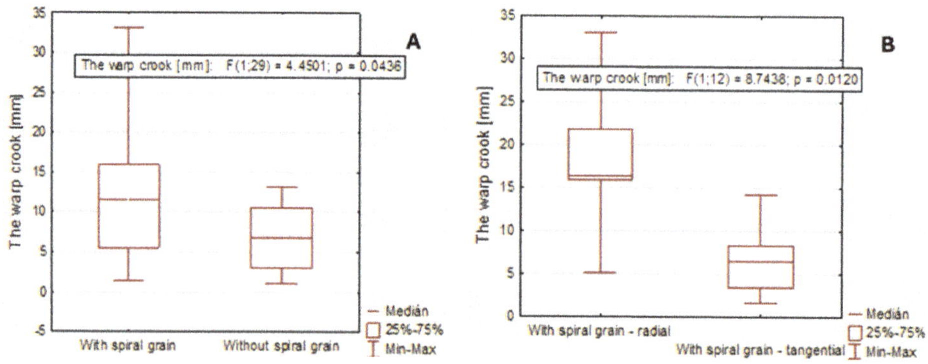

Figure 14. Statistical evaluation. Warp crook (**A**) difference between the log with and without spiral grain; (**B**) difference between tangential and radial lumber.

A significant difference was confirmed after comparing tangential and radial lumber, also indicated by the resulting value from the STATISTICA program, which was 0.012. Therefore, the warp depends on the initial position of the log during the first stage of wood processing (Figure 14B).

4. Conclusions

- Our research confirmed that the spiral grains remarkably affect shape stability. A warped bow with spiral grain values increased by almost 50% compared with lumber without spiral grain. This statement was confirmed even with statistical results where the set limit value of $p < 0.05$ was not exceeded, either with warp crook ($p = 0.0436$) or with warp bow ($p = 0.00001$).

- Regarding warp crook, statistical significance was also confirmed for different types of lumber (radial or tangential), where the significance value reached $p = 0.0120$. This result can be explained mainly by the original position of tangential lumber, which was closer to the girth of the log, than radial lumber, which was closer to the pith, which could lead to different shape stability.
- According to the measured values, the warps during the warp crook depend on the sizes of the spiral grain and the original position of the lumber, which means whether the lumber is closer to the girth of the log or closer to the pith.
- According to DIN standard (4074-5) and sizes of shape change for warp bow, only one piece of lumber would be obtained in the case of lumber with spiral grain and up to twelve pieces of lumber in the case of sawn lumber without spiral grain. This result confirms that the spiral grain remarkably influences the resulting quality of the lumber.

This research could broaden the possibilities of using hornbeam lumber (*Carpinus betulus* L.) as a construction material.

Author Contributions: Conceptualization, I.K. and P.V.; methodology, H.M.U. and I.K.; software, P.V. and I.K.; validation P.V. and I.K.; formal analysis, P.V.; investigation P.V.; resources, T.V.; data curation, I.K.; writing—original draft preparation, P.V.; writing—review and editing, T.V.; visualization, H.M.U.; supervision, I.K.; project administration, P.V. and T.V.; funding acquisition, I.V and P.V. All authors have read and agreed to the published version of the manuscript.

Funding: This work was funded by the Slovak Research and Development Agency under contract no. APVV-21-0049.

Institutional Review Board Statement: Not applicable.

Informed Consent Statement: Not applicable.

Data Availability Statement: The data presented in this study are available on request from the corresponding author.

Acknowledgments: This work was supported by the Scientific Grant Agency of the Ministry of Education, Science, Research and Sport of the Slovak Republic and the Slovak Academy of Sciences—project VEGA no. 1/0063/22.

Conflicts of Interest: The authors declare no conflicts of interest.

References

1. Sühlfleisch, K. Wie Krümmung und Drehwuchs die Holzqualität beeinflussen. *Waldarbeit Holzmerkmale Magazine*, April 2020; pp. 38–39.
2. Mania, P.; Siuda, F.; Roszyk, E. Effect of Slope Grain on Mechanical Properties of Different Wood Species. *Materials* **2020**, *13*, 1503. [CrossRef] [PubMed]
3. Pope, D.; Marcroft, J.; Whale, L. The effect of global slope of grain on the bending strength of scaffold boards. *Holz Als Roh Werkst.* **2005**, *63*, 321–326. [CrossRef]
4. Säll, H. Spiral Grain in Norway Spruce. *Acta Wexionensia* **2002**, *22*, 171.
5. Pyszyński, W. Mechanism of formation of spiral grain in Aesculus stems: Dissymmetry of deformation of stems caused by cyclic torsion. *Acta Soc. Bot. Pol.* **1977**, *46*, 501–522. [CrossRef]
6. Mattheck, G.C. *Case Studies*; Springer: Berlin/Heidelberg, Germany, 1991; pp. 20–115. ISBN 978-3-540-54276-6. [CrossRef]
7. Eklund, L.; Säll, H. The influence of wind on spiral grain formation in conifer trees. *Trees* **2000**, *14*, 324–328. [CrossRef]
8. Leelavanichkul, S.; Cherkaev, A. Why the grain in tree trunks spirals: A mechanical perspective. *Struct. Multidisc. Optim.* **2004**, *28*, 127–135. [CrossRef]
9. Zagórska-Marek, B.; Little, C.H.A. Control of fusiform initial orientation in the vascular cambium of Abies balsamea stems by indol-3-ylacetic acid. *Can. J. Bot.* **1986**, *64*, 1120–1128. [CrossRef]
10. Kubler, H. Function of spiral grain in trees. *Trees* **1991**, *5*, 125–135. [CrossRef]
11. Chester, T.; Stith, D.; Morse, K.; Fidler, W. Spiral of Wood Grain in Tree Trunks at San Jacinto Mountain/Possible Causes for Spiral Grain. 2014. Available online: http://tchester.org/sj/analysis/tree_wood_grain_helicity.html (accessed on 10 June 2024).
12. Wellner, C.A.; Lowery, D.P. *Spiral Grain: A Cause of Pole Twisting*; Intermountain Forest a Range Experiment Station, Forest Service, U.S. Department of Agriculture: Ogden, UT, USA, 1967; Volume 38. [CrossRef]

13. Fodor, F.; Lankveld, C.; Németh, R. Testing common hornbeam (*Carpinus betulus* L.) acetylated with the Accoya method under industrial conditions. *iFor. Biogeosci. For.* **2017**, *10*, 948–954. [CrossRef]
14. Ormarsson, S.; Dahlblom, O.; Petersson, H. A numerical study of the shape stability of sawn timber subjected to moisture variation—Part 1: Theory. *Wood Sci. Technol.* **1998**, *32*, 325–334. [CrossRef]
15. Forsberg, D.; Warensjö, M. Grain Angle Variation: A Major Determinant of Twist in Sawn *Picea abies* (L.) Karst. *Scand. J. For. Res.* **2001**, *16*, 269–277. [CrossRef]
16. Ekevad, M. Twist of wood studs: Dependence on spiral grain gradient. *J. Wood Sci.* **2005**, *51*, 455–461. [CrossRef]
17. Ormarsson, S. Numerical Analysis of Moisture-Related Distortion in Sawn Timber. Ph.D. Thesis, Chalmers University of Technology, Department of Structural Mech, Gothenburg, Sweden, 1999; p. 213.
18. Ormarsson, S.; Dahlblom, O.; Petersson, H. A numerical study of the shape stability of sawn timber subjected to moisture variation—Part 2: Simulation of drying board. *Wood Sci. Technol.* **1999**, *33*, 407–423. [CrossRef]
19. Ormarsson, S.; Dahlblom, O.; Petersson, H. A numerical study of the shape stability of sawn timber subjected to moisture variation—Part 3: Influence of annual ring orientation. *Wood Sci. Technol.* **2000**, *34*, 207–219. [CrossRef]
20. Ormarsson, S.; Dahlblom, O.; Johansson, M. Finite element study of growth stress formation in wood and related distortion of sawn timber. *Wood Sci. Technol.* **2009**, *43*, 387–403. [CrossRef]
21. Straže, A.; Kliger, R.; Johansson, M.; Gorišek, Ž. The influence of material properties on the amount of twist of spruce wood during kiln drying. *Eur. J. Wood Prod.* **2011**, *69*, 239–246. [CrossRef]
22. *STN EN 1309-3 (49 1013)*; Gul'atina a Rezivo, Metódy Merania Rozmerov, Časť 3: Znaky a Biologické Poškodenie. (Logs and lumber, Methods of Measuring Dimensions, Part 3: Signs and Biological Damage). Slovak Office of Standards, Metrology and Testing: Bratislava, Slovakia, 2018.
23. *DIN 4074-5:2008-12*; Sortierung von Holz nach der Tragfähigkeit—Teil 5: Laubschnittholz. (Sorting Wood According to Load-Bearing Capacity—Part 5: Hardwood Lumber). German Institute for Standardization: Berlin, Germany, 2008.
24. *ASTM D245*; Standard Practice for Establishing Structural Grades and Related Allowable Properties for Visually Graded Lumber. ASTM: West Conshohocken, PA, USA, 2000.
25. Sandberg, D. Radially sawn timber. *Holz Als Roh Werkst.* **1997**, *55*, 175–182. [CrossRef]
26. Bratasz, L.; Kozłowska, A.; Kozłowski, R. Analysis of water adsorption by wood using the Guggenheim-Anderson-de Boer equation. *Eur. J. Wood Prod.* **2012**, *70*, 445–451. [CrossRef]

applied sciences

MDPI

Article

New Bio-Based Binding Systems for Plywood Panels

Electra Papadopoulou [1,*], Dimitrios Moutousidis [1], Sotirios Kountouras [1], Aikaterini Argyropoulou [2], Panagiotis Stathopoulos [3], Alexios Leandros Skaltsounis [3], Rafail O. Ioannidis [4], Eleftheria Xanthopoulou [4], Lamprini Malletzidou [5,6] and Konstantinos Chrissafis [5]

1 Chimar Hellas S.A., 15 km National Road Thessaloniki—Polygyros, 57001 Thessaloniki, Greece; dimitris.moutousidis@ari.gr (D.M.); kountouras@ari.gr (S.K.)
2 PharmaGnose S.A., 57th km Athens-Lamia National Road, 32011 Oinofyta, Greece; kargyropoulou@pharmagnose.com
3 Division of Pharmacognosy and Natural Products Chemistry, Department of Pharmacy, National and Kapodistrian University of Athens, Panepistimiopolis Zografou, 15771 Athens, Greece; stathopan@pharm.uoa.gr (P.S.); skaltsounis@pharm.uoa.gr (A.L.S.)
4 Laboratory of Chemistry and Technology of Polymers and Dyes, Department of Chemistry, Aristotle University of Thessaloniki, 54124 Thessaloniki, Greece; rafahlioannidis@gmail.com (R.O.I.); elefthxanthopoulou@gmail.com (E.X.)
5 Laboratory of Advanced Materials and Devices, School of Physics, Faculty of Sciences, Aristotle University of Thessaloniki, 54124 Thessaloniki, Greece; labrinim@auth.gr (L.M.); hrisafis@physics.auth.gr (K.C.)
6 Laboratory of Archaeometry and Physicochemical Measurements, "Athena" Research Centre, Kimmeria University Campus, 67100 Xanthi, Greece
* Correspondence: papadopoulou@ari.gr; Tel.: +30-6946-463660

Abstract: This study investigated the use of non-formaldehyde binders in the production of plywood panels, focusing on mixtures containing 70% poly 4,4′-methylene diphenyl isocyanate (pMDI) and 30% soy flour (SF), along with blends of soy flour and agricultural residues (olive by-products—with and without extraction of their bioactive ingredients—and defatted hemp seeds). The basic properties of these biomaterials, such as moisture content, pH, and buffering capacity, were determined with laboratory analysis. Adhesive mixtures were characterized using attenuated total reflectance Fourier transform infrared spectroscopy (ATR-FTIR) and thermogravimetric analysis (TGA). The adhesive's bonding ability was evaluated by manufacturing plywood panels on a laboratory scale, simulating industrial practices. The glue lines were visually inspected with a stereomicroscope. Micro-ATR-FTIR study of the cross-sections of plywood panels showed the full consumption of isocyanate groups indicating effective curing of the adhesive. Mixtures containing olive residues, particularly olive skin and stones, showed improved thermal stability in the TGA study. The mechanical properties of the plywood panels were assessed with three-point bending tests, while their shear strength and wood failure performance were tested according to the European standards used in the relevant industry (EN 314.1:2004 and EN 314.2:1993). In terms of flexural properties, the adhesive with non-extracted (NE) residual olive skin (ROS) showed the highest flexural strength of around 17 MPa and a flexural modulus of 650 MPa. The formulations containing extracted materials from hemp seeds (HSs) and residual olive skin (ROS) showed the best overall performance with wood failure values of 85% and 75% after the most severe cyclic test (EN314.1:2004-Pretreatment 5.1.3). Overall, the results showed that binders prepared with residual olive skin and defatted hemp seeds have promising performance and can be used in the manufacture of plywood panels.

Keywords: plywood; bio-based binder; by-products; olive; hemp; ATR-FTIR; TGA

1. Introduction

Although polymeric methylene diphenyl diisocyanate (pMDI) resins have been successfully used as non-formaldehyde binders in the manufacture of wood panel products such as particle board (PB) and medium density fiberboard (MDF), their use in the manufacture of plywood panels (PW) has been limited for a number of reasons. These include the difficulty of cleaning with water the equipment used to apply pMDI resins to veneers, the reaction of the NCO group in PMDI with the hydroxyl groups in wood, resulting in premature curing prior to hot pressing, the tendency of pMDI resin to stick to the caul plates during pressing making the plywood difficult to remove, and the incompatibility of pMDI with water-based adhesives commonly used in plywood production due to its high reactivity with water. A solution to overcome these drawbacks is to mix pMDI adhesives with starch to control their reactivity by forming urethane bonds between starch and pMDI, making them easier to apply to plywood [1]. The starch used industrially to make plywood is usually derived from wheat. However, in the context of a sustainable economy, it is desirable to use other lower value biomasses or wastes for this application.

In this study, residues of soybeans, hemp seeds, and olive fruit and leaves were investigated as pMDI resin fillers in the manufacture of plywood panels. All these materials are found in large quantities as wastes, and their use in this application gives them added value.

Soybean cultivation in Europe has grown steadily in recent years, reaching a record high of 11.5 million tonnes in 2023 [2], and is expected to increase by a further 5–10% in 2024 [3]. Soybeans are rich in protein and also contain fiber and oligosaccharides, making them a valuable food source for both humans and animals. However, soy has been studied for many applications beyond food, such as health products [4], dietary supplements, cosmetics [5–7], and chemical formulations. In the case of adhesives for plywood panels, He et al., 2012 [8], investigated a mixture of soy protein hydrolysate, different resins (phenol-formaldehyde, melamine-formaldehyde, phenol-melamine-formaldehyde, methylenediphenyl diisocyanate, and epoxy) and an amine polymer-epichlorohydrin adduct. It was found that the plywood panels were not successful in all cases, with the degree of hydrolysis of the soy playing a significant role. Lei et al., 2014 [9], improved the water resistance of a soy-based adhesive by mixing it with epoxy resin or melamine-formaldehyde resin or a mixture thereof, making it suitable for plywood production. Other researchers have investigated the performance of non-formaldehyde mixtures of soybean flour with cardanol-based epoxy resin or citric acid [10,11]. The results of these studies have shown the potential of such binders for plywood panel production. Alawode et al., 2022 [12], evaluated the potential of using soy flour (among other materials) as a partial replacement of pMDI resin for the production of both hardwood and softwood plywood panels. The results showed that replacing 30% of the pMDI resin with soy flour significantly improved its adhesion properties. Following this, it has been considered that the partial substitution of soy flour in pMDI should not exceed that of 30% because of the poor waterproof properties of soy flour itself [13,14].

Industrial hemp is another crop that is attracting increasing research interest due to its rapid growth and diverse applications in commercial products such as textiles, paper, medicine, food, animal feed, paints, biofuel, biodegradable plastics, and construction materials. Industrial hemp seeds, which are the fruits of *Cannabis sativa*, represent another valuable but underutilized biomass resource after the extraction of nutritional ingredients. Specifically, hemp seeds are covered by a thin, two-layered pericarp and contain an endosperm and two cotyledons. They contain approximately 25–30% oil, 25–30% protein, 30–40% fiber, and 6–7% moisture [15–17]. After extraction, the remaining biomass is considered waste. Although its unprecedented production growth (expected to grow at a CAGR of 22.44% over the next 10 years) will inevitably lead to large amounts of defatted residual material, there has been limited research into the use of defatted hemp seed residues. Kawalerczyk et al., 2020 [18], investigated the potential of incorporating hemp meal as a filler in melamine urea formaldehyde (MUF) adhesives for plywood production. They found that the incorporation of 20–25 pbw (parts by weight) hemp flour facilitated the production of plywood panels with mechanical properties comparable to the reference plywood bonded with MUF adhesive mixed with rye flour. More recently, in 2024, Barretto et al. [19] used extracted hemp proteins, both modified and unmodified, in combination with glutaraldehyde to synthesize wood adhesives. The adhesiveness of these resins was tested on cherry wood samples of specific dimensions. The results showed that hemp protein has significant potential as a renewable and environmentally friendly alternative to plant protein-based adhesives.

Regarding the olive tree, it is another crop whose cultivation produces waste that causes environmental problems and whose management entails significant economic costs. Olive growing and olive oil production processes generate significant solid residues, both from the processing of the fruit and from the waste olive leaves. An average mature olive tree can yield between 10 and 70 kg of olive oil, which is equivalent to roughly 2500 to 17,500 pieces of olive fruit, and a corresponding amount of residues [20]. It is therefore clear that a huge amount of olive leaf and fruit waste such as pomace, exhausted olive cake, pomace skins, and stones is produced every year. Of course, these residues are a source of valuable bioactive compounds such as simple phenols and phenolic acids, secoiridoids, flavonoids, lignans, and triterpenic acids, which are used in several applications including pharmaceuticals, dietary supplements, biofunctional foods, and cosmetics. Their extensive use is now the focus of commercial interest in the global market, but this also creates other waste in large quantities that usually remain unused.

Several scientists have investigated the use of olive residues—olive stones in particular—in various applications, such as fillers in thermoplastic resin composites [21–23], the recovery of various chemicals from them [24,25], as ingredients in skin care formulations, as a natural mulch for plants, and as a fuel [26,27]. However, there are only a few publications on wood adhesives, and none of them used olive residues after removal of the valuable bioactive compounds. Specifically, Benhamou et al., 2021 [28], compared cactus waste seeds and olive stones as fillers in phenol-formaldehyde (PF) to produce wood composites. The study showed that incorporation of up to 20 per cent cactus waste seeds and olive stones in PF resins resulted in higher viscosity, gel time, and solids content than the control PF, while the corresponding plywood panels had lower formaldehyde emissions than the control. Elsahli et al., 2016 [29], used olive stone/unsaturated polyester resin compositions at various ratios to produce particleboards that were evaluated for their mechanical and physical properties according to European standards. It was found that such panels have acceptable performance.

In this context, we investigated the possibility of replacing wheat flour, an edible resource, with biomass waste in the manufacture of pMDI-based adhesives for plywood panels, with the aim of producing a non-competitive and environmentally friendly product. Specifically, we examined the replacement of wheat flour with a combination of commercial defatted soybean flour and various biomass residues. The biomass residues investigated included olive leaves (OLs), exhausted olive pomace (EOP), residual olive skin (ROS), and olive stones (OSs), both with and without the removal of their bioactive compounds. Additionally, oil-free hemp seeds (HSs) were also studied.

Previous research has investigated the partial replacement of pMDI with up to 30% soybean meal [12,30]. Building on this, the present study investigates the partial replacement of soy flour (50%) with olive and hemp residues. In particular, the pMDI-based adhesives were prepared by mixing 70% of pMDI with 15% of defatted soybean flour and 15% of other biomass residues (by weight). Additionally, control formulations were prepared with only soybean flour and wheat/soybean flour mixtures. The adhesive mixtures were characterized by infrared spectroscopy to examine the chemical structure of the adhesives and by thermogravimetric analysis (TGA) to study their thermal stability. To test the properties of the novel pMDI-based adhesives, they were applied in the preparation of plywood panels according to conventional industrial practices. Finally, these plywood panels were further characterized by infrared spectroscopy, mechanical tests according to European standards, and TGA to verify, respectively, the curing of the adhesive, the mechanical performance of the plywood panels, and their thermal stability. Promising results were obtained, suggesting that the studied biomass residues can indeed be used to replace wheat flour in pMDI-based adhesives.

2. Materials and Methods

2.1. Materials

pMDI is a bulk material. Defatted soy flour (SF) with the trade name "Prolia" was provided by Cargill company (Wayzata, MN, USA). Hemp seeds (HSs) of the *Felina 32* technical variety were provided by the Centre for Renewable Energy Sources and Saving (CRES) in Attiki, Greece. Regarding the olive residues, 3 categories of olive pomace waste (OPW) were collected and studied: (a) exhausted olive pomace (EOP); (b) residual olive skin (ROS); (c) olive stone (OS). The raw materials were obtained from a collaborating pomace oil factory located in Messinia region of Greece. For reference, commercial wheat flour (WF) was also included in the study.

2.2. Preparation of Materials

The exhausted olive pomace (EOP) by-product originated after chemically extracting the residual oil from two- and three-phase olive pomace waste. Subsequently, the raw material was dried in a laboratory oven at 140 °C, until most of its moisture was removed. Then, the moisture content of the EOP was less than 5%, which was sufficient for their pulverization with a standard laboratory blender. The raw materials of residual olive skin (ROS) and olive stone (OS) resulted from the exhausted olive pomace after processing it through sieves with a diameter of 1 mm. All dried materials were kept in sealed bags suitable for human use, at room temperature, prior to their dispatch to the laboratories for analysis and further separation of bioactive ingredients (Hydroxytyrosol, Tyrosol, Maslinic, and Oleanolic acids) by extraction using methanol as a solvent.

In parallel, waste olive leaves were collected, dried, and powdered, and then oleuropein was separated by ethanol/water (80/20) extraction. The remaining biomass was dried afterwards. Regarding hemp seeds, the oil was extracted by cold pressing using a screw press.

All the above materials before and after the extraction of oil and bioactive ingredients were used in adhesive mixtures with pMDI for plywood panel production (Table 1), after being grounded and sieved to obtain particle size below 200 μm.

Table 1. Materials used in the adhesive mixtures with pMDI for plywood panels.

Material	Characteristic
Soy flour Prolia FLR 200/90	Non-extracted (NE)
Exhausted Olive Pomace (EOP)	Non-extracted (NE)
Exhausted Olive Pomace (EOP)	Extracted (E)
Residual olive skin (ROS)	Non-extracted (NE)
Residual olive skin (ROS)	Extracted (E)
Olive stone (OS)	Non-extracted (NE)
Olive stone (OS)	Extracted (E)
Olive Leaves (OLs)	Non-extracted (NE)
Olive Leaves (OLs)	Extracted (E)
Hemp seeds (HSs)	Non-oil-extracted (NE)
Hemp seeds (HSs)	Extracted (Oil-free) (E)

2.3. Preparation of Adhesive Mixtures

The adhesive mixtures were prepared by mechanical stirring of the materials at ambient temperature. The pMDI was 70% of the mixture and the flours 30%.

Details of the formulation of the mixtures are given in Table 2 below.

Table 2. Composition of adhesive mixtures.

a/a	Materials	Weigh (g)
1	pMDI	70
	Soy flour	30
2	pMDI	70
	Soy flour	15
	Wheat flour	15
3	pMDI	70
	Soy flour	15
	EOP-non-extracted (NE)	15
4	pMDI	70
	Soy flour	15
	EOP-extracted (E)	15
5	pMDI	70
	Soy flour	15
	ROS-non-extracted (NE)	15
6	pMDI	70
	Soy flour	15
	ROS-extracted (E)	15
7	pMDI	70
	Soy flour	15
	OS-non-extracted (NE)	15
8	pMDI	70
	Soy flour	15
	OS-extracted (E)	15
9	pMDI	70
	Soy flour	15
	OL-non-extracted (NE)	15
10	pMDI	70
	Soy flour	15
	OL-extracted (E)	15
11	pMDI	70
	Soy flour	15
	HS-extracted (E) (oil-free HSs)	15
12	pMDI	Unsuitable for use *
	Soy flour	
	HS-non-extracted (NE)	

* Hemp seeds without oil removal could not be used, as their pulverization resulted in a paste-like mass that could not be sieved.

The ingredients were agitated for 2–3 min and then applied on veneers.

2.4. Preparation of Plywood Panels

Plywood panels are typically constructed by gluing wood veneers together at right angles and then subjecting them to high temperatures and pressures. The adhesive used is usually a mixture of polymer resin (commonly PF resin), water and a filler, often wheat flour. This process produces strong boards that are widely used in construction and various other applications. When pMDI is used as the adhesive, the mixture is anhydrous to prevent any reaction between pMDI and water. The amount of adhesive used ranges from 0.1 to 0.25 kg/m^2. Hot pressing ensures good and sufficient contact between the boards and facilitates the curing of the adhesive for a strong bond. Special hydraulic presses with heated plates are used for this process, with temperatures ranging from 120 to 150 °C. The curing time depends on the thickness of the construction and the moisture, thermal conductivity and specific heat of the wood. In practice, the time required ranges from 2 to 15 min, depending on the thickness of the veneers, the type of adhesive used, and the degree of hardening required.

In this study, three-layer plywood panels were produced on a laboratory scale using typical production parameters [31,32] (Table 3). After pressing, the panels were stacked and left to allow the adhesive to fully cure. Finally, the boards were trimmed to their final dimensions, squared, and tested to determine their properties.

Table 3. Typical parameters of plywood panel production.

Wood of Veneers:	Softwood (Scots Pine)
Dimensions of veneers:	50 × 50 cm
Moisture of veneers:	3.5–6.5%
Assembling time:	10–20 min
Waiting time before hot pressing:	>1 h
Hot press temperature:	130–150 °C
Pressing time:	2–7 min
Specific pressure:	10–25 kg/cm^2

2.5. Characterization Methods

The biomasses of this study were characterized for their moisture content, pH, and buffer capacity.

The adhesive mixtures with pMDI were studied with Fourier transform infrared (FTIR) spectroscopy and thermogravimetric analysis (TGA).

All plywood panels were tested for their mechanical properties according to the European standards in force as well as with 3-point bend tests. Their thermal properties were determined via thermogravimetric analysis (TGA), while they were subjected to surface analysis with stereomicroscope. The various measurements were conducted as described below.

2.5.1. Moisture Content (MC) of Biomasses of Olive and Hemp Biomasses

The moisture content (MC) of the various biomasses was determined with the gravimetric method, also known as direct method, by comparing the weight of biomasses before and after drying them in an oven [2]. According to this method, firstly, a quantity of these samples is weighed using a precision balance. An oven dryer is then used to remove moisture from the chips. When the chips have reached a constant weight without further loss, the dry weights are measured again using the precision balance. The moisture content (MC) is then determined using the following equation:

$$\mathrm{MC}(\%) = \frac{(\mathrm{w} - \mathrm{d})}{w} * 100 \tag{1}$$

where w = wet (initial) weight, and d = dry (final) weight.

2.5.2. pH and Buffer Capacity of Biomasses of Olive and Hemp Biomasses

Understanding the pH and buffering capacity of biomasses is essential for their effective use in gluing processes [33]. For pH measurements, 2 g (oven-dried basis) of ground biomass from each type were mixed with 30 mL of cold, freshly boiled distilled water (pH 7.3). The samples were stirred periodically at room temperature for 72 h, after which the pH was recorded. The measurement was repeated twice to obtain the average pH value [34]. The pH was measured by a CRISON pH meter (Crison Instruments, s.a., Alella, Spain) calibrated at three points (pH 4, pH 7, and pH 10) for high accuracy. The samples were then titrated to pH 3 with nominal 0.05 N HCl (for alkalinity determination). The alkalinity was expressed as ml of HCl consumed after correction for the volume of titrant required to adjust the distilled water to the target pH [34].

2.5.3. Fourier Transform Infrared (FTIR) Spectroscopy Study of Adhesive Mixtures and Plywood Panels

All the adhesive mixtures as mentioned in Table 2 were studied after being cured for 2 h in 120 °C. Spectra of pMDI alone and mixtures of pMDI/soy flour (70/30) and pMDI/soy flour/wheat flour (70/15/15) were studied for comparison reasons.

FTIR spectra of and the adhesive mixtures were recorded with a Cary 670 spectroscope (Agilent Technologies, Palo Alto, CA, USA) using a diamond attenuated total reflectance (ATR) accessory, (GladiATR, Pike Technologies, Madison, WI, USA). The spectra were collected in the mid IR area (4000–400 cm^{-1}), with 32 scans and a resolution of 4 cm^{-1}. The specimens were not manipulated for the measurements; they were placed on the ATR diamond and pressured by the instrument's click-stop pressure clamp.

Furthermore, micro-ATR-FTIR measurements were performed on cross-sections of the manufactured plywood panels. The measurements were employed with a Cary 620 FTIR microscope (Agilent Technologies, Palo Alto, CA, USA)—coupled with the aforementioned Cary 670 spectroscope—equipped with a germanium micro-ATR accessory attached on the 15× objective lens of the microscope, using a 64 × 64 focal plane array (FPA) detector (Agilent Technologies, Palo Alto, CA, USA). Each FPA image consists of 4096 data points/individual spectra, each one of them collected from a 1.1 × 1.1 μm^2 independent observed area. The measurements were performed in the range of 3950–700 cm^{-1}, with a resolution of 4 cm^{-1} and 64 co-added scans. For the preparation of the cross-sections, specimens from the panels were cut mechanically using a cutting machine to a thickness of 0.5 cm, and then polished with SiC papers up to 4000 mesh.

2.5.4. Visual Assessment of the Quality of Glue Lines in Plywood Panels

The glue lines of the panels were examined with a SteREO Discovery V.20, stereomicroscope (Zeiss, Oberkochen, Germany), equipped with a GRYPHAX Altair camera (Jenoptik, Jena, Germany).

2.5.5. Plywood Panels Performance Tests

The plywood panels were tested according to the European standards used by the relevant industry and refer to their mechanical strength and bonding quality. These standards are as follows:

- EN 314-1:2004 [35]. Plywood—Bonding quality—Part 1: Test methods.
- EN 314-2:1993 [36]. Plywood—Bonding quality—Part 2: Requirements.

The standard EN 314-1 outlines the methods for testing plywood panels, while EN 314-2 classifies the panels and establishes threshold values for their shear strength (measured with a Zwick/Roell machine) and wood failure performance. According to the EN314-1 standard, the panels are subjected to various pre-treatments before their testing for shear strength and wood failure evaluation. They are as follows:

- Pre-treatment 5.1.1 (EN314-1): Immersion in water of (20 ± 3) °C for 24 h.

- Pre-treatment 5.1.3 (EN314-1): Immersion in boiling water for 4 h, followed by drying in ventilated oven for 16 h to 20 h at (60 ± 3) °C and subsequent immersion in boiling water for 4 h followed by cooling in water at (20 ± 3) °C for at least 1 h.

In order for plywood panels to be suitable for various applications, they must meet certain combined shear strength and wood failure limits as shown in Table 4 below.

Table 4. EN314-2:1993 standard bonding performance requirements.

Mean Shear Strength fv, N/mm^2	Mean Apparent Cohesive Wood Failure w, %
0.2 < fv < 0.4	≥80
0.4 < fv < 0.6	≥60
0.6 < fv < 1.0	≥40
1.0 < fv	No requirement

Although the above standards apply to industrial applications, they provide a reliable method of evaluation and laboratory boards as well.

In this study, the plywood panels were evaluated as boards for use in outdoor applications, and therefore, their properties were measured after pretreatments 5.1.1 and 5.1.3, which are the severest for plywood products. Shear force was applied to the central glue line of each test piece. After shearing, the test pieces were analyzed for the percentage of wood failure. The results were compared to those of panels prepared with the reference adhesive mixture. All evaluations were conducted according to the European standard EN 314-2: 1993, which correlates mean values of shear strength with mean values of wood failure (Table 4).

2.5.6. Three-Point Bending Tests of Plywood Panels

Three-point bending tests were performed using a Shimadzu EZ Flexural Tester Model EZ-LX (Kyoto, Japan), with a 2 kN load cell. Samples (26.5 mm wide and ca 9 mm thick) were tested flatwise on the support span according to ASTM D790-17 [37]. The support span-to-depth ratio was 16:1 (tolerance ±1), and a constant crosshead speed of 1.0 mm/min was applied.

2.5.7. Thermogravimetric Analysis (TGA) of Adhesive Mixtures and Plywood Panels

Thermogravimetric analysis was performed to the adhesive mixtures and the manufactured panels with a SETSYS TG-DTA 16/18 (Setaram, Lyon, France). In every case, the analysis was performed on samples of (3.0 ± 0.2) mg placed in alumina crucibles in dynamic conditions, from room temperature up to 800 °C for the adhesive mixtures and up to 500 °C for the plywood panels, with a heating rate of 20 °C/min, using nitrogen flow (50 mL/min).

The pMDI-based soy/flour mixture was measured (a) immediately after mixing its components and (b) after 2 h at room temperature to compare the progress of reactions between the components of the mixture. All the adhesive mixtures—as mentioned in Table 2—were studied after being cured for 2 h in 120 °C. For the study of the plywood panels, material was scraped from the surface of the middle veneer after their detachment, without further manipulation.

3. Results and Discussion

The aim of this study was to investigate the feasibility of replacing wheat flour with biomass residues in pMDI-based adhesives. To this end, a series of pMDI adhesives were prepared consisting of 70% pMDI and 30% biomass residues, including 15% defatted soybean flour and 15% other biomass residues (olive and hemp). The biomass materials used in this study were characterized for moisture content, pH, and buffer capacity. The adhesives were analyzed using FTIR and TGA and then applied to the production of plywood panels. These panels were produced by gluing wood veneers together at right angles and subjecting them to high temperature and pressure. The resulting plywood panels were characterized to evaluate the properties of the adhesive blends. The test results are presented below.

3.1. Properties of Studied Biomaterials

The properties of the biomasses are shown in Table 5.

Table 5. Properties of biomasses olive and hemp biomasses.

Material	Solids, %	pH at 21 °C, [ ]	Buffer Capacity, mL
EOP-non-extracted (NE)	92.14	5.95	63
EOP-extracted (E)	91.76	7.10	18
ROS-non-extracted (NE)	92.28	5.50	44
ROS-extracted (E)	90.95	7.01	12
OS-non-extracted (NE)	88.9	7.27	12
OS-extracted (E)	89.87	7.60	11
OL-non-extracted (NE)	94.58	5.59	46
OL-extracted (E)	89.61	5.95	18
HS-extracted (E)	96.00	5.85	10

The above results show that non-extracted (NE) materials are more acidic than extracted biomass. This also affects their buffer capacity as non-extracted materials present a higher buffer capacity than the extracted ones.

3.2. ATR-FTIR Analysis of the Adhesive Mixtures

Figure 1a shows the ATR-FTIR spectrum collected from the neat and uncured pMDI together with the spectra of cured pMDI adhesive mixtures with soy, and with soy and wheat. Neat and uncured pMDI presents all the characteristic bands as supported in numerous references [38–40]. The difficulty to observe FTIR bands in the 2300–2200 cm^{-1} region is noted, due to the diamond ATR accessory that was employed for the collection of the spectra. For this reason, the isocyanate band at 2278 cm^{-1}—according to the bibliography—is biased and located at 2242 cm^{-1}.

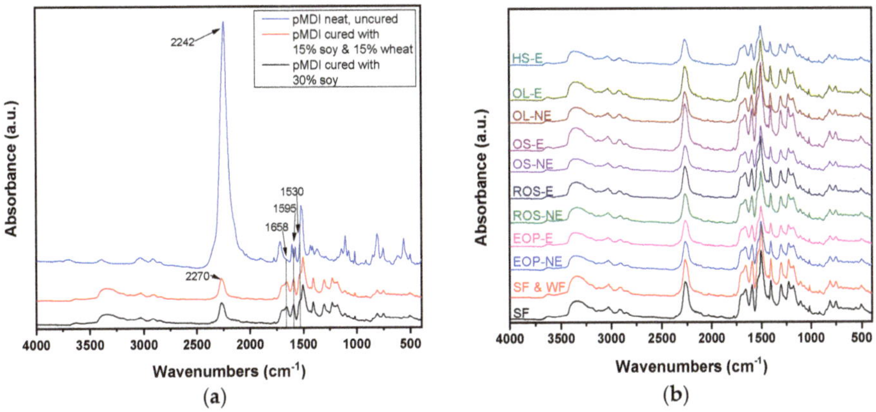

Figure 1. (**a**) ATR-FTIR spectra of neat and uncured pMDI, and its mixtures with soy, and soy/wheat; (**b**) ATR-FTIR spectra of all the cured adhesive mixtures.

Both the cured adhesive mixtures—with plain soy flour and soy/wheat flours—present the same spectral characteristics (Figure 1a). The isocyanate band decreases and shifts to 2270 cm^{-1}, but is not fully consumed; this is indicative that the pMDI polymerization with the additives is not full but isocyanates remain to react with the wood of the veneers. A new band of C=O at 1663 cm^{-1} is observed, indicative of NH-CO-NH linkages [38]. The bands at 1610 and 1577 cm^{-1} of neat and uncured pMDI disappear and a new band at 1595 cm^{-1} rises, indicating the formation of secondary urea linkages. Finally, the shoulder at 1530 cm^{-1} is attributed to CHN groups of secondary urethane formation.

Figure 1b shows the ATR-FTIR spectra of the cured adhesive mixtures, as they are described in Table 2. In Figure 1a, the glue mixtures with soy flour and with both soy and wheat flours are again presented for comparative reasons. In every case, all the spectra of the cured glue mixtures present similar characteristics in comparison with the cured mixtures of soy flour, and of soy and wheat flour, as already discussed; the band of the isocyanate group at 2270 cm^{-1} is decreased in comparison with the neat and uncured pMDI, but the group is not fully consumed, as isocyanates remain to react with the wood of the veneers.

3.3. TGA Analysis of the Adhesive Mixtures

Figure 2 shows the TGA curves of mass loss and its derivative for the adhesive mixture of pMDI with 30% soy flour. The adhesive mixture was studied immediately after combining its ingredients (uncured), two hours after its preparation (prepolymer) and after its curing (cured). The uncured adhesive mixture presents a 7% mass loss up to 160 °C, which is attributed to humidity. Several local maxima of the mass loss rate can be observed, which correspond to its unreacted ingredients, such as the degradation of pMDI at 245 °C, the degradation of MDI monomers at 430 °C, and, finally, the degradation of the remaining isomers of MDI at 500 °C [41]. Moreover, and due to the soy flour, two additional local mass loss rate maxima are reported, at ~350 °C, due to starch and protein decomposition [30,42]. Regarding the cured adhesive mixture, its decomposition presents a mass loss rate maximum at 344 °C, in accordance with the literature [43]. Finally, the prepolymer presents a mass loss of 0.5% due to humidity up to 160 °C, which indicates that the moisture content of soy flour is immediately consumed as pMDI starts to cure. This is supported from the major step of degradation that the three analyzed samples present; their corresponding maxima of mass loss rate are located at 250, 322, and 343 °C for the uncured, the prepolymer, and the cured mixture, respectively.

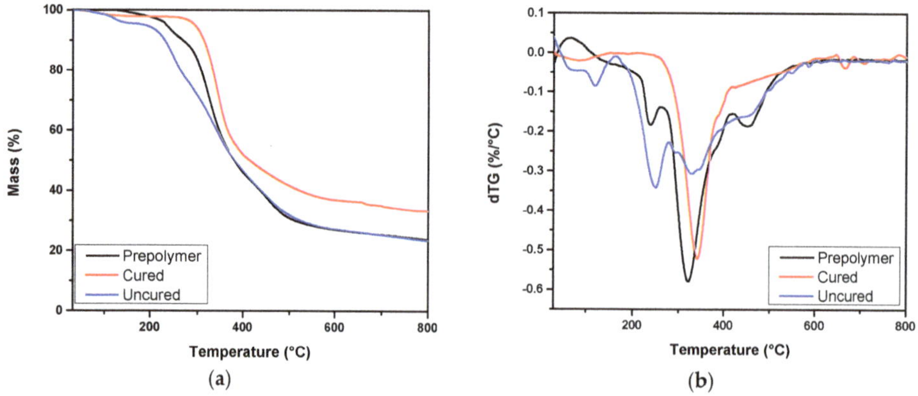

Figure 2. Thermogravimetric curves of the adhesive mixture pMDI/soy flour 70/3: (**a**) mass loss; (**b**) dTG.

Figure 3 shows the TGA results of the various studied adhesive mixtures of pMDI with 15% soy flour, and 15% flour of the agricultural wastes, according to Table 2, while the mixture of pMDI with 30% soy flour—as aforementioned—is also presented for comparative reasons. All the specimens were studied after being cured. The adhesive mixtures present a major degradation step, with a local mass loss rate maximum at 330–360 °C, which is in total agreement with cured pMDI mixtures according to the literature [43]. As shown by the dTG curves (Figure 3b), the glue mixtures with residual olive skin (ROS), olive stone (OS), and olive leaves (OLs)—all extracted—present a better thermal stability of all the other mixtures, while the glue mixtures with wheat flour (WF), extracted (EOP-E) and non-extracted exhausted olive pomace (EOP-NE), residual olive skin non-extracted (ROS-NE), and extracted hemp seeds (HS-E) have a lower thermal stability.

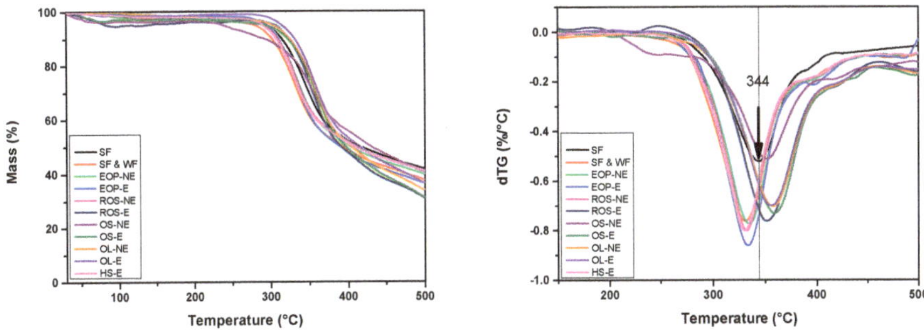

(a) (b)

Figure 3. Thermogravimetric curves of the adhesive mixtures of pMDI with soy flour, wheat flour, and agricultural residues: (**a**) mass loss; (**b**) dTG.

3.4. Plywood Test Results

All of the adhesive formulations in Table 2 were used in the production of three-layer plywood panels using the process described in Section 2.4. The results of their evaluations are given below.

3.4.1. Plywood Visual Observations

The quality of the glue lines was also assessed by visual observation (Figure 4).

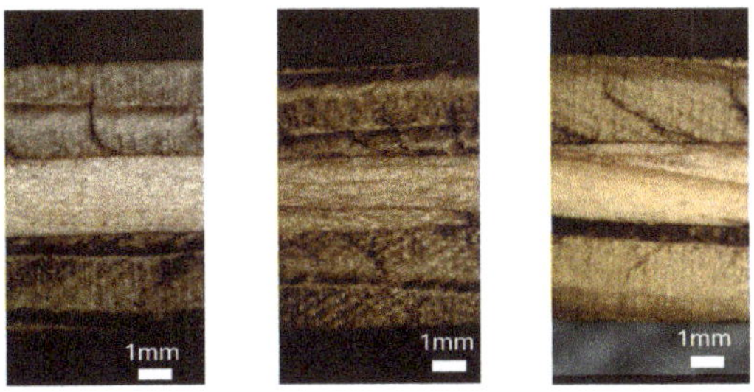

(1) (2) (3)

Figure 4. *Cont.*

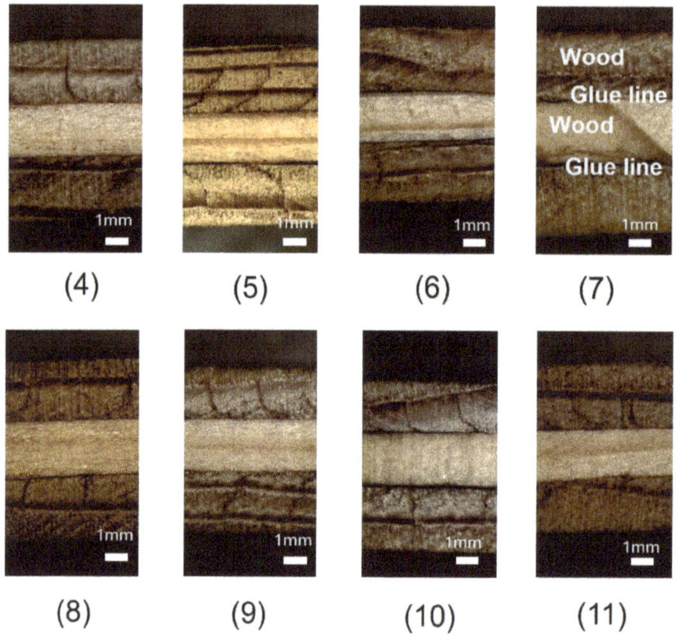

Figure 4. Stereomicroscope photographs of plywood panels made with the following: (**1**) pMDI/SF (control 1), (**2**) pMDI/SF/WF (control 2), (**3**) pMDI/SF/EOP-NE, (4) pMDI/SF/EOP-E, (**5**) pMDI/SF/ROS-NE, (**6**) pMDI/SF/E, (7) pMDI/SF/OS-NE, (**8**) pMDI/SF/OS-E, (**9**) pMDI/SF/OL-NE, (**10**) pMDI/SF/OL-E, and (**11**) pMDI/SF/HS-E. Indicatively, each part of the plywood panel is shown in photo no. 7. They are the same for all photos.

Figure 4 above shows the cross-sections of the panels and allows a qualitative assessment of the glue lines. This study has shown the successful and uniform coating of the wood surfaces by the adhesive mixture, which is a critical parameter for their good bonding.

3.4.2. Micro-ATR-FTIR Analysis of Plywood Panels

Figure 5 presents the FPA micro-FTIR analysis of the cross-section of the plywood panel manufactured with the heartwood non-extracted adhesive mixture. The analysis was performed in micro-ATR mode, using a germanium crystal which does not affect the 2400–2000 cm^{-1} area as a diamond one. The FPA chemical image of the area of interest is indicated in Figure 5a and presented in Figure 5b, showing the intensity of the band at 1710 cm^{-1} in a color scale from blue (absence of band) to red (most intense). The band at ~1710 cm^{-1} is attributed to urethane linkages (R-NH-COOR) that are formed, due to the reaction of the isocyanate groups with hydroxyls [38]. The glue mixture is distributed in the bonding line, within a 20 μm thickness. The micro-ATR spectrum of the glue mixture (Figure 5c) shows that the isocyanate group is fully consumed, as no band is noted at ~2240 cm^{-1}, except from the bands at 2400–2300 cm^{-1} due to atmospheric CO_2.

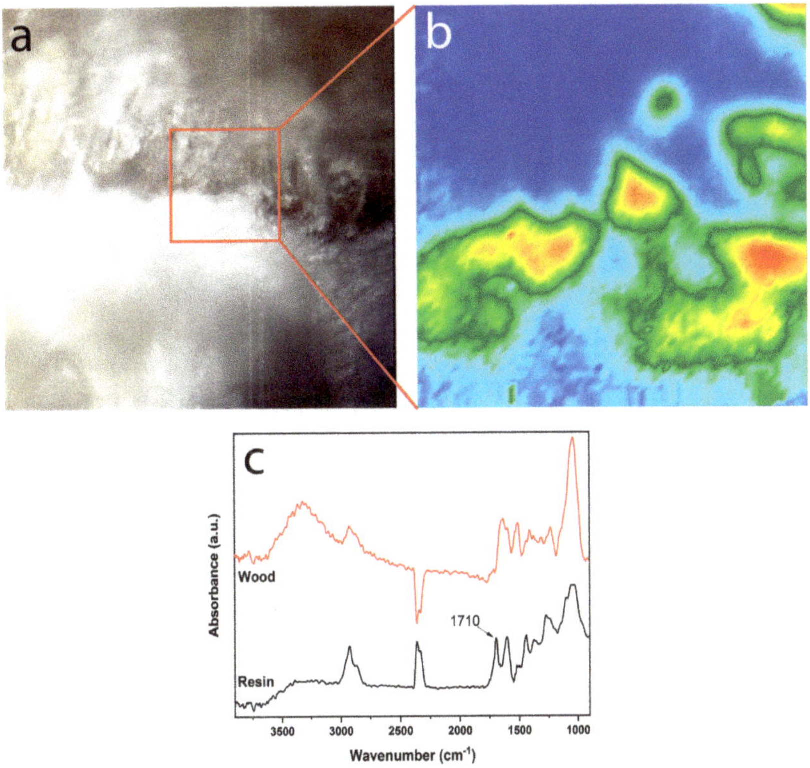

Figure 5. FPA micro-ATR-FTIR analysis or the bonding line of the plywood manufactured using the adhesive mixture with non-extracted residual olive skin (ROS-NE), as representative of the plywood panels. (**a**) Cross-section image (350 μm × 350 μm) in visible region. The area of interest is indicated with the red rectangle. (**b**) FPA chemical image representing the spatial distribution of the band at ~1710 cm^{-1} (70 μm × 70 μm). (**c**) ATR-FTIR spectra collected from single 1.1 μm × 1.1 μm pixels of the wood veneer (blue area in (**b**)) and the glue mixture (red area in (**b**)).

3.4.3. Mechanical Properties of Plywood Panels

Plywood Panels Evaluation According to the European Standards

The shear strength and wood failure performance of panels was evaluated according to the European standards EN314.1 and EN 314.2 as described in Section 2.5.4. The results are shown in the following Table 6.

Table 6. Properties of plywood panels made with pMDI/SF/various OPW and HS.

Formulation:	**1 (Control 1)**	**2 (Control 2)**	**3**	**4**	**5**	**6**	**7**	**8**	**9**	**10**	**11**
Resin:						pMDI/SF					
Biomaterial:	-	WF	EOP	EOP	ROS	ROS	OS	OS	OL	OL	HS
Type	-		NE	E	NE	E	NE	E	NE	E	E
EN314.1											

Table 6. *Cont.*

Formulation:	1 (Control 1)	2 (Control 2)	3	4	5	6	7	8	9	10	11
Pre-treatment 5.1.1	Immersion in water of 20 °C for 24 h										
Shear strength, N/mm^2	1.65	1.80	1.66	2.03	1.90	1.75	1.78	1.76	1.72	1.75	1.70
Wood failure, %	75	70	40	55	70	85	45	0	40	25	85
Pre-treatment 5.1.3	4 h in boiling water-16 h drying at 60 °C-4 h in boiling water-1 h in cool water										
Shear strength, N/mm^2	1.56	1.32	1.54	1.60	1.65	1.64	1.69	1.83	1.67	1.36	1.56
Wood failure, %	75	40	20	35	55	75	40	50	25	20	85

The above results show that all adhesive formulations have a shear strength greater than 1 N/mm^2, and therefore, according to the specifications of the EN314-2 standard for bonding performance requirements (Table 4), there is no requirement for a specific wood failure performance. In practice, however, the industry requires a wood failure of more than 50%. In this light, the mixtures with ROS and especially the extracted material (ROS-E) together with hemp seeds (HS-E) have the best overall performance after both pre-treatments (5.1.1. and 5.1.3.)—even from the reference formulations (formulations 1 and 2 of Table 6).

Plywood Flexural Test Results

The flexural properties of the panels were studied by subjecting them to three-point bending tests. This measurement shows the resistance of the panels to bending as well as the maximum strength that the panels can withstand. Figure 6 shows the flexural stress–strain curves of the examined panels, and the corresponding flexural data are presented in Table 7.

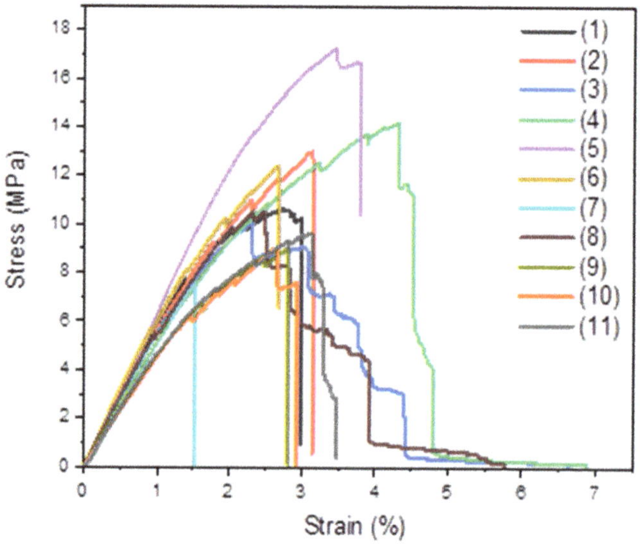

Figure 6. Flexural stress–strain curves of the plywood panels made with the following: (1) pMDI/SF (control 1), (2) pMDI/SF/WF (control 2), (3) pMDI/SF/EOP-NE, (4) pMDI/SF/EOP-E, (5) pMDI/SF/ROS-NE, (6) pMDI/SF/E, (7) pMDI/SF/OS-NE, (8) pMDI/SF/OS-E, (9) pMDI/SF/OL-NE, (10) pMDI/SF/OL-E, and (11) pMDI/SF/HS-E.

Table 7. Flexural data of the prepared plywood panels.

Sample	Flexural Stress at Break (MPa)	Flexural Strain (%)	Flexural Modulus (MPa)	Flexural Strength (MPa) *
pMDI/SF	10.20	3.00	604.6	10.60
pMDI/SF/WF	12.60	3.16	580.2	12.97
pMDI/SF/EOP-NE	10.08	2.31	592.1	10.17
pMDI/SF/EOP-E	14.10	4.33	557.0	14.20
pMDI/SF/ROS-NE	16.52	3.81	655.2	17.25
pMDI/SF/E	12.10	2.70	647.3	12.41
pMDI/SF/OS-NE	7.45	1.54	563.9	7.60
pMDI/SF/OS-E	10.10	2.49	583.1	10.55
pMDI/SF/OL-NE	8.95	2.83	494.2	9.05
pMDI/SF/OL-E	7.57	2.95	467.8	9.01
pMDI/SF/HS-E	9.20	3.11	536.8	9.63

Flexural strength (MPa) *: It is the maximum flexural stress sustained by the samples during the three-point bending tests.

With some exceptions, most plywood panels with olive and hemp residue binding systems exhibited similar mechanical behavior compared to the reference panel, indicating that the novel biobased adhesives under study could potentially replace the conventional ones in terms of mechanical behavior. The panels with pMDI/SF/ROS-NE resin exhibited the best mechanical performance among all other panels, meaning it exhibited the highest endurance to bending deformation. Specifically, it could withstand maximum flexural stress (flexural strength) and flexural modulus close to 17 MPa and 650 MPa, respectively. Chen et al., 2024 [44], investigated starch-based gels as green alternatives adhesives for plywood applications providing flexural data comparable to the present bio-based binders.

3.4.4. TGA of the Plywood Panels

The TGA performed on material removed from the middle veneer of each plywood panel manufactured with the adhesive mixtures under study is presented in Figure 7. In every case, the thermal degradation of the adhesive mixtures follows that of the wood, as it was impossible to extract just the adhesive mixture that was cured between the veneers; following this, the mass loss and dTG curves of the plain wood veneer are shown for comparative reasons [45]. A slight mass loss due to humidity that does not exceed 3.5% is reported up to 160 °C. All the samples present a local maximum in the mass loss rate at about 375–390 °C due to wood degradation predomination and, in particular, the cellulose decomposition. Plain wood is considered as the most thermally stable one from all the samples under study, as its thermal decomposition maximum is located at 390 °C. The samples containing adhesive mixtures with olive tree residues, and especially that with residual olive skin (ROS) and olive stone (OS), present a somewhat better thermal stability in comparison with the other glue mixtures, with local degradation maxima above 380 °C. It should be mentioned that although all the adhesive mixtures—as collected from the wooden panels after their curing—present similar thermal characteristics, the one composed of extracted hemp seeds (HS-E) differs, as it presents two more local degradation maxima, at 310 and 439 °C. The first one is also observed in other adhesive mixtures, mainly the one with olive leaves (OL-NE), and can be related with the depolymerization of hemicellulose, cellulose, lignin, and pectins, which are associated with the agriculture residues under study [46]. The latter could be associated with lignin pyrolysis or the decomposition of char residues [47].

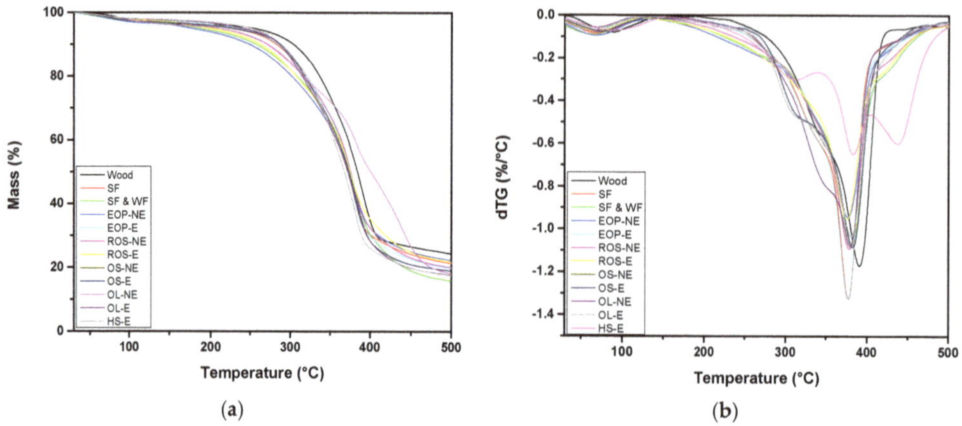

Figure 7. (**a**) Mass loss curves and (**b**) dTG curves of the plywood panels manufactured with the various adhesive mixtures.

Summarizing all the above results, it can be seen that the removal of bio-components from biomasses increases their pH, resulting in less acidic materials and a reduction in their buffer capacity (Table 5). The FTIR analysis of the adhesive mixtures with the agricultural residues before and after their curing shows that they exhibit the exact characteristics with those of the glue mixture with soy flour. Furthermore, it is noted that in the cured adhesive mixtures, the peak corresponding to the isocyanate group is not totally consumed, and isocyanate groups still remain to react during the adhesion stage of the manufacturing of the plywood panels. This is supported by the FPA micro-FTIR analysis of the plywood panels, where the isocyanate groups were totally consumed. Visual observation of the glue lines showed good adhesion in all cases, and mechanical tests revealed that all the panels made from all studied biomaterials exhibit excellent mechanical properties and high wood failure. Notably, the panels with ROS and HSs demonstrated mechanical properties and wood failure performance that are comparable to, or even superior to the reference compositions (Tables 6 and 7). The TGA of the adhesive mixtures showed that the ones containing ROS, OS, and OL had better thermal stability compared to all the other blends. In the case of the material collected from the bonding lines of the plywood panels, the TGA indicated that their thermal behavior closely follows that of plain wood.

4. Conclusions

This study demonstrated that agricultural biomass residues, such as soybean meal, and olive and hemp cultivation wastes can serve as effective substitutes for the traditionally used wheat flour in pMDI-based formulations for plywood panels. Of the various residues tested, the bio-ingredient-free olive fruit skin residues and oil-free (defatted) hemp seeds showed the best overall combined performance in terms of shear strength and wood failure, which are the most important properties for the relevant industrial sector. Similarly, the adhesive system with non-extracted olive skin (ROS-NE) showed the highest resistance in the three-point flexure tests, with a flexural strength and modulus of around 17 MPa and 650 MPa, respectively, while all ROS-based formulations were among the most thermally stable formulations. The use of such waste materials, which have no other use, in this application was investigated for the first time in this work, and the results of the study pave the way for their use in this value-added application.

Author Contributions: Conceptualization, E.P. and K.C.; methodology, E.P., D.M., L.M. and K.C.; validation, S.K., R.O.I., E.X. and L.M.; formal analysis, A.L.S. and P.S.; investigation, E.P., S.K., L.M. and R.O.I.; resources, A.A., P.S. and A.L.S.; data curation, L.M.; writing—original draft preparation, E.P., P.S., A.A., L.M. and R.O.I.; writing—review and editing, E.P., D.M., L.M., A.A., P.S. and E.X.; supervision, K.C.; project administration, K.C. and E.P.; funding acquisition, K.C. All authors have read and agreed to the published version of the manuscript.

Funding: This research has been co-financed by the European Regional Development Fund of the European Union and Greek national funds through the Operational Program Competitiveness, Entrepreneurship and Innovation, under the call Special Actions Aquaculture-Industrial Materials-Open Innovation in Culture (project code: T6YBP-00161).

Institutional Review Board Statement: Not applicable.

Informed Consent Statement: Not applicable.

Data Availability Statement: The original contributions presented in the study are included in the article, further inquiries can be directed to the corresponding author.

Conflicts of Interest: Authors Electra Papadopoulou, Dimitrios Moutousidis and Sotirios Kountouras were employed by the company Chimar Hellas S.A. Author Aikaterini Argyropoulou was employed by the company PharmaGnose S.A. The remaining authors declare that the research was conducted in the absence of any commercial or financial relationships that could be construed as a potential conflict of interest.

References

1. Gopal, G.T.; Nath, S.; Sujatha, D.; Rangaraju, T.S. Starch based PMDI (Polymeric methyl diphenyl diisocyanate) Adhesives for Plywood. *IOSR J. Appl. Chem. IOSR-JAC* **2016**, *9*, 57–61. [CrossRef]
2. Rahman, A.; Marufuzzaman, M.; Street, J.; Wooten, J.; Gude, V.G.; Buchanan, R.; Wang, H. A comprehensive review on wood chip moisture content assessment and prediction. *Renew. Sustain. Energy Rev.* **2024**, *189 Pt A*, 113843. [CrossRef]
3. Available online: https://www.donausoja.org/record-high-for-soybean-output-in-europe-11-5-million-tonnes/ (accessed on 3 January 2024).
4. Guo, M. Soy food products and their health benefits. In *Functional Foods Principles and Technology*; Woodhead Publishing Series in Food Science, Technology and Nutrition; Woodhead Publishing: Sawston, UK, Chapter 7; 2009; pp. 237–277. [CrossRef]
5. Na Takuathung, M.; Teekachunhatean, S.; Chansakaow, S.; Klinjan, P.; Inpan, R.; Kongta, N.; Tipduangta, P.; Tipduangta, P.; Dukaew, N.; Sakuludomkan, C.; et al. The effects of SOY extract nutraceuticals on postmenopausal women's health: A randomized, double-blind, placebo-controlled trial. *J. Funct. Foods* **2024**, *113*, 106055. [CrossRef]
6. Leite, B.; Bezerra, P.; Stocco, B.; Abichabki, N.; Andrade, L.; Fonseca, M.; Torqueti, M. Biotransformed soybean cream as a new nutraceutical for skin care: Collagen stimulation in vitro and ex vivo. *Braz. J. Med. Biol. Res.* **2023**, *56*, e12781. [CrossRef]
7. Hydrolyzed Soy Protein, SpecialChem, The Material Selection Platform. Available online: https://cosmetics.specialchem.com/inci-ingredients/hydrolyzed-soy-protein (accessed on 12 January 2024).
8. He, G.; Feng, M.; Dai, C. Development of soy-based adhesives for the manufacture of wood composite products. *Holzforschung* **2012**, *66*, 857–862. [CrossRef]
9. Lei, H.; Du, G.; Wu, Z.; Xi, X.; Dong, Z. Cross-linked soy-based wood adhesives for plywood. *Int. J. Adhes. Adhes.* **2014**, *50*, 199–203. [CrossRef]
10. Zhu, Z.; Zhang, E.; Tu, Y.; Ye, M.; Chen, N. An Eco-Friendly Wood Adhesive Consisting of Soybean Protein and Cardanol-Based Epoxy for Wood Based Composites. *Polymers* **2022**, *14*, 2831. [CrossRef]
11. Hao, Z.; Xi, X.; Hou, D.; Lei, H.; Li, C.; Xu, G.; Du, G. A fully bio-based soy protein wood adhesive modified by citric acid with high water tolerance. *Int. J. Biol. Macromol.* **2023**, *253 Pt 6*, 127135. [CrossRef]
12. Alawode, A.O.; Owofadeju, F.K.; Musah, M.; Asafu-Adjaye, O.; Peng, Y.; Via, B. Performance of Different Fillers in Polymeric Methylene Diphenyl Diisocyanate (pMDI) Resin. *For. Prod. J.* **2022**, *72*, 14–21. [CrossRef]
13. Asafu-Adjaye, O.; Alawode, A.O.; Via, B.; Banerjee, S. Soy Flour Substitution in Polymeric Diphenylmethane Diisocyanate Resin Reduces Press Platen Sticking. *For. Prod. J.* **2022**, *72*, 5–7. [CrossRef]
14. Zheng, X.; Li, P.; Lin, Y.; Li, X. Study on Performance of Flame Retardant and Smokeless Reed/Magnesite Cement Inorganic Particleboard. *For. Prod. J.* **2021**, *71*, 224–232. [CrossRef]
15. Callaway, J.C. Hemp Seed Production in Finland. *J. Ind. Hemp* **2004**, *9*, 97–103. [CrossRef]
16. Farinon, B.; Molinari, R.; Costantini, L.; Merendino, N. The seed of industrial hemp (*Cannabis sativa* L.): Nutritional quality and potential functionality for human health and nutrition. *Nutrients* **2020**, *12*, 1935. [CrossRef] [PubMed]
17. Dunford, N. Hemp Seed Oil Properties, Extension. OKSTATE.edu, 2020. Available online: https://extension.okstate.edu/fact-sheets/hemp-seed-oil-properties.html (accessed on 12 January 2024).

18. Kawalerczyk, J.; Siuda, J.; Mirski, R.; Dziurka, D. Hemp flour as a formaldehyde scavenger for melamine-urea-formaldehyde adhesive in plywood production. *BioResources* **2020**, *15*, 4052–4064. [CrossRef]
19. Barretto, R.; Qi, G.; Xiao, R.; Jones, C.; Sun, X.S.; Li, Y.; Griffin, J.; Wang, D. Hempseed protein is a potential alternative source for plant protein-based adhesives. *Int. J. Adhes. Adhes.* **2024**, *133*, 103740. [CrossRef]
20. Petar, D. How Many Olives Are Harvested on an Average Olive Tree? 26 November 2023. Available online: https://oliveknowledge.com/how-many-olives-are-harvested-on-olive-tree/ (accessed on 12 February 2024).
21. Koutsomitopoulou, A.; Bénézet, J.; Bergeret, A.; Papanicolaou, G. Preparation and characterization of olive pit powder as a filler to PLA-matrix bio-composites. *Powder Technol.* **2014**, *255*, 10–16. [CrossRef]
22. Farag, E.; Alshebani, M.; Elhrari, W.; Klash, A.; Shebani, A. Production of particleboard using olive stone waste for interior design. *J. Build. Eng.* **2020**, *29*, 101119. [CrossRef]
23. Valvez, S.; Maceiras, A.; Santos, P.; Reis, P.N.B. Olive Stones as Filler for Polymer-Based Composites: A Review. *Materials* **2021**, *14*, 845. [CrossRef]
24. Rodríguez, G.; Lama, A.; Rodríguez, R.; Jiménez, A.; Guillén, R.; Fernández-Bolaños, J. Olive stone an attractive source of bioactive and valuable compounds. *Bioresour. Technol.* **2008**, *99*, 5261–5269. [CrossRef]
25. Bolek, S. Olive stone powder: A potential source of fiber and antioxidant and its effect on the rheological characteristics of biscuit dough and quality. *Innov. Food Sci. Emerg. Technol.* **2020**, *64*, 102423. [CrossRef]
26. Olive Pit: Beyond the Olive: Lesser Known Uses of Olive Pits, Faster Capital. Available online: https://fastercapital.com/content/Olive-pit--Beyond-the-Olive--Lesser-Known-Uses-of-Olive-Pits.html (accessed on 12 February 2024).
27. What Do Olive Stones Have in Common with Barbecue and Soccer Fields? LEROMA, 2021. Available online: https://leroma.de/blog/id/olive_stones.html (accessed on 20 February 2024).
28. Benhamou, A.A.; Boussetta, A.; Kassab, Z.; Nadifiyine, M.; Salim, M.H.; Grimi, N.; EL Achaby, M.; Moubarik, A. Investigating the characteristics of cactus seeds by-product and their use as a new filler in phenol formaldehyde wood adhesive. *Int. J. Adhes. Adhes.* **2021**, *110*, 102940. [CrossRef]
29. Elsahli, E.; Elhrari, W.; Klash, A.; Shebani, A. The use of olive stone waste for production of particleboard using commercial polyster sealer as a binding agent. In Proceedings of the 1st International Conference on Chemical, Petroleum, and Gas Engineering, Alkhoms, Libya, 25–30 January 2016.
30. Hand, W.G.; Ashurst, W.R.; Via, B.; Banerjee, S. Curing behavior of soy flour with phenol-formaldehyde and isocyanate resins. *Int. J. Adhes. Adhes.* **2018**, *87*, 105–108. [CrossRef]
31. Muniz, G.I.B.D.; Iwakiri, S.; Viana, L.C.; Andrade, M.; Weber, C.; Almeida, V.C. Production of plywood panels from Pinus taeda using veneers of different densities and phenol-formaldehyde resin with high and low molecular weights. *Cerne* **2013**, *19*, 315–321. [CrossRef]
32. Lengowski, E.C.; Bonfatti, E.A.; Dallo, R.; Nisgoski, S.; de Mattos, J.L.M.; Prata, J.G. Nanocellulose-reinforced phenol-formaldehyde resin for plywood panel production. *Maderas Cienc. Tecnol.* **2020**, *23*. [CrossRef]
33. Johns, W.E.; Niazi, K.A. Effect of pH and buffering capacity of wood on the gelation time of urea-formaldehyde resin. *Wood Fiber Sci.* **1980**, *12*, 255–263.
34. Wang, X.; Huang, Z.; Cooper, P.; Wang, X.M.; Zhang, Y.; Casilla, R. The ability of wood to buffer highly acidic and alkaline adhesives. *Wood Fiber Sci.* **2010**, *42*, 398–405.
35. *EN 314-1:2004*; Plywood—Bonding Quality—Part 1: Test Methods. British Standards Institution: London, UK, 2004.
36. *EN 314-2:1993*; Plywood—Bonding Quality—Part 2: Requirements. British Standards Institution: London, UK, 1993.
37. *ASTM D790-17*; Standard Test Methods for Flexural Properties of Unreinforced and Reinforced Plastics and Electrical Insulating Materials. ASTM: West Conshohocken, PA, USA, 2017.
38. Li, Z.; Zhao, S.; Wang, Z.; Zhang, S.; Li, J. Biomimetic water-in-oil water/pMDI emulsion as an excellent ecofriendly adhesive for bonding wood-based composites. *J. Hazard. Mater.* **2020**, *396*, 122722. [CrossRef]
39. Liu, Z.-M.; Wang, F.-H.; Wang, X.-M. Spectroscopic Analysis of the Interface for Wheat Straw Specimen Glued with PMDI. *Wood Fiber Sci.* **2007**, *39*, 109–119.
40. El-Shekeil, Y.A.; Sapuan, S.M.; Khalina, A.; Zainudin, E.S.; Al-Shuja'A, O.M. Influence of chemical treatment on the tensile properties of kenaf fiber reinforced thermoplastic polyurethane composite. *Express Polym. Lett.* **2012**, *6*, 1032–1040. [CrossRef]
41. Mohamed, A.; Finkenstadt, V.L.; Gordon, S.H.; Palmquist, D.E. Thermal and mechanical properties of compression-molded pMDI-reinforced PCL/gluten composites. *J. Appl. Polym. Sci.* **2010**, *118*, 2778–2790. [CrossRef]
42. Soares, R.M.D.; Scremin, F.F.; Soldi, V. Thermal stability of biodegradable films based on soy protein and corn starch. *Macromol. Symp.* **2005**, *229*, 258–265. [CrossRef]
43. Hosseinpourpia, R.; Adamopoulos, S.; Echart, A.S.; Eceiza, A. Polyurethane films prepared with isophorone diisocyanate functionalized wheat starch. *Eur. Polym. J.* **2021**, *161*, 110826. [CrossRef]
44. Chen, Y.; Rao, Y.; Liu, P.; Wu, L.; Zhang, G.; Zhang, J.; Xie, F. High-amylose starch-based gel as green adhesive for plywood: Adhesive property, water-resistance, and flame-retardancy. *Carbohydr. Polym.* **2024**, *339*, 122247. [CrossRef] [PubMed]
45. Malletzidou, L.; Papadopoulou, E.; Papadopoulos, L.; Markessini, C.; Bikiaris, D.N.; Chrissafis, K.; Vourlias, G. Novel Biobased Polyester Resins for Wood Applications: Morphological, Thermal, and Chemical Characterization. *Macromol. Symp.* **2022**, *405*, 2100229. [CrossRef]

46. Asimakidou, T.; Chrissafis, K. Thermal behavior and pyrolysis kinetics of olive stone residue. *J. Therm. Anal. Calorim.* **2022**, *147*, 9045–9054. [CrossRef]
47. Rachini, A.; Le Troedec, M.; Peyratout, C.; Smith, A. Comparison of the thermal degradation of natural, alkali-treated and silane-treated hemp fibers under air and an inert atmosphere. *J. Appl. Polym. Sci.* **2008**, *112*, 226–234. [CrossRef]

MDPI AG
Grosspeteranlage 5
4052 Basel
Switzerland
Tel.: +41 61 683 77 34

Applied Sciences Editorial Office
E-mail: applsci@mdpi.com
www.mdpi.com/journal/applsci

www.ingramcontent.com/pod-product-compliance
Lightning Source LLC
Chambersburg PA
CBHW061436210926
51850CB00008B/1524